Principles and Applications of Clinical Mass Spectrometry

PRINCIPLES and APPLICATIONS of CLINICAL MASS SPECTROMETRY

Small Molecules, Peptides, and Pathogens

EDITED BY

NADER RIFAI, PhD
Professor of Pathology
Harvard Medical School
Louis Joseph Gay-Lussac Chair in
  Laboratory Medicine
Director of Clinical Chemistry
Boston Children's Hospital
Boston, MA, United States

ANDREA RITA HORVATH, MD, PhD
Professor, Department of Clinical Chemistry &
  Endocrinology
New South Wales Health Pathology
School of Medical Sciences
University of New South Wales
Sydney, Australia

CARL T. WITTWER, MD, PhD
Professor of Pathology
University of Utah School of Medicine
Medical Director, Immunologic Flow Cytometry
ARUP Laboratories
Salt Lake City, UT, United States

ASSOCIATE EDITOR

ANDREW N. HOOFNAGLE, MD, PhD
Professor
Head, Division of Clinical Chemistry
Department of Laboratory Medicine
University of Washington
Seattle, Washington, United States

Elsevier
Radarweg 29, PO Box 211, 1000 AE Amsterdam, Netherlands
The Boulevard, Langford Lane, Kidlington, Oxford OX5 1GB, United Kingdom
50 Hampshire Street, 5th Floor, Cambridge, MA 02139, United States

Copyright © 2018 Elsevier Inc. All rights reserved.

No part of this publication may be reproduced or transmitted in any form or by any means, electronic or mechanical, including photocopying, recording, or any information storage and retrieval system, without permission in writing from the publisher. Details on how to seek permission, further information about the Publisher's permissions policies and our arrangements with organizations such as the Copyright Clearance Center and the Copyright Licensing Agency, can be found at our website: www.elsevier.com/permissions.

This book and the individual contributions contained in it are protected under copyright by the Publisher (other than as may be noted herein).

Notices
Knowledge and best practice in this field are constantly changing. As new research and experience broaden our understanding, changes in research methods, professional practices, or medical treatment may become necessary.

Practitioners and researchers must always rely on their own experience and knowledge in evaluating and using any information, methods, compounds, or experiments described herein. In using such information or methods they should be mindful of their own safety and the safety of others, including parties for whom they have a professional responsibility.

To the fullest extent of the law, neither the Publisher nor the authors, contributors, or editors, assume any liability for any injury and/or damage to persons or property as a matter of products liability, negligence or otherwise, or from any use or operation of any methods, products, instructions, or ideas contained in the material herein.

Library of Congress Cataloging-in-Publication Data
A catalog record for this book is available from the Library of Congress

British Library Cataloguing-in-Publication Data
A catalogue record for this book is available from the British Library

ISBN: 978-0-12-816063-3

For information on all Elsevier publications visit our website at https://www.elsevier.com/books-and-journals

  Working together to grow libraries in developing countries

www.elsevier.com • www.bookaid.org

Publisher: Susan Dennis
Acquisitions Editor: Kathryn Morrissey
Editorial Project Manager: Carly Demetre
Production Project Manager: Paul Prasad Chandramohan
Cover Designer: Miles Hitchen

Cover image: Reproduced with permission from TransTech Publications A. Jaworek et al., Electrospray Nanocoating of Microfibres, *Solid State Phenomena* 2008;**140**:127−132.

Typeset by TNQ Technologies

CONTENTS

Contributors, vii
Preface, ix

1 **Chromatography,** 1
 David S. Hage
2 **Mass Spectrometry,** 33
 Alan L. Rockwood, Mark M. Kushnir, and Nigel J. Clarke
3 **Sample Preparation for Mass Spectrometry Applications,** 67
 David A. Wells
4 **Mass Spectrometry Applications in Infectious Disease and Pathogens Identification,** 93
 Phillip Heaton and Robin Patel
5 **Development and Validation of Small Molecule Analytes by Liquid Chromatography-Tandem Mass Spectrometry,** 115
 Russell P. Grant and Brian A. Rappold
6 **Proteomics,** 181
 Andrew N. Hoofnagle and Cory Bystrom

Index, 203

CONTRIBUTORS

Cory Bystrom, BS, MS, PhD
Vice President, Research and Development
Cleveland HeartLab
Cleveland, Ohio

Nigel J. Clarke, BSc(Hons), PhD
Vice President, Advanced Technology
Quest Diagnostics Nichols Institute
San Juan Capistrano, California

Russell P. Grant, PhD
Vice President
Research and Development
Laboratory Corporation of America
Burlington, North Carolina

David S. Hage, PhD
Professor of Chemistry
University of Nebraska
Lincoln, Nebraska

Phillip Heaton, MS, PhD
Technical Director of Microbiology and Molecular Diagnostics
Pathology and Laboratory Medicine
Children's Hospitals and Clinics of Minnesota
Minneapolis, Minnesota

Andrew N. Hoofnagle, MD, PhD
Professor
Head, Division of Clinical Chemistry
Department of Laboratory Medicine
University of Washington
Seattle, Washington

Mark M. Kushnir, PhD
Adjunct Assistant Professor of Pathology
University of Utah School of Medicine
Senior Scientist
ARUP Institute for Clinical and Experimental Pathology
Salt Lake City, Utah

Robin Patel, MD, FRCP(C), D(ABMM), FIDSA, FACP, F(AAM)
Professor of Medicine and Microbiology
Chair, Division of Clinical Microbiology
Department of Laboratory Medicine and Pathology
Mayo Clinic
Rochester, Minnesota

Brian A. Rappold, BS
Scientific Director
Essential Testing
Collinsville, Illinois

Alan L. Rockwood, PhD, DABCC
Professor of Pathology
University of Utah School of Medicine
Scientific Director of Mass Spectrometry
ARUP Laboratories
Salt Lake City, Utah

David A. Wells, PhD
Founder and Principal Consultant
Wells Medical Research Services
Laguna Beach, California

PREFACE

Over the last several decades, automated instruments in the clinical laboratory have morphed from "open systems," in which users can develop and implement novel tests, to "closed systems," in which users are dependent on manufacturers for the test menu. Mass spectrometers are an exception to this trend—instruments that have been widely used for decades in research laboratories but only relatively recently have become part of the armamentarium of the clinical lab.

Over the last few decades, the sensitivity, specificity, and reliability of mass spectrometers have increased and costs have decreased. In addition, legacy instruments requiring substantial laboratory space and knowledge in computer programming have evolved into moderate-throughput, user-friendly, bench-top workstations. As a result, their use has become widespread. According to the Web of Science, the number of published articles that use mass spectrometry since 2000 has increased threefold (Fig. 1).

Initially, mass spectrometry was used in the clinical laboratory for the confirmation of the presence of drugs of abuse and the quantification of small molecules such as steroids and drugs. However, the clinical utility of mass spectrometry has now extended to pathogen identification as well as the measurement of proteins, peptides, and nucleic acids.

The development of a clinical assay on a mass spectrometer is not trivial. It requires knowledge and expertise not only of the instrument itself but also in pre-and postanalytical processes. Appropriate sample preparation, a good understanding of chromatographic principles, and a solid understanding of mass spectral and chromatographic data are essential for the successful development of these assays.

In this handbook "Principles and Applications of Clinical Mass Spectrometry: Small Molecules, Peptides, and Pathogens," world-renown clinical laboratorians address the issues mentioned above. This compilation of chapters was initially published as part of the Tietz Textbook of Clinical Chemistry and Molecular Diagnostics. The editors believe that this handbook will be useful to any investigator interested in the clinical applications of mass spectrometry, including those in the pharmaceutical industry and those involved in biomarker research.

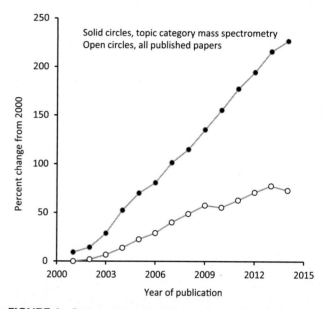

FIGURE 1 Percent change in the number of publications from 2000 for the topic category "mass spectrometry." *From Clin Chem 2016;62:1, with permission.*

Chromatography

*David S. Hage**

ABSTRACT

Background
Clinical tests often involve the use of one or more separation steps to isolate, enrich, or separate a target compound from other chemicals in the sample. Chromatography is one of the most common methods for achieving this type of separation. In this method, the components of a mixture are separated based on their differential interactions with two chemical or physical phases: a mobile phase and a stationary phase that is held in place by a supporting material. There are many forms of chromatography based on the different mobile phases, stationary phases, and supports that can be used in this method, which has led to a wide range of applications for this technique.

Content
This chapter describes the basic principles of chromatography and discusses various forms of this method that are used for chemical analysis or to prepare specimens for analysis by other techniques. The methods of gas chromatography and liquid chromatography are discussed, as well as the techniques of planar chromatography, supercritical fluid chromatography, and multidimensional separations. The mobile phases, stationary phases, and supports that are used in each of these methods are described. The instrumentation and detection schemes that are employed in these methods are also discussed.

Biological fluids are complex mixtures of chemicals. This means that clinical tests for specific components in these fluids often involve the use of one or more separation steps to isolate, enrich, or separate the target compound of interest from other chemicals in the sample. Chromatography is one of the most common methods for achieving this type of separation. This chapter describes the basic principles of chromatography and discusses various forms of this method that are used for chemical analysis or to prepare specimens for analysis by other techniques.

BASIC PRINCIPLES OF CHROMATOGRAPHY

General Terms and Components of Chromatography

Chromatography is a method in which the components of a mixture are separated based on their differential interactions with two chemical or physical phases: a mobile phase and a stationary phase.[1-4] The basic components and operation of a typical chromatographic system are illustrated in Fig. 1.1. The mobile phase travels through the system and carries sample components with it once the sample has been applied or injected. The stationary phase is held within the system by a support and does not move. As a sample's components pass through this system, the components that have the strongest interactions with the stationary phase will be more highly retained by this phase and move through the system more slowly than components that have weaker interactions with the stationary phase and spend more time in the mobile phase. This leads to a difference in the rate of travel for these components and their separation as they move through the chromatographic system.

The type of chromatographic system that is shown in Fig. 1.1 uses a column (or a tube) to contain the stationary phase and support, while also allowing the mobile phase and sample to pass through the system. This approach was first described in 1903 by Mikhail Tswett, who used this method to separate plant pigments into colored bands by using a column that contained calcium carbonate as both the support and stationary phase.[5] Tswett gave the name *chromatography* to this method. This name is derived from Greek words *chroma* and *graphein,* which mean "color" and "to write." This term is still used to describe this technique, even though most modern chromatographic separations do not involve colored sample components.

The type of chromatography that was used by Tswett, in which the stationary phase and support are held within a column, is known as "column chromatography." In chromatography, the stationary phase may be the surface of the support, a coating on this support, or a chemical layer that is cross-linked or bonded to the support.[2,6,7] In column chromatography, the support may be the interior wall of the column or it may be a material that is placed or packed into the column. A column is the most common format for chromatography. However, it is also possible to use a support and stationary phase that are present on a plane or open surface. This second format is known as "planar

---

*The author gratefully acknowledges the contributions of Drs. Glen L. Hortin, Bruce A. Goldberger, M. David Ullman, Carl A. Burtis, and Larry D. Bowers to this chapter in previous editions.n

chromatography," as will be discussed in more detail later in this chapter.[2,7]

One way of classifying chromatographic methods is based on the type of support that they employ; two examples are the techniques of column chromatography and planar chromatography. Chromatographic methods also can be classified based on the mobile phase that is present. For instance, a chromatographic method that uses a mobile phase that is a gas is called gas chromatography (GC),[8] and a chromatographic method that uses a liquid mobile phase is known as liquid chromatography (LC).[9] It is also possible to divide chromatographic methods according to the type of stationary phase that is present or the way in which this stationary phase is interacting with sample components. Examples of these classifications include the GC methods of gas-solid chromatography (GSC) or gas-liquid chromatography (GLC) and the LC methods of adsorption chromatography, partition chromatography, or ion-exchange chromatography (IEC). Each of these categories, as well as others, will also be discussed later in this chapter.

The instrument that is used to perform a separation in chromatography is known as a chromatograph.[7,10] For instance, in GC the instrument is a gas chromatograph, and in LC the instrument used to carry out this method is a liquid chromatograph. These instruments can provide a response that is related to the amount of a compound that is exiting (or eluting) from the column as a function of the elution time or the volume of mobile phase that has passed through the system. The resulting plot of the response versus time or volume is known as a chromatogram,[7,10] as is illustrated in Figs. 1.1 and 1.2.

The average time or volume that is required for a particular chemical to pass through the column is known as that chemical's retention time (t_R) or retention volume (V_R). These values both increase with the strength and degree to which the chemical is interacting with the stationary phase. The elution time or volume for a compound that is nonretained or that does not interact with the stationary phase is known as the void time (t_M) or void volume (V_M). If the retention time or retention volume is corrected for the void time or void volume, the resulting measure of retention is known as the adjusted retention time (t_R', where $t_R' = t_R - t_M$) or the adjusted retention volume (V_R', where $V_R' = V_R - V_M$). For two chemicals to be separated by chromatography, it is necessary for these chemicals to have different values for t_R and V_R (or t_R' and V_R').[2,7,10]

Most separations that are used for chemical analysis in column chromatography are carried out by injecting a relatively small volume or amount of sample onto the chromatographic system. This situation results in a chromatogram that consists of a series of peaks that represent the different compounds in the sample as they each elute from the column. The retention time or retention volume of each peak can be used to help identify the eluting compound, whereas the area or height of the peak can be used to measure the amount of the compound that is present.

The width of each peak is also of interest in a chromatogram. The peak width reflects the separating performance or efficiency of the chromatographic system. The width of a peak in a chromatogram is often represented by its baseline width (W_b) or its half-height width (W_h) (Fig. 1.3).[2,7,10] As the widths for the peaks in a chromatogram become sharper, it becomes easier for the chromatographic system to separate two peaks with similar interactions with the system and to separate more peaks in a given amount of time. Sharper peaks are also easier to measure than broader peaks and tend to produce better limits of detection.

Retention and Selectivity

For two chemicals to be separated by chromatography, these chemicals need to have some differences in how they are

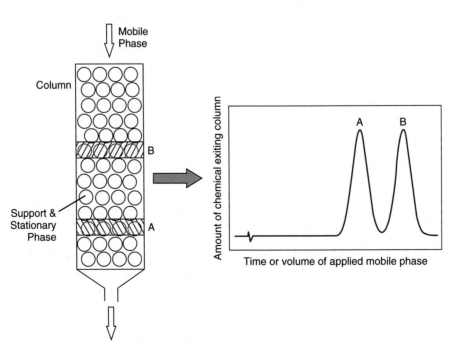

FIGURE 1.1 The general components of a chromatographic system, as illustrated here by using a column to separate two chemicals, A and B.

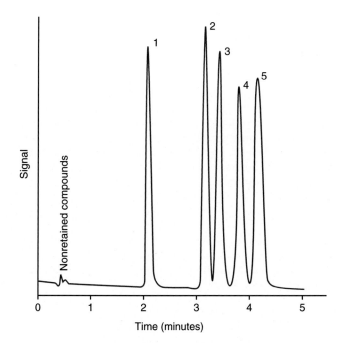

FIGURE 1.2 Chromatogram from a separation of tricyclic antidepressants based on reversed-phase chromatography and high-performance liquid chromatography. Detection was based on the use of an absorbance detector that monitored the column eluent at 215 nm. NOTE: the signal is displayed at 0.1 absorbance units-full scale (AUFS). (Courtesy Vydac/Grace Materials Technologies, Columbia, Maryland.)

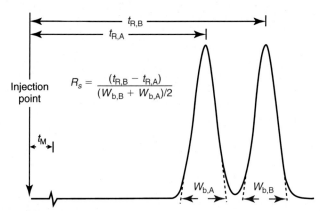

FIGURE 1.3 An example of a general chromatogram that may be obtained when using a column. In this example, compound B is eluted later than compound A. R_s, Resolution; t_M, void time; $t_{R,A}$ and $t_{R,B}$, retention times for solutes A and B; $W_{b,A}$ and $W_{b,B}$, baseline peak widths for compounds A and B.

interacting with the stationary phase versus the mobile phase. Besides using the retention time and retention volume (or adjusted retention time and adjusted retention volume) to describe these differences, another way of representing retention in chromatography is by using the retention factor (k). This term is also sometimes represented as k' or called the capacity factor.[7,10] The retention factor is a measure of the average time a chemical resides in the stationary phase versus the time it spends in the mobile phase. This value can be calculated from experimental data by using any of the following equivalent relationships.[2,7,10]

$$k = (t_R - t_M)/t_M = (t_R')/t_M$$

or

$$k = (V_R - V_M)/V_M = (V_R')/V_M$$

As these equations suggest, the retention factor is a unitless number where a value of 0 indicates that no binding or interactions are occurring between a chemical and the stationary phase or that this compound is eluting from the system at the void time. As the chemical undergoes greater interactions with the stationary phase, this will result in longer retention times and an increased value of k. In practice, it is desirable to have a value for k that is between 1 and 10 to provide reasonable separations between compounds without the need for excessive lengths of time for their elution from the column.

The retention factor is useful in describing a compound's retention in chromatography for several reasons. First, the value of k should be independent of the flow rate and column size. Also, k can be directly related to the strength of the interactions that are occurring between a chemical and the stationary phase or mobile phase, as well as the relative amount of stationary phase versus mobile phase that is present in the column. This last feature is illustrated by the following equation for a chromatographic system in which a chemical is separated based on its ability to partition between the mobile phase and stationary phase. Similar relationships can be written for other types of separation mechanisms.[2]

$$k = K_D(V_S/V_M)$$

In this relationship, the value of k is directly related to (1) the distribution equilibrium constant (K_D) for partitioning of the analyte into the stationary phase versus the mobile phase and (2) the relative amount of stationary phase in the column (as represented here by V_S) versus the amount of mobile phase that is present (as represented by V_M, the void volume). The value of k in this situation will increase if there is either an increase in K_D, which reflects the tendency of the chemical to enter the stationary phase over the mobile phase, or the ratio (V_S/V_M), which is a term also known as the *phase ratio*.[2,7]

Any separation in chromatography requires that there be some difference in retention for the chemicals that are to be separated from each other. One way of describing this difference in retention is by using the separation factor or selectivity factor (α).[7,10] The separation factor for two compounds (A and B) is equal to the ratio of their retention factors (k_A and k_B),

$$\alpha = k_B/k_A$$

where the retention factor for the later eluting component is given in the numerator. If two chemicals have the same retention in a chromatographic system, the value of α will equal 1 and no separation will be possible. If the peaks for A and B do have different retention, the value of α will be greater than 1 and will increase as the degree of separation increases.

The values of both the retention factor and selectivity factor are determined by the chemicals that are being separated, as well as the stationary phase and mobile phases that are present in the chromatographic system. A large difference in retention and a large separation factor are desirable when the goal is to selectively isolate one chemical from others in a sample. However, smaller differences in retention and in separation factors are often used when the chromatographic system is used to separate several chemicals and peaks from the same sample. In this second situation, a value for α of 1.1 or greater represents an adequate separation in many common types of chromatography. However, chromatographic methods that result in broad peaks may need even larger values of α to produce a good separation between two chemicals.

Band-Broadening and Efficiency

Besides needing a difference in retention for a separation to occur, the peaks for two neighboring chemicals must be sufficiently narrow to allow this difference to be observed. The injection of even a sample with a small volume will experience some increase in width, or band-broadening, as this peak travels through the chromatographic system. This broadening of peaks is produced by various processes related to the rate of movement or diffusion of the applied chemicals as they pass around or within the support and within or between the mobile phase and stationary phase. These band-broadening processes, in turn, are affected by factors such as the diameter or type of support within the chromatographic system, the flow rate, the diffusion coefficient of the chemical in the mobile phase and stationary phases, and the degree of retention of the chemical in the column (Box 1.1).

> **BOX 1.1 Factors That Can Affect Chromatographic Efficiency**
>
> - Column length (affecting the number of theoretical plates, *N*, but not the plate height, *H*)
> - Particle size of support (packed bed column) or tube diameter (open tubular column)
> - Uniformity in size, shape, and packing of the support
> - Flow rate and linear velocity
> - Temperature and rate of solute diffusion
> - Mobile phase viscosity
> - Degree of compound retention
> - Initial injection volume
> - Volume of connecting tubing, detector, and system components besides the colum

Together, these processes and factors determine the overall efficiency or extent of band-broadening obtained.

The efficiency and degree of band-broadening in a chromatographic system are related experimentally to the final observed width of a chemical's peak. This width can be described by measures such as the baseline width (W_b), the half-height width (W_h), or the standard deviation (σ) of the peak. These values, in turn, can be used to find another measure of chromatographic efficiency known as the number of theoretical plates, or plate number (*N*). The value of *N* for any type of chromatographic peak can be calculated by using the following formula:

$$N = (t_R/\sigma)^2$$

where t_R is the retention time for the peak and σ is the standard deviation of the peak in the same units of time as t_R.[7,10] This equation takes on the following two equivalent forms for a Gaussian-shaped peak[2,7]:

$$N = 16(t_R/W_b)^2 \text{ or } N = 5.545(t_R/W_h)^2$$

These last two equations make use of the fact that a Gaussian peak has a baseline width, as measured by the intersection of the baseline with tangents along either side of the peak, that is equal to 4 σ, and a half-width width that is equal to 2.355 σ.

The value of *N* can be thought of as representing the effective number of times that a chemical has been distributed between the mobile phase and stationary phase as this chemical has passed through the chromatographic system. A larger value for *N* represents many such steps, which makes it easier to distinguish between two chemicals that have only small differences in their retention. Experimentally, a large value of *N* results in a high chromatographic efficiency and sharp peaks, which are both desirable for either separating chemicals with similar retention or quickly separating many chemicals in the same sample.

There are several other ways in which the efficiency of a chromatographic system can be described. One way is by using the number of theoretical plates (*N*) per unit length of the chromatographic system (*L*), as given by the ratio (*N/L*). This ratio helps in comparing systems with different lengths, because the value of *N* increases in direct proportion to the length of the column or support bed that is used in a separation for chromatography. Although this means that a longer chromatographic system will always lead to a larger value for *N* and greater efficiency, the use of a longer system also results in longer separation times.

Another way of describing column efficiency is the height equivalent of a theoretical plate or plate height (HETP, or *H*).[7,10] The value of *H* is found by dividing the length of the chromatographic system by the number of theoretical plates for this system.

$$H = L/N$$

The value of *H* represents the length of the column or chromatographic system that makes up one theoretical plate or one distribution step for a chemical between the mobile phase and stationary phase. Although a large value of *N* (or N/L) represents a chromatographic system with high efficiency, the same system would be represented by a small value for H (or L/N).

A valuable feature of using H to describe chromatographic efficiency is that this term can be related directly to the parameters and processes that affect band-broadening. A common example of this is the *van Deemter equation*,[11] which shows how the overall value of H is affected by the linear velocity of the mobile phase (u), which is directly related to the flow rate (F) through the relationship $u = (F \times L)/V_M$.[10,12]

$$H = A + B/u + C u$$

The terms A, B, and C in this equation are constants that represent the contributions of several types of band-broadening processes. For instance, the A term represents the contributions of band-broadening processes that are independent of the linear velocity and flow rate, such as eddy diffusion and mobile phase mass transfer. The B term is the contribution to the plate height by longitudinal diffusion, which is a process that becomes more important as the flow rate and linear velocity are decreased. Finally, the C term represents the contributions from processes that lead to an increase in H as the flow rate or linear velocity is increased. The processes that make up the C term are stagnant mobile phase mass transfer and stationary phase mass transfer. The van Deemter equation predicts that the combined effect of these band-broadening processes will be an optimum range of flow rates and linear velocities over which the lowest plate heights, and best efficiencies, will be obtained.[11] In practice, the usual goal in varying the flow rate in chromatography is to identify those conditions that provide the most rapid separation times while still providing adequate resolution of all peaks that are of interest in the samples being separated.

Several factors that affect chromatographic efficiency are listed in Box 1.1. For instance, efficiency can be improved by using longer columns, which increases the value of N but does not alter H. It is also possible to change the flow rate to its optimum value, to use smaller diameter support particles, to use nonporous or pellicular particles instead of fully porous support particles, or to use a relatively narrow-diameter coated capillary instead of a packed bed column. All of these latter factors help to lower the value of H, which in turn increases the value of N for a given length of column or chromatographic bed. However, there are practical limits to how much some of these experimental parameters can be changed. As an example, a reduction in the diameter of the support particle will lead to greater efficiency, but it will also result in higher back pressures across the chromatographic system, require the use of lower flow rates, or both.

Resolution and Peak Capacity

The overall extent to which two peaks are separated in chromatography can be described by using a term known as the resolution (R_s), as is illustrated in Fig. 1.3. The resolution between two neighboring peaks can be found by using the following formula.[7,10]

$$R_s = \frac{(t_{R,B} - t_{R,A})}{(W_{b,B} + W_{b,A})/2}$$

In this equation, $t_{R,A}$ and $t_{R,B}$ are the average retention times for compounds A and B (where B elutes after A), while $W_{b,A}$ and $W_{b,B}$ are the baseline widths for the peaks of these compounds (in time units, in this case). An equivalent equation can be written in terms of the retention volumes of A and B and their baseline widths in volume units. The use of either approach will give a unit-less value for R_s that represents the average number of baseline widths that separate the centers of the two peaks.

Fig. 1.4 shows how the separation of two neighboring peaks changes as the value of R_s increases for these peaks. An R_s value of 0 is obtained when there is no separation between the peaks and they have exactly the same retention times or retention volumes. The degree of peak separation increases as the value of R_s increases. An R_s value of 1.5 or greater is often said to represent a complete separation between two equally sized peaks, or baseline resolution. However, for many separations, resolution values between 1.0 or 1.25 and 1.5 also may be adequate, especially if the peaks are about the same size and are to be measured using their peak heights rather than their peak areas.

Several approaches can be used to alter or improve the resolution between two peaks in chromatography (Fig. 1.5). These approaches are indicated by the following expression, which is sometimes known as the resolution equation of chromatography.[13]

$$R_s = \left[\left(N^{1/2}\right)\big/4\right] \times [(\alpha - 1)/\alpha] \times [k/(1 + k)]$$

In this equation, k is the retention factor for the second of two neighboring peaks, α is the separation factor between the first and second peaks, and N is the number of theoretical plates for the chromatographic system. This relationship indicates that resolution of two peaks in chromatography

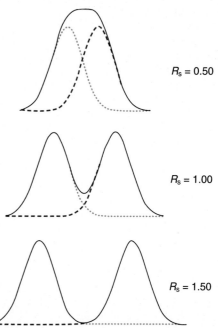

FIGURE 1.4 Degree of separation obtained for two chromatographic peaks that are present in a 1:1 area ratio as the resolution between these peaks (R_s) is varied.

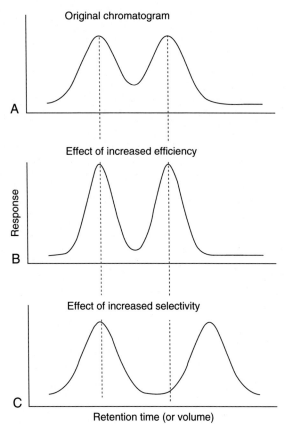

FIGURE 1.5 Effects of selectivity and efficiency on the resolution of peaks in chromatography. These three situations represent cases in which there is **(A)** poor or moderate resolution between two neighboring peaks, **(B)** good resolution between the peaks as a result of high column efficiency, or **(C)** good resolution between the peaks as a result of good column selectivity.

can be changed in three ways: (1) by altering the efficiency of the system, as represented by N; (2) by changing the overall degree of peak retention, as represented by k; or (3) by changing the selectivity of the column for one compound versus another, as represented by α. An increase in N, such as can be obtained through the use of a longer column, will lead to an increase in R_s that is proportional to $N^{1/2}$. An increase in the retention factor (k) or selectivity (α) will also lead to a nonlinear increase in resolution.

Another way of describing a chromatographic separation is in terms of the *peak capacity*. The peak capacity is the maximum number of peaks (or sample components) that can be separated, in theory, during a single chromatographic separation.[14-16] The value of the peak capacity can be found by assuming there is a continuous distribution of peaks that are separated by an average baseline width (or 4 standard deviations). In practice, the number of components that can be separated in a single run by a given system will be lower than the theoretical peak capacity because the retention times of their peaks will not be evenly distributed. The peak capacity of a system that is used for high-performance liquid chromatography (HPLC) is usually limited to several hundred peaks, whereas higher values can be obtained in methods such as capillary GC. Factors that can be used to increase the peak capacity include increasing the efficiency of the system (eg, by using a longer column) and using gradient elution or extended run times. Another approach for increasing the peak capacity is to use a multidimensional separation, as will be discussed later in this chapter.

> **POINTS TO REMEMBER**
>
> **General Ways to Improve Peak Resolution in Chromatography**
> - Increase the efficiency of the system
> - Increase the overall degree of peak retention
> - Increase the selectivity of the column for the peak of one compound versus another

Analyte Identification and Quantification

Chromatography is often used as an analytical tool to qualitatively identify analytes in a sample and to measure the concentrations of these analytes. For example, the retention time, retention volume, and retention factor for an analyte are all characteristic values that reflect how this chemical is interacting within a particular chromatographic system. These retention values can be compared to those for a known sample of the same compound to help confirm its identity. However, other confirmation also may be needed because other compounds may have similar retention characteristics.

One way additional confirmation can be obtained is if the unknown compound and reference compound have the same retention under several types of chromatographic conditions, such as on different columns or column/mobile phase combinations. In the case of capillary GC or LC columns, it is possible to simultaneously introduce samples onto two columns that contain different stationary phases and that are connected to separate detectors. If the unknown compound and a reference compound match in their retention properties on the two columns, this greatly enhances the chance for correctly identifying the unknown analyte. An alternative and even more reliable approach for identification is to use a detection method that provides structural information on the analyte, such as mass spectrometry (see Chapters 2 and 4 to 6).

The peak area or peak height can be used to produce quantitative information on an analyte that is separated from other sample components by chromatography. Peak areas tend to provide a more precise means for measuring an analyte, whereas peak heights are easier to use if there is not complete resolution between the analyte and its neighboring peaks. Both external and internal calibration techniques can be used in chromatography for such measurements.[2,17] In external calibration, standard solutions containing known quantities of the analytes are processed and separated in the same manner as samples that contain one or more of these analytes (Fig. 1.6). A calibration curve is then constructed by plotting the peak height or peak area (or the spot density, in the case of planar methods) versus the concentration or mass of analyte that was applied in the standard solutions. This curve can then be used with the peak area or peak height that is determined for the same analyte in the samples to find the concentration or amount of this analyte present.

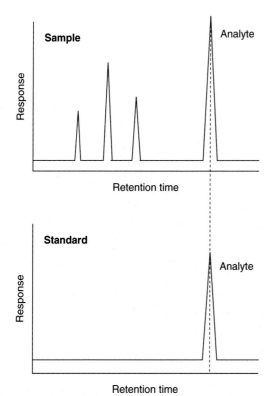

FIGURE 1.6 Use of external calibration and standards to quantify an analyte based on its peak height or area in a chromatogram for an injected sample.

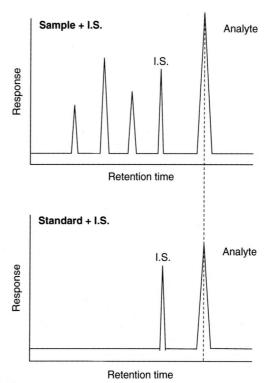

FIGURE 1.7 Use of internal calibration and samples or standards containing an internal standard (I.S.) to quantify an analyte based on its peak height or area in a chromatogram for an injected sample.

In the method of internal calibration (also called internal standardization), standard solutions of the analyte are again prepared; however, a constant amount of a different compound known as the internal standard is also now added to each standard solution and sample (Fig. 1.7). The internal standard should be a chemical that was not originally present in either the sample or the standard, is similar in its chemical and physical properties to the analyte, and can be measured independently from the analyte. This internal standard is typically added to the samples and standards before they are processed by any pretreatment steps, such as extraction or derivatization. The addition of this agent can help normalize the results for any variations that may occur during the pretreatment steps or during sample/standard injection onto the chromatographic system. This normalization is made by constructing a calibration curve in which the y-axis is based on the ratio of the peak height or peak area for the analyte in a given standard or sample divided by the peak height or peak area for the internal standard in the same standard or sample. This ratio is plotted versus the concentration or amount of analyte in each standard. This calibration plot can then be used to find the concentration or amount of the analyte that was present in each sample.[2,17]

GAS CHROMATOGRAPHY

GC is a common type of chromatography often used in chemical separations and analysis. GC can be defined as a chromatographic method in which the mobile phase is a gas.[7] The first modern GC system was developed in the mid-1940s by Cremer[18,19] and became popular after work by James and Martin in 1952, who used this method to separate fatty acids.[20]

In GC, a gaseous mobile phase is used to pass a mixture of volatile solutes through a column containing the stationary phase.[7,8] The mobile phase is typically an inert gas such as nitrogen, helium, or argon or a low mass gas such as hydrogen. Because of the low densities of gases under typical GC operating conditions, the compounds injected onto a GC column do not have any appreciable interactions with the gaseous mobile phase. Instead, this gas acts to merely carry samples through the column. As a result, the term carrier gas is commonly used to refer to the mobile phase in GC.[7]

Solute separation in GC is based on differences in the vapor pressures of the injected compounds and in the different interactions of these compounds with the stationary phase. For instance, a more volatile chemical will spend more time in the gaseous mobile phase than a less volatile solute and will tend to elute more quickly from the column. In addition, a chemical that selectively interacts with the stationary phase more strongly than another chemical will tend to stay longer in the column. The overall result is a separation of these chemicals based on their volatility and interactions with the stationary phase.

Types of Gas Chromatography

There are several ways of classifying GC methods based on the type of stationary phase present. These categories include GSC, GLC, and bonded phase GC.

TABLE 1.1 Stationary Phases Commonly Used in Gas-Liquid Chromatography and as Bonded Phases in Gas Chromatography

Composition	Polarity	Commercial Examples	Typical Applications
100% Methylpolysiloxane	Nonpolar	OV-1, SE-30	Drugs, amino acid derivatives
5% Phenyl–95% methylpolysiloxane	Nonpolar	OV-23, SE-54	Drugs
50% Phenyl–50% methylpolysiloxane	Intermediate polarity	OV-17	Drugs, steroids, glycols
50% Cyanopropylmethyl–50% phenylmethylpolysiloxane	Intermediate polarity	OV-225	Fatty acid methyl esters, carbohydrate derivatives
Polyethylene glycol	Polar	Carbowax 20M	Acids, alcohols, glycols, ketones

Gas-Solid Chromatography

GSC is a type of GC in which the same material acts as both the stationary phase and the support.[7] In this method, chemicals are retained by their adsorption to the surface of the support. This support is often an inorganic material such as silica or alumina. Other supports that can be used in this method are molecular sieves, which are porous materials that are made from a mixture of silica and alumina, or organic polymers such as porous polystyrene.[2,12,21]

The retention of an analyte on a GSC support will be affected by several factors. These factors include the surface area of the support, the size of the pores in the support, and the types of functional groups that are present on the surface of the support. Using a support with a high surface area will lead to higher retention than a support with a lower surface area. The selection of an appropriate pore size may be important if the analytes are large enough to be able to access the surface within only some of these pores. The functional groups and polarity of the support and its surface will also determine which types of analytes will have the strongest adsorption to this surface. Polar materials such as silica, alumina, and molecular sieves will usually have strong binding to polar compounds and to those that can form hydrogen bonds. Polystyrene and other less polar supports will have weaker and less selective interactions with chemicals and tend to give separations that are based more on the volatility of the components in an applied sample.

Gas-Liquid Chromatography and Bonded Phases

In GLC, the stationary phase is a liquid that is placed as a coating or layer on the support.[7] This is the most common type of GC for chemical analysis. Various types of liquids can be used for this purpose (see examples in Table 1.1). All of these liquids must have a low volatility to allow them to stay within the column at the high temperatures that are often used in GC separations. Many GLC stationary phases are based on polysiloxanes, which have the basic structure shown in Fig. 1.8.[12] The molar mass of the —Si-O-Si- chain in a polysiloxane can range in size from a few thousand to over a million grams per mole. The side chains that are attached to the silicon atoms in this chain can have structures that range from nonpolar methyl groups to polar cyanopropyl groups. These side chains also can be present in various ratios as mixtures. The overall polarity and types of chemicals that will be retained the most by this type of stationary phase will be determined by the amounts and types of side chains that are present.

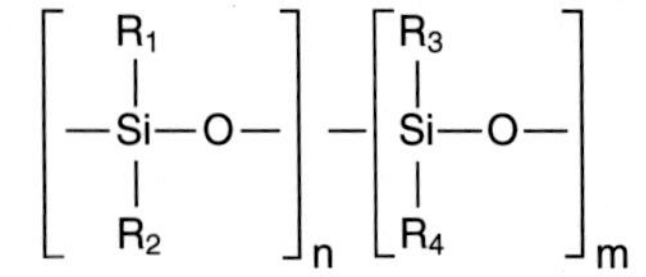

$$\left[\begin{array}{c} R_1 \\ | \\ -Si-O- \\ | \\ R_2 \end{array} \right]_n \left[\begin{array}{c} R_3 \\ | \\ -Si-O- \\ | \\ R_4 \end{array} \right]_m$$

FIGURE 1.8 General structure of a polysiloxane. The side groups are represented by R_1 through R_4, while n and m represent the relative lengths (or amounts) of each type of segment in the overall polymer.

One issue in using a liquid as a stationary phase in GC is that some of this liquid will eventually leave the column over time. This loss of the stationary phase is known as column bleed.[12,22] This process is not desirable because it will result in a change in the amount of stationary phase present and a change in the ability of the GC system to retain chemicals. This process also may cause the signal of the GC detector to have a high background or to be noisy as the liquid stationary phase leaves the column and passes through the detector.

Column bleed can be minimized by using a bonded phase instead of a liquid as the stationary phase in the GC column. The resulting method is sometimes known as bonded phase GC. A bonded phase can be produced by reacting functional groups on a stationary phase such as a polysiloxane with silanol groups on the surface of silica. Alternatively, the stationary phase can be cross-linked to make it less volatile and more stable. Besides providing a stationary phase that is more stable, a bonded phase also can provide a stationary phase that has a thinner and more uniform coating than a stationary phase based on a liquid coating. Although bonded phases are more expensive than liquid stationary phases, bonded phases are often preferred for analytical work because of their better thermal stability and better efficiencies.[12,22]

POINTS TO REMEMBER

Types of Gas Chromatography Based on the Stationary Phase
- GSC
- GLC
- Bonded phase GC

Gas Chromatography Instrumentation

The typical components of a gas chromatograph are illustrated in Fig. 1.9.[21] The first major component is the source of the gaseous mobile phase, which is used to supply the

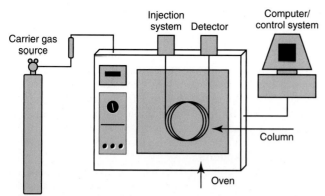

FIGURE 1.9 General design of a gas chromatograph. (Modified from a figure courtesy Restek Corporation, Bellefonte, Pennsylvania.)

carrier gas at a controlled pressure and flow rate. Next, there is the injection system, through which samples are placed into the gas chromatograph and converted into a volatile form. This is followed by the column, which contains the support and the stationary phase. This column is held in an oven for temperature control. The fourth part of the GC system is a detector that monitors sample components as they leave the column. Finally, there is a computer or control system that acquires data from the detector and allows control of the GC system.[21]

Carrier Gas Sources and Flow Control

The function of the carrier gas source is to provide the gas that will be used as the mobile phase for the GC separation. The carrier gas is usually supplied by a standard gas cylinder. However, the carrier gas is sometimes provided by using a gas generator that is connected to the GC system. Such a generator can be used to isolate nitrogen from air or produce hydrogen gas through the electrolysis of water.[2]

Good flow control is needed in GC to provide a constant or well-defined flow of the carrier gas. This control makes it possible to maintain good column efficiency and obtain reproducible elution times. Systems that are used to provide constant flow rates may use a simple mechanical device, such as a pressure regulator, or a more sophisticated electronic control device. Methods in GC such as temperature programming, as will be discussed later, require electronic pressure control to regulate the carrier gas flow rate and pressure during a chromatographic run. Such a controller may be operated in a constant-flow or constant-pressure mode. In the constant-flow mode, the pressure required to provide a flow rate that is independent of the carrier gas viscosity is determined and maintained by the system through the use of a pressure transducer and pressure regulator.

The magnitude of the carrier gas flow rate will depend on the type of column being used. For example, packed columns require typical flow rates that range from 10 to 60 mL/min. Capillary columns use much lower flow rates (eg, 1 to 2 mL/min). Because of the greater efficiencies of capillary columns versus packed columns, operating at a consistent flow rate is even more critical for the operation of the capillary columns.

Various gases can be used as the mobile phase in GC. The choice of carrier gas will depend on factors such as the type of column and detector used, as well as the expense, purity, and chemical or physical properties of the gas. Hydrogen and helium are the carrier gases of choice with capillary columns. Only high-purity hydrogen and helium should be used for this purpose. For packed columns, the most frequently used carrier gas is nitrogen.

Carrier gas impurities such as water, oxygen, and hydrocarbons can (1) harm or alter the column, (2) negatively influence the performance of some detectors, and (3) adversely affect the measurement of chemicals. The carrier gas should be as pure as possible to avoid such problems. The carrier gas should be dry, and the tubing used to connect the gas source to the GC system should be free from contamination. Molecular sieve beds and specialized inline traps are often used to remove water, hydrocarbons, oxygen, and particulate matter that may be present in the carrier gas.[23]

Many GC detectors work best with certain types of carrier gases. For instance, work with packed columns often involves the use of nitrogen as the carrier gas when working with a flame ionization detector (FID), electron capture detector (ECD), or thermal conductivity detector (TCD), which are each described in more detail later. Helium is often used with capillary columns and in work with a FID or TCD, whereas nitrogen/argon-methane mixtures are used with an ECD.

Injection Systems and Sample Derivatization

The injection of a sample into a GC system has to be done with minimal disruption of gas flow into the column. Most clinical GC methods make use of liquid-phase samples, for which the sample components are first extracted into or dissolved in a nonaqueous liquid or adsorbed onto a microextraction fiber (see Chapter 3). This liquid or microextraction fiber is then placed into the chromatographic system by using a precise and rapid online injector (eg, an autosampler or automated injection system). With packed columns, a glass microsyringe is used to inject a 1- to 10-µL portion of the sample through a septum, which serves as the interface between the injector and the chromatographic system. On the other side of the septum is located a heated injection port. Volatile chemicals in the sample and the solvent are flash-vaporized in this heated port and swept into the column by the carrier gas. To ensure rapid and complete volatilization, the temperature of the heated injection port is usually maintained at a temperature that is at least 30 to 50° C higher than the column temperature.

Common problems during injection include septum leaks and the adsorption of sample components onto the septum. In addition, because the injection port is heated, thermal decomposition products may be produced here from the sample and enter into the column. This process can result in spurious peaks, or "ghost" peaks, in the chromatogram. This type of contamination is most likely to occur at high injection temperatures. A Teflon-coated septum, or low-bleed septum, can be used to minimize this problem. In addition, the inner surface of the septum can be purged with the carrier gas and vented before the purge gas passes into the column. This approach is especially effective in reducing septum-related problems, and most commercial injectors are equipped with continuous-purge capabilities. The septum is a consumable component of the gas chromatograph and should be replaced at least once every 100 injections.

Because of the low sample capacities and slow carrier gas flow rates that are used with capillary columns, split and splitless injection techniques are used to introduce samples into such columns.[2,22] In the method of split injection, only a small portion of the vaporized sample enters the column, with the remainder being passed through a side vent. In splitless injection, most of the sample enters the column.[4] The split flow injection mode is used for samples that contain relatively high concentrations of the target analytes, whereas the splitless mode is used for samples that contain relatively low concentrations of the analytes.

Temperature-programmable injection ports are available and may be used in either the split or splitless injection mode. In this type of port, the sample is injected at a temperature slightly higher than the boiling point of the solvent that contains the sample. Under these conditions, most of the sample components will condense on a glass or fused silica wool insert that is present in the injector, while the solvent is vaporized and removed. The injector is then rapidly heated at rates of up to 100°C/min. The rapid heating vaporizes the analytes, which then move into the column. This rapid heating is advantageous because any thermally labile compounds that may be present in the sample are exposed to the high temperatures for only a short time. The ability of this approach to provide separate steps for solvent removal and analyte vaporization can allow the injection of sample volumes of up to hundreds of microliters. This ability can improve analyte detection when the amount of sample that is available is not a limiting factor.

Headspace analysis is a sample introduction technique that can be used with aqueous solutions or samples that contain some nonvolatile components.[22] In this method, a portion of the vapor phase (or "headspace") that is above a liquid or solid sample is used for the analysis. This vapor phase contains a portion of some of the more volatile components of the sample and can be directly injected onto a GC system for analysis. Headspace analysis can be carried out using either a static method or a dynamic method. In the static method, the sample is placed in an enclosed container and allowed to reach equilibrium for the distribution of its components between the sample and the vapor phase above the sample. A portion of the vapor phase is then injected onto the GC system for analysis. In the dynamic method, an inert gas is passed through the sample and used to sweep away the volatile components. These components are then captured by a solid adsorbent or a cold trap and later injected onto the GC system for analysis.

Although a fairly large number of low-mass chemicals can be injected directly onto a GC system, many more are not sufficiently volatile or thermally stable for their direct application to a GC system. A common way of making a chemical more volatile and thermally stable is to alter its structure through derivatization.[2,24] This usually involves replacing one or more polar groups on the analyte with less polar groups. This change tends to make the chemical more volatile by reducing dipole-related interactions or hydrogen bonding and also often makes the chemical more thermally stable. Various types of reactions can be used for this purpose in GC. A common example is the replacement of an active hydrogen on an alcohol, phenol, amine, or carboxylic acid group with a trimethylsilyl (TMS) group, producing a TMS derivative. Other examples include the use of alkylation (eg, the formation of a methyl ester through the esterification of a carboxylic acid) or acylation (eg, the production of an acetate derivative from an alcohol or amine).[24] Along with increasing the volatility and thermal stability of a compound, some of these derivatization reactions also can be used to change the response of the analyte to certain detectors, such as an ECD through the addition of halogen atoms to a compound's structure.

Columns and Supports

Both packed columns and capillary columns are used in GC.[2,7,21,25] Packed GC columns are filled with support particles that are based on either uncoated supports, as used in GSC, or that have liquid coatings or bonded stationary phases, as used in GLC and bonded phase GC. These packed columns vary from 1 to 4 mm in inner diameter and have typical lengths of 1 to 2 m, with the outside of the column being fabricated from tubes of glass or stainless steel. Packed GC columns are useful when it is necessary to apply a relatively large amount of a sample onto the GC system. However, packed columns also tend to have lower efficiencies than capillary columns. This last factor results in packed columns being mainly used for separations in which a relatively small number of compounds are to be separated.

Capillary columns, which are also known as open-tubular columns, consist of a column that has the stationary phase attached to or coated on its interior surface. Capillary columns have typical inner diameters of 0.1 to 0.75 mm and lengths that often range from 10 to 150 m. The capillary columns with narrow bores are more efficient, and the wider bore columns have greater sample capacities. Capillary GC columns are usually made from fused silica capillaries that have a polyimide or aluminum coating on the outside to give the capillary sufficient strength and flexibility for use in a GC system. Although capillary columns have lower sample capacities than packed columns, they also provide better peak resolution and higher efficiencies. These properties make capillary columns the most common type of support used in GC for analytical applications.

There are several types of capillary columns. Three common types are (1) porous-layer open tubular (PLOT) columns, (2) support-coated open tubular (SCOT) columns, and (3) wall-coated open tubular (WCOT) columns.[2,7,21,25] In PLOT columns, a porous layer is placed on the inner wall of the capillary columns. This porous layer is made by either chemical means (eg, etching) or by depositing a layer of porous particles on the wall from a suspension. The porous layer serves as a support and/or stationary phase for use in GSC. PLOT columns are primarily used for analysis of gases and separation of low-mass hydrocarbons.

SCOT columns have an inner wall with a thin layer of a support onto which a stationary phase is coated or attached. This type of column is used with liquid stationary phases or bonded phases. WCOT columns consist of a capillary tube whose inner wall is coated directly with a liquid stationary phase or a bonded phase. WCOT columns tend to be more efficient than SCOT columns but also have a smaller sample capacity.

In addition to traditional packed columns and capillary columns, research has been carried out in the development of GC columns on microchips.[26] These devices have great potential for use in high-speed GC and miniaturized GC systems.[27]

Temperature Control

All types of GC columns and systems require careful control of temperature. The accurate, precise control of the column and injector temperatures is required to obtain optimal performance and reliable results in a GC system. Control of the column temperature is achieved by using a column oven, in which the column is heated directly by resistive heating.[21,28] The temperatures of the injection system and detectors are also usually controlled by resistance heating. Temperature control of the column is especially important, particularly in applications in which the retention times or volumes of eluting peaks are compared with those of standards for compound identification. For instance, a change of only 1°C in column temperature can lead to a 5% change in retention time.

Depending on the type of GC separation being carried out and the complexity of the sample, the column may be maintained at a constant temperature during the separation (ie, a method known as isothermal elution) or the temperature may be varied as a function of time (ie, a technique known as temperature programming).[7] Temperature programming is used for most clinical applications. In temperature programming, the sample components with the lowest boiling points and weakest interactions with the GC column will elute first, followed by chemicals that have higher boiling points and/or stronger interactions with the column. As a result, it is possible with temperature programming to separate a complex mixture of chemicals with a wide range of boiling points and volatilities. Temperature programming also usually provides sharper and more distinct peaks in less time than can be obtained with isothermal elution. The main advantage of isothermal elution is that it can be faster for simple samples that do not contain a wide range of chemicals with different volatilities. Also, it is essential with temperature programming to use computer control to provide a reproducible and well-defined temperature gradient during the analysis.

The thermal stability of the stationary phase is important to consider during the development of a GC method. Because each stationary phase has a specific temperature range over which it is stable, it is necessary to keep the column temperature within this usable range. For nonpolar stationary phases in silica capillary columns, the upper temperature limit is often determined by the stability of the polyimide coating on the capillary. The introduction of aluminum clad columns has broadened this usable temperature range. Oxidation reactions that may occur at high temperatures tend to limit the operating temperature of stationary phases that have an intermediate polarity or that have a higher polarity.

Before any GC column is used for routine analysis, it must be "thermally conditioned" by heating the column at various temperatures and for different lengths of time. This process helps to remove volatile contaminants, including residual monomers from a polymeric stationary phase that may be initially present in the column. Furthermore, the thermal conditioning of used columns can remove nonvolatile contaminants that have accumulated on this column and that can lead to unstable baselines.

To thermally condition a column, the column should be disconnected from the detector and purged for at least 5 minutes with pure carrier gas. The column should then be heated to a temperature that is above 50°C. The column temperature is then passed through a normal temperature program for three or four cycles. Alternatively, the column can be maintained at the maximum operating temperature for 12 to 24 hours. Thermal conditioning at lower temperatures can prolong the life of the column, but longer conditioning times are required under these conditions to achieve good baseline stability. Preconditioned capillary columns are also available to minimize such problems.

Gas Chromatography Detectors

A variety of detectors can be used in GC systems (Table 1.2). These include universal detectors that can detect a broad range of analytes and more selective devices that may detect only specific groups of analytes. Examples that will be examined in this section include the (1) FID, (2) nitrogen-phosphorus detector, (3) ECD, (4) photoionization detector (PID), (5) TCD, and (6) mass spectrometric detectors.[12,21,22] Many other types of detectors have been used in GC, and it has become common to place two or more detectors in series to enhance the specificity and sensitivity of GC systems.

Flame Ionization Detector. An FID is a common detector used for GC in clinical laboratories.[12,21,22] This type of detector is often used during GC analysis of ethanol and other volatiles in blood or other aqueous samples. Typical chromatograms are shown in Fig. 1.10 of volatile compounds that have been examined by using headspace analysis and a GC system equipped with an FID. During the operation of an FID, the carrier gas that is leaving the column is mixed with hydrogen, and the eluting compounds are burned by a flame that is surrounded by air and an oxygen-rich environment. Approximately one organic molecule in 10,000 results in the production of a gas-phase ion. These ions are detected by a collector electrode that is positioned above the flame. The magnitude of the current that is generated by these ions is related to the mass of carbon that was delivered to the detector. This signal can then be used for both the detection and quantification of organic compounds that are eluting from the column.

The advantages of an FID include its simplicity, reliability, versatility, and ease of operation. Another advantage of using an FID is that this detector gives little or no signal for common carrier gases (eg, He, Ar, or N_2) or typical contaminants in such gases (eg, O_2 and H_2O). An FID is easy to use with temperature programming and is a good general detector for the routine clinical analysis of organic compounds. One disadvantage of the FID is its destructive nature, so it cannot be connected directly to other GC detectors. However, an FID still can be used in combination with another detector if part of the carrier gas stream is split between the FID and the other detector.

Nitrogen-Phosphorus Detector. The nitrogen-phosphorus detector (NPD) is also known as a thermionic selective detector (TSD). This detector is similar to an FID but instead of a flame uses an electrically heated alkali bead, which is generally made of rubidium. This heated bead is placed directly above where the mixture of the carrier gas and hydrogen enter the detector.[12,21,22] Ions are generated at or above the surface of the heated alkali bead, which

TABLE 1.2 Examples of Detectors Used in Gas Chromatography

Type of Detector	Principle of Operation	Selectivity	Approximate Limit of Detection
Flame ionization detector (FID)	Production of gas phase ions from combustion of organic compounds	General: Organic compounds	10^{-12} g carbon
Nitrogen-phosphorus detector (NPD; thermionic detector, TSD)	Heated alkali bead selectively ionizes nitrogen- or phosphorus-containing compounds	Nitrogen- or phosphorus-containing compounds	10^{-14}–10^{-13} g nitrogen or phosphorus
Electron capture detector (ECD)	Capture of electrons by chemicals with electronegative groups	Chemicals with electronegative groups	10^{-15}–10^{-13} g
Mass spectrometry (MS)	Production of gas phase ions, followed by separation/analysis of these ions based on their mass-to-charge ratios	Universal: Full-scan mode Selective: Selected ion monitoring mode (SIM)	10^{-10}–10^{-9} g Full-scan mode 10^{-12}–10^{-11} g SIM mode
Thermal conductivity detector (TCD)	Measurement of change in thermal conductivity of carrier gas as compounds elute from the column	Universal	10^{-9} g
Photoionization detector (PID)	Measurement of gas phase ions that are produced due to chemical ionization with light	General: Organic compounds	10^{-12}–10^{-11} g
Flame photometric detector (FPD)	Phosphorus- and sulfur-containing compounds emit light when burned in a flame; emitted light is detected	Phosphorus and sulfur-containing compounds	10^{-12} g phosphorus 10^{-11} g sulfur
Infrared (IR) spectroscopy	Absorption of IR light	IR-absorbing compounds	10^{-9} g

Portions of this table are based on data from Hage, DS, Carr, JD. *Analytical chemistry and quantitative analysis.* New York: Pearson; 2011, and references cited therein.

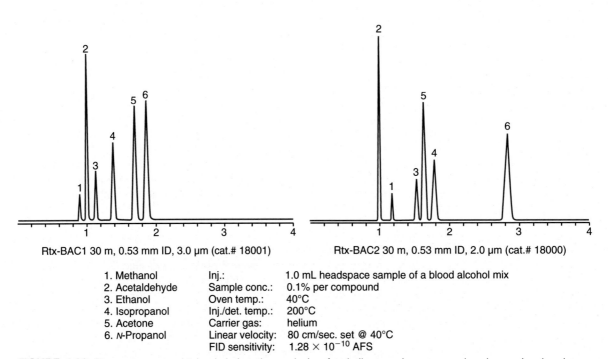

Rtx-BAC1 30 m, 0.53 mm ID, 3.0 μm (cat.# 18001)   Rtx-BAC2 30 m, 0.53 mm ID, 2.0 μm (cat.# 18000)

1. Methanol
2. Acetaldehyde
3. Ethanol
4. Isopropanol
5. Acetone
6. N-Propanol

Inj.: 1.0 mL headspace sample of a blood alcohol mix
Sample conc.: 0.1% per compound
Oven temp.: 40°C
Inj./det. temp.: 200°C
Carrier gas: helium
Linear velocity: 80 cm/sec. set @ 40°C
FID sensitivity: 1.28×10^{-10} AFS

FIGURE 1.10 Chromatograms obtained during the analysis of volatile organic compounds when using headspace analysis and gas chromatography. (Courtesy Restek Corporation, Bellefonte, Pennsylvania.)

supplies electrons to electronegative compounds that surround the bead and leads to the formation of negatively charged ions. These ions are then collected at an electrode and generate a current that is used to detect and quantify the eluting compounds.

Nitrogen- or phosphorus-containing compounds are especially good at creating ions in an NPD. This feature makes the NPD particularly useful for monitoring low concentrations of analytes that have nitrogen or phosphorus in their structures. The NPD is frequently used in GC for detection of organic bases and acids. This type of detector does not respond to common GC carrier gases or their impurities, and several types of carrier gases can be used with this detector. However, it is necessary to have the alkali bead in this detector changed on a regular basis because this material will slowly degrade over time.

Electron Capture Detector. The ECD is another example of a selective GC detector. The operation of an ECD is based on the capture of secondary electrons by electronegative compounds that are eluting from the column.[12,21,22] High-energy electrons, or beta particles, are provided in an ECD by a radioactive source such as ^{63}Ni or ^{3}H that is housed in the detector. As the beta particles are produced, they collide with the carrier gas and lead to the release of a large number of secondary electrons. When only the carrier gas is passing through this detector, a consistent supply of the secondary electrons is created. These secondary electrons are collected at a positive electrode and measured. When a chemical with electronegative groups elutes from the column, some of these secondary electrons are captured and fewer reach the electrode. The resulting change in current is used to detect and measure the amount of analyte that was eluting from the column.

An ECD can provide both selective and sensitive detection for chemicals that contain electronegative groups. This includes chemicals that contain halogen atoms (I, Br, Cl, and F) or nitro groups ($-NO_2$). It also includes chemicals that are polynuclear aromatic hydrocarbons, anhydrides, or conjugated carbonyl compounds, along with many among others. Derivatization with reagents containing polychlorinated or polyfluorinated groups can be used with some chemicals to also allow them to be monitored with an ECD.

Argon and nitrogen are usually employed as the carrier gases for a GC system with an ECD because their relatively large size makes it easy for them to collide with beta particles and produce secondary electrons. Some methane is also usually combined with these carrier gases to produce a steady stream of secondary electrons and provide a stable detector response. It is important that these gases be pure and dry because the presence of oxygen and water can foul this type of detector. Because an ECD uses a radioactive source, this source needs to be replaced on a regular basis by a certified technician.

Mass Spectrometry. Mass spectrometers are also used as detectors for GC. This combination is known as gas chromatography/mass spectrometry (GC/MS).[12] GC/MS is a powerful method for identifying analytes and quantifying them as they elute from a GC column. Ionization methods that are often used in GC/MS include electron impact ionization and chemical ionization, which are both discussed in Chapter 2. Some mass analyzers that are commonly used in GC/MS are quadrupole mass analyzers and ion traps (see Chapter 2), although other types of mass analyzers can be used as well.

In the full-scan mode of GC/MS, information is acquired by the MS system on a wide range of ions. This mode is used when the goal is to detect many compounds in a single run or provide data that can be used to identity an unknown compound from its mass spectrum. In this mode, the mass spectrometer acts as a general detector for the GC system. GC/MS also can monitor specific analytes by using selected ion monitoring (SIM). This mode uses the mass spectrometer to monitor only a few ions that are representative of the analytes of interest. SIM is used when selective detection and low detection limits are desired, but does require that information be available in advance on the types of ions that will be generated from the analytes.

The response of a GC/MS system can be represented in several ways. For instance, in the full-scan mode a plot can be made of the number of ions measured at each elution time. This type of graph is known as a mass chromatogram, or total ion chromatogram, and can be used to show the overall response of the system to the eluting analytes. It is also possible in the full-scan mode to use all of the collected data to show the mass spectrum that is acquired at a given elution time. This plot can be used to help identify a compound that is eluting at that time from the GC/MS system. Finally, a plot can be made of the number of ions that are detected at only specific mass-to-charge ratios as a function of the elution time. This last plot is called a selected ion chromatogram and is used in the SIM mode or the full-scan mode when looking for specific compounds that may be eluting from the GC/MS system.

Other Gas Chromatography Detectors. A variety of other detectors can be part of a GC system. One example is the TCD. A TCD is a general detector that can monitor both inorganic and organic compounds. It detects and measures these compounds based on their ability to change how the carrier gas/analyte mixture will conduct heat away from a hot wire filament (ie, a property referred to as thermal conductivity).[12,21,22] The carrier gas used with a TCD is often helium or hydrogen, which have the greatest differences in their thermal conductivities from most organic or inorganic compounds. However, nitrogen or other carrier gases also can be used with a TCD.

The primary advantage of a TCD is its ability to detect many types of chemicals, as long as they are present at sufficient quantities to be detected. This detector is nondestructive and can be easily combined with a second detector. However, a TCD also can respond to contaminants in the carrier gas and can give a change in the background response during temperature programming. In addition, the TCD tends to have much higher detection limits than other common GC detectors.

Another group of GC detectors make use of the interactions of chemicals with light. One example is the PID.[12,21] The PID is similar in design and operation to an FID in that both use electrodes to detect and measure ions produced from chemicals eluting from the column. However, in the PID these ions are produced through interaction of the eluting chemicals with ultraviolet (UV) radiation rather than through the use of a flame. Another GC detector that makes use of light is a flame photometric detector (FPD).[12] An FPD is a selective detector for phosphorus- or sulfur-containing compounds. Like an FID, an FPD passes the eluting chemicals into a flame, but with the FPD measuring the release of light from excited-state phosphorus- or sulfur-containing species rather than the production of ions. It is also possible to combine GC with infrared (IR) spectroscopy, giving a method known as gas chromatography/infrared spectroscopy (GC/IR).[12] This combination can be used for both chemical measurement and identification by looking at the absorption of IR radiation by chemicals as they elute from a GC column.

Data Acquisition and System Control

As with most modern analytical instruments, computers are used to both control and automate GC systems, as well as collect and process data from these systems. With regard to system control, the computer can regulate parameters such as (1) the carrier gas composition and flow rate; (2) the column back pressure; (3) the column and detector temperatures, including temperature programming; (4) the sample injection process; (5) detector selection and operation; and (6) the timing steps that are used during system operation and a chemical analysis.

In terms of data acquisition and processing, the computer can monitor the signals generated by the GC system's detectors, including the acquisition and storage of data at specified time intervals. From this information, the area or height of each chromatographic peak can be measured, and this information can be used to determine the analyte concentration represented by each peak. Algorithms are available for this process in modern GC systems that allow for the generation of calibration curves or conversion factors based on either internal or external calibration methods. The computer system also can be used to search databases to aid in identification of analytes based on their retention times or response at the detector (eg, their mass spectra).[21,29] If desired, the data acquisition system can then be used to generate a report on the results for each chromatographic run. Alternatively, these data can be stored for later examination or reprocessing.

Laboratory Safety in Gas Chromatography

Standard safety precautions should be followed when placing and securing the gas cylinder or mobile phase source that is used for GC. If hydrogen is used as the carrier gas, extra precautions should be taken in training laboratory personnel in the handling and use of this flammable and potentially explosive gas. Proper ventilation facilities should be available in the work area to deal with carrier gas that has passed through the GC system. All samples, reagents, and solutions that are used during a GC analysis should also be handled and stored using appropriate laboratory procedures.

LIQUID CHROMATOGRAPHY

LC is a type of chromatography in which the mobile phase is a liquid. Separations in LC are based on the distribution of chemicals between a liquid mobile phase and a stationary phase.[9] This was the type of chromatography used by Tswett when he first began to practice column chromatography in 1903. Although LC was mainly a preparative tool until the 1960s, it is now the dominant type of chromatography used for chemical analysis in clinical or biomedical laboratories. A key advantage of this method over GC is the ability of LC to work directly with liquid samples, such as those encountered with clinical or biological specimens.

Before the mid-1960s, the supports used in LC columns were based on large and irregularly shaped particulate supports such as those used in packed columns for GC. These supports were useful in preparative work but were not suitable for many analytical applications because they tended to result in broad peaks and separations with low resolution. In addition, these supports generally had limited mechanical stability and could be used only at relatively low operating pressures.

Developments began to occur in the 1960s to produce smaller, more mechanically stable and more efficient supports for LC, along with the instrumentation that could be used with such materials.[30] This resulted in a method that is now known as HPLC.[10,30-32] The use of these more efficient supports made it possible to obtain narrower peaks, better separations, and lower limits of detection in LC. These reasons, along with the ability of HPLC to be used as an automated method, have made this technique the method of choice for most routine chemical separations and analysis methods in modern laboratories, including those in a clinical setting. Other advantages of HPLC and LC include the wide range of separation mechanisms, stationary phases, solvents, and detectors that can be employed in such methods.

In HPLC, particulate supports with relatively small diameters are often used to hold the stationary phase within the column. Because the pressure drop across a packed bed column is related to the square of a support particle's diameter, relatively high pressures can be required to pump liquids through HPLC columns with even moderate lengths. As a result, this technique has also been referred to as high-pressure liquid chromatography, although HPLC is the preferred name. Most modern HPLC systems can work at pressures up to 5000 to 6000 psi. Specialized systems that can operate at even higher pressures have recently been developed for a method known as ultra-high-performance liquid chromatography or ultra-high-pressure liquid chromatography."[33-36]

One important difference between LC and GC is that the retention of chemicals in LC can depend on the interactions of these chemicals with both the mobile phase and stationary phase. This means the composition and nature of the mobile phase are important to consider when adjusting the retention of a chemical in an LC system. The term strong mobile phase is used to describe a mobile phase that leads to weak retention for an analyte on a given type of stationary phase. The weakest retention for a chemical will occur when this substance favors staying in the mobile phase instead of the stationary phase. The term weak mobile phase refers to the opposite situation in which a chemical favors the stationary phase versus the mobile phase and has its highest retention within a given column.[2]

POINTS TO REMEMBER

Mobile Phase Strength in Liquid Chromatography
- The mobile phase is important in LC in determining chemical retention.
- A strong mobile phase is a mobile phase that leads to weak retention for an analyte on a given type of stationary phase and column.
- A weak mobile phase is a mobile phase in which a chemical favors the stationary phase and has its highest retention within a given column.

Types of Liquid Chromatography

Liquid chromatographic methods are classified according to the chemical or physical mechanisms by which they separate chemicals (Fig. 1.11). The five main types of LC based on the separation mechanism are (1) adsorption chromatography, (2) partition chromatography, (3) IEC, (4) size-exclusion chromatography (SEC), and (5) affinity chromatography.[2,7] Most clinical applications use LC separations that are based on partition chromatography or IEC; however, the other types of LC also have valuable clinical applications.

Adsorption Chromatography

Adsorption chromatography is a type of LC in which chemicals are retained based on their adsorption and desorption at the surface of the support, which also acts as the stationary phase (see Fig. 1.11). This method is also sometimes referred to as liquid-solid chromatography.[10] Retention in this method is based on the competition of the analyte with molecules of the mobile phase as both bind to the surface of the support. The degree of a chemical's retention in adsorption chromatography will depend on (1) the binding strength of this chemical to the support, (2) the surface area of the support, (3) the amount of mobile phase displaced from the support by the chemical, and (4) the binding strength of the mobile phase to the support.[12] Electrostatic interactions, hydrogen bonding, dipole-dipole interactions, and dispersive interactions (ie, van der Waals forces) all may affect retention in this type of chromatography.[37,38]

The binding strength of the mobile phase with the support in adsorption chromatography is described by the mobile phase's elutropic strength.[10,12,39] A liquid or solution that has a large elutropic strength for a given support will act as a strong mobile phase for that material because this mobile phase will tend to bind tightly to the support and cause the analyte to elute more quickly as it spends more time in the mobile phase. As an example, a relatively polar solvent such as methanol will have a higher elutropic strength for a polar support such as silica than a nonpolar solvent such as carbon tetrachloride. In the same manner, a liquid or solution that has a low elutropic strength for a support would represent a weak mobile phase for that support in adsorption chromatography (eg, carbon tetrachloride on silica).

Three types of adsorbents are generally used in adsorption chromatography: (1) polar acidic supports, (2) polar basic supports, and (3) nonpolar supports. The most common polar and acidic support used in adsorption chromatography is silica. The surface silanol groups on this support tend to adsorb polar compounds and work particularly well for basic substances. Alumina is the main type of polar and basic adsorbent that is used in adsorption chromatography. Like silica, alumina retains polar compounds, but alumina works especially well for polar acidic substances. Florisil is an alternative polar and basic support that can be used in place of alumina, such as when catalytic decomposition of an analyte is observed with this latter material. Other types of supports that can be used in adsorption chromatography are nonpolar adsorbents such as charcoal and polystyrene.

Partition Chromatography

The second major type of LC based on the separation mechanism is partition chromatography. Partition chromatography is an LC method in which solutes are separated based on their partitioning between a liquid mobile phase and a stationary phase that is coated or bonded onto a solid support (see Fig. 1.11).[40-42] The support in most types of partition chromatography is silica, although other types of supports also can be employed. This method originally involved coatings of liquid stationary phases that were immiscible with the desired mobile phase. However, most current columns used in partition chromatography employ stationary phases that

Adsorption chromatography
Separation based on adsorption of chemicals to the surface of a support

Partition chromatography
Separation based on partitioning of chemicals into a layer of the stationary phase

Ion-exchange chromatography
Separation of ions based on their binding to fixed charges on a support

Size-exclusion chromatography
Separation of chemicals based on their size and ability to enter a porous support

Affinity chromatography
Separation of chemicals based on their interactions with a biologically related binding agent

FIGURE 1.11 Main types of liquid chromatography based on their separation mechanisms.

are bonded to the support. These bonded phases are more stable than the coated layers of stationary phases that were initially used in partition chromatography and provide better column efficiencies.

The two main types of partition chromatography based on the polarity of the stationary phase are normal-phase chromatography and reversed-phase chromatography.[7,10,12,39] Normal-phase chromatography is a type of partition chromatography in which a polar stationary phase is used.[7,10,43] This is the first type of partition chromatography that was developed, and it is also known as normal-phase liquid chromatography. The stationary phase in this method typically contains groups that can form hydrogen bonds or undergo dipole-related interactions. Examples of bonded stationary phases for normal-phase chromatography are those that contain aminopropyl groups, cyano groups, and diol groups. Because this method has a polar stationary phase, it will have its highest retention for polar compounds. A weak mobile phase in this method will be a nonpolar liquid. A strong mobile phase in normal-phase chromatography is a polar liquid, such as methanol or water.

Normal-phase chromatography can be used in many of the same applications as separations in adsorption chromatography that use silica or alumina supports. These applications usually involve the separation or analysis of chemicals that are present in organic solvents and of substances that contain one or more polar functional groups. Examples of chemicals that are of clinical interest and for which normal-phase chromatography has been used include steroids and sugars.[12,31,32,44]

The second major type of partition chromatography is reversed-phase chromatography, which is also known as reversed-phase liquid chromatography. Reversed-phase chromatography is a type of partition chromatography that uses a nonpolar stationary phase.[7,10] It is the most popular type of liquid chromatography and the most common type found in clinical laboratories.[12,31,32,45] One reason for this is that the weak mobile phase in reversed-phase chromatography is a polar solvent, such as water. This property makes this type of LC convenient for the analysis and separation of chemicals in aqueous-based systems, such as serum, urine, and blood.[12,31,32,39] A strong mobile phase in this method is a liquid that is less polar than water, such as acetonitrile or methanol. Because of the presence of a nonpolar stationary phase, nonpolar compounds will have the highest retention in reversed-phase chromatography.

Reversed-phase chromatography has many applications in the areas of clinical chemistry and biomedical research. Examples of chemicals that have been separated or analyzed by this method include drugs, drug metabolites, amino acids, peptides, proteins, carbohydrates, lipids, and bile acids. A separation of antidepressant drugs by reversed-phase chromatography was shown earlier in Fig. 1.2. Compounds representing the greatest challenge for reversed-phase separations are highly polar compounds such as sugars or amino acids, which tend to be weakly retained by reversed-phase columns, and basic compounds, which may exhibit peak tailing as the result of their interactions with silica. Derivatization of some compounds (eg, amino acids) has been employed to improve their retention on reversed-phase columns.[3] It is also important to consider both the type of analyte and support that are being used in these separations. For instance, large chemicals such as peptides or proteins will require reversed-phase supports with larger pore sizes than those routinely used for the separation of small molecules.[46-48]

A relatively wide range of stationary phases and supports are available for reversed-phase separations.[6,49-51] The most common stationary phases used in reversed-phase chromatography are those based on octadecyl (C_{18}), octyl (C_8), phenyl, or butyl (C_4) groups that are attached to a support such as silica. Similar materials are commonly used in solid-phase extraction (see Chapter 3). The retention characteristics of these silica-based columns will depend on (1) the nature of bonded phase, (2) the amount of the bonded phase (often expressed as the percent of carbon load), (3) the surface area of the support, (4) the pore size of the support, and (5) the quantity of accessible groups on the support (eg, silanol groups on silica) that can be used to prepare the bonded phase. Alternative reversed-phase materials such as porous graphite, fluorinated hydrocarbons, and hydrophobic stationary phases with embedded polar groups offer different selectivities than C_{18}- or C_8-silica. Silica tends to dissolve slowly at a pH greater than 8.0 or at a pH below 2.0, so separations that make use of silica supports are usually done in this pH range unless the silica has been stabilized by surface treatment.[49] Some of the other supports that are available for reversed-phase chromatography, such as polystyrene or porous graphite, are stable over a broader pH range (eg, pH 2.0 to 13.0).

During the bonding of a reversed-phase stationary phase to silica it usually is not possible to cover all of the available silanol groups. These remaining silanol groups may interact with some analytes and lead to mixed-mode interactions that produce broad peaks and result in a decrease in peak resolution. For instance, the peak tailing that may occur for some basic compounds on silica is caused by coulombic interactions of these compounds with the conjugate base form of silanol groups. These interactions can be minimized by reacting many of the silanol groups with a small organosilane such as trimethylchlorosilane in a method known as endcapping.[10] In addition, the pH of the mobile phase can be lowered to decrease the amount of silanol groups that are present in their charged form. Additives such as trifluoroacetic acid or triethylamine can be added to the mobile phase to minimize interactions of the silanol groups with analytes.[12,39,44]

The strength of a mobile phase in both normal-phase chromatography and reversed-phase chromatography can be described by using the solvent polarity index. A weak mobile phase for normal-phase chromatography will be a solvent or solvent mixture that has a low value for the solvent polarity index, whereas a strong mobile phase in this method would be one with a high solvent polarity index. The opposite trend occurs in reversed-phase chromatography, in which a weak mobile phase will have a high solvent polarity index, and a strong mobile phase will have a low solvent polarity index. Some large aliphatic stationary phases in reversed-phase chromatography, such as C_{18}-silica, may undergo phase collapse if they are used in only an aqueous mobile phase;

this process probably represents the folding of the aliphatic groups down onto the surface to decrease their exposure to water. This effect can be minimized by including a small amount of organic modifier in the mobile phase or by using a bonded phase with a shorter chain length.

Samples are usually applied or injected onto a reversed-phase column in the presence of an aqueous solution or water that contains a low concentration of an organic solvent such as methanol or acetonitrile. The partitioning of chemicals that are weak acids or weak bases can be adjusted in reversed-phase chromatography by changing the pH to minimize the charge of these solutes. Because most acid-base reactions are fast, this situation will usually result in only one observed peak with a retention time that is the weighted average of what would be seen for the acid or base forms of the compound. The same type of effect and shifts in retention can occur for chemicals that undergo other rapid reactions, such as complex formation with mobile phase additives.[12] The mobile phase strength in normal-phase chromatography and reversed-phase chromatography is often changed by using mixtures of solvents or solutions and gradients in which the proportions of solvents or solutions are changed during an analysis (see Fig. 1.12).[52] It is also possible to modify the polarity of aqueous solutions in reversed-phase chromatography by changing the salt concentration of the mobile phase.

This last approach is employed in a variation of reversed-phase chromatography that is known as hydrophobic interaction chromatography (HIC).[12] This method is applied mainly in the separation of large biomolecules such as proteins. HIC makes use of a weakly hydrophobic stationary phase that is made up of small nonpolar groups such as phenyl or butyl residues. The weak mobile phase that is used with this type of column, and which promotes the binding of proteins or related biomolecules to the stationary phase, is an aqueous solution that contains a high salt concentration. The retained biomolecules are eluted by using a strong mobile phase that is less polar, which in this case is an aqueous solution that has a lower salt concentration.

Ion-Exchange Chromatography

The third type of liquid chromatography is IEC. IEC is a type of liquid chromatography in which ions are separated by their adsorption onto a support that contains fixed charges on its surface.[7,10] This method relies on the interaction (or exchange) of ions in the sample or mobile phase with fixed ionic groups of the opposite charge that are bound to the support and act as the stationary phase (see Fig. 1.11). Depending on the charge of the groups that make up the stationary phase, the types of ions that bind to the column may be either cations (ie, positively charged ions) or anions (ie, negatively charged ions). These two methods are referred to as cation-exchange chromatography and anion-exchange chromatography, respectively.[12]

Supports for cation-exchange chromatography contain negatively charged functional groups. These groups may be the conjugate bases of strong acids, such as sulfonate ions that are formed by the deprotonation of sulfonic acid, or the conjugate bases of weak acids, such as those produced from carboxyl or carboxymethyl groups. The supports used in anion-exchange chromatography are usually the conjugate acids of strongly basic quaternary amines, such as triethylaminoethyl groups, or the conjugate acids of weak bases, such as aminoethyl or diethylaminoethyl groups. Supports that can be modified to contain these charged groups for use in IEC include silica and polystyrene, as well as carbohydrate-based materials such as agarose, dextran, or cellulose.[10,12,44] The carbohydrate-based supports are particularly useful in preparative work with biological agents, which can have strong binding to materials such as underivatized silica or polystyrene. The large pore size of supports such as agarose also makes these materials valuable in separations involving biological macromolecules such as proteins and nucleic acids.[10,12,44]

A strong mobile phase in IEC is usually a mobile phase that contains a high concentration of competing ions. The presence of these competing ions will make it more difficult for a charged analyte to bind to the fixed charges that act as the stationary phase. A weak mobile phase in IEC is one that contains few or no competing ions or that otherwise promotes binding by charged analytes to the column. Changing the competing ion concentration is the most common approach for adjusting the retention of analyte ions in IEC. The retention of ions in this method also may be affected by (1) pH, (2) the type of competing ion used, (3) the type of fixed charges used as the stationary phase, and (4) the density of these fixed charges on the support. Many stationary phases in IEC can exhibit mixed-mode retention through a combination of coulombic interactions and adsorption. As an example, ion-exchange resins that are used for amino acid analysis are able to separate amino acids with virtually the same charge because of differences in the adsorption of these amino acids onto the stationary phase.

IEC has a number of clinical applications. Common examples are the use of this method in the separation and analysis of amino acids and hemoglobin variants. IEC is

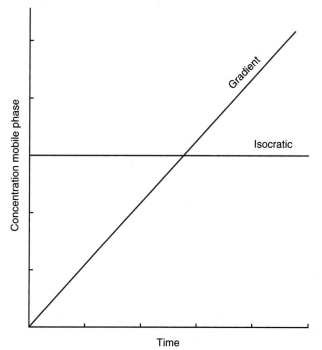

FIGURE 1.12 Examples of isocratic elution (ie, constant mobile phase composition) or gradient elution based on solvent programming (ie, varying mobile phase composition).

also frequently used as a preparative tool in biomedical research for purifying proteins, peptides, and nucleotides. A modified form of IEC, known as ion chromatography, can be used with a conductivity detector to analyze small inorganic and organic ions.[12] The water purification systems that are used in many laboratories are another important application of IEC. In these purification systems, supports containing a mixture of cation- and anion-exchange groups are used to remove anions and cations from water, in which hydrogen ions are exchanged for other cations and hydroxide ions are exchanged for other anions. Most of these hydrogen ions and hydroxide ions then combine to form deionized water.[2]

Size-Exclusion Chromatography

SEC is an LC technique that separates molecules or other particles based on size (Figs. 1.11 and 1.13).[2,12,44] In this method, a porous support is used that has an inert surface and few or no interactions with the injected sample components. This support also should have a range of pore sizes that approach, or are similar to, the sizes of the compounds that are to be separated. As a sample travels through a column that contains this support, small components of the sample can enter all or most of the pores and larger components may enter only a few or none of the pores. The result is a separation based on size or molar mass, in which the larger components elute first from the column.

In SEC, all of the injected components will elute in a fairly narrow volume range. This range extends from the volume of mobile phase that is outside of all the pores of the support (also known as the excluded volume, V_E) to the total amount of mobile phase in the column, as represented by the void volume (V_M). The stationary phase in SEC can be thought of as the volume of the mobile phase in the pores of the support that can be entered by a given solute. The extent of retention in this method can be described by using the measured t_R or retention volume V_R for an injected component or by using the ratio K_o, which is calculated by using the following equation[44]:

$$K_o = \frac{(V_R - V_E)}{(V_M - V_E)}$$

The value of K_o represents the fraction of the volume between V_M and V_E in which a given sample component elutes. Small components will have a value for K_o that is equal to or approaches 1, whereas large components will have a K_o value that is equal to or approaches 0. Components with intermediate sizes will have values for K_o between 0 and 1.

Many types of porous supports have been used in SEC. Cross-linked carbohydrate-based supports such as dextran and agarose are often used in this method for work with aqueous-based samples and biological compounds such as proteins or nucleic acids. Polyacrylamide gel and silica or glass beads that have been modified into a diol-bonded form also can be used for aqueous samples and biological compounds. Polystyrene is usually employed as the support when SEC is to be used with synthetic polymers and samples that are present in organic solvents.[12,31,44] For each of these supports, the range of pore sizes that are present will determine the range of sizes for the injected compounds that can be separated. In the case of carbohydrate-based supports

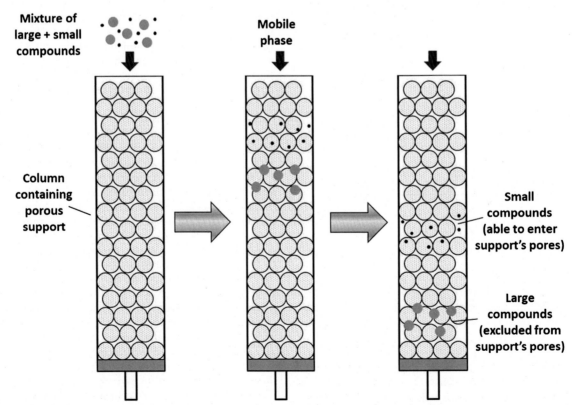

FIGURE 1.13. General principle of size-exclusion chromatography. This method separates compounds based on their size and by using a column that contains a porous support.

and polymeric supports, this range will become smaller as the degree of cross-linking of the support is increased.

The mobile phase in SEC can be a polar solvent, such as water, or a nonpolar solvent, which is usually tetrahydrofuran. Because the stationary phase is based on a physical difference in the accessible pore volume for the solutes, rather than on chemical interactions, there is no weak mobile phase or strong mobile phase in this method. The choice of the mobile phase is instead determined by the solubility of the desired analytes and the stability of the support and the column. If water or an aqueous mobile phase is used in SEC (which typically involves a carbohydrate-based support, polyacrylamide or diol-bonded silica or glass), the resulting method is often called gel filtration chromatography. If an organic mobile phase is used (which generally involves the use of polystyrene as the support), the size-exclusion method is referred to as gel permeation chromatography.[10] Other names that are sometimes used for SEC are steric exclusion, molecular exclusion, and molecular sieve chromatography.

SEC in the gel filtration mode allows the separation of molecules under physiologic salt conditions. This feature is useful for identifying intact complexes of agents such as lipoproteins, antibody-antigen complexes, and the binding of proteins with their target compounds. SEC is often used as a rapid preparative technique to exchange buffers or to remove salts from large sample components. In addition, this method can be used to remove small molecules from large biomolecules, such as in the isolation of drugs, fatty acids, and peptides from proteins.

SEC also can be used to estimate the molecular weight of a biomolecule, such as a protein or nucleic acid, or to characterize the distribution of molecular weights for a polymer. This is done by first calibrating the size-exclusion column with compounds that are similar to the desired analytes in structure but that have known molecular weights.[53] A calibration curve can be made by plotting the logarithm of the molecular weight versus the measured retention time, retention volume, or calculated value of K_o for each standard compound. This plot is then used to determine the molecular weight of the unknown compounds based on their measured retention on the same column. This approach can be used to provide a good estimate of the molecular weight; however, it is a relatively low-resolution technique that does require substantial differences in molecular weight to create significant shifts in retention. For example, the diameters of globular proteins change in proportion to the cube root of their molecular weights, so roughly an eightfold difference in molecular weight is required to yield a twofold change in diameter.[53] Linear polymers, such as DNA and proteins that have been denatured and treated with sodium dodecyl sulfate or guanidine hydrochloride, have a much larger diameter for the same molecular weight, and their diameter changes approximately in proportion to the square root of molecular weight. This latter effect allows for smaller differences in molecular weight to be observed by SEC for linear molecules compared to globular molecules (eg, nondenatured globular proteins).

Affinity Chromatography and Chiral Separations

The fifth main type of LC is affinity chromatography. Affinity chromatography is an LC method that makes use of biologically related interactions for the retention and separation of chemicals (see Figs. 1.11 and 1.14).[7,10,54-56] This method uses the selective, reversible interactions found in many biological systems, such as the binding of an antibody with an antigen, the interactions of an enzyme with a substrate or inhibitor, the binding of a hormone with a receptor, and the interactions of a lectin with a carbohydrate. These selective interactions are used in affinity chromatography by immobilizing one of a pair of interacting compounds onto a support for use as the stationary phase. This immobilized binding agent is called the affinity ligand,[55] and it is used to create a column that can selectively bind and capture the complementary compound from applied samples.

The stationary phase for affinity chromatography is usually prepared by covalently immobilizing the affinity ligand to the support.[55] This is often done through the reaction of amine, sulfhydryl, carboxyl, or carbonyl groups on the affinity ligand with activated sites on the support, although other types of groups can be employed. During this process, the orientation of the affinity ligand and its accessibility to its target compound are both important to consider in providing good activity for the binding agent. Use of a spacer between the surface of the support and the affinity ligand also may be needed with smaller binding agents. Besides covalent immobilization, it is also sometimes possible to adsorb the affinity ligand to the support. This can be done in a general manner, such as through polar or ionic interactions; it also can be done through biospecific adsorption to a secondary binding agent. Examples of this latter approach include the biospecific adsorption of antibodies to an immobilized immunoglobulin-binding protein such as protein A or protein G, and the biospecific adsorption of biotin-labeled agents to immobilized avidin or strepavidin.[55] Another alternative approach for preparing affinity ligands is to form a molecularly imprinted polymer in the presence of the desired target or a target analog.[55,57] In this case, the shape and structure of the binding pockets that remain in the polymer after release

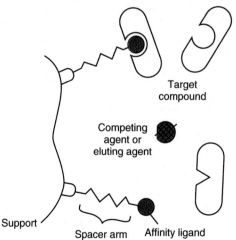

FIGURE 1.14 General mechanism of separation in affinity chromatography. The target analyte is first allowed to bind to the immobilized affinity ligand in the column. After the nonretained sample components have been washed from the column, the retained analyte is later released by using an elution buffer. This elution may be based on a nonspecific method (eg, the addition of a chaotropic agent or a change in pH), or it may be accomplished by adding a biospecific competing agent to the mobile phase to displace the analyte from the column.

of the target template can be used to selectively bind to the same or similar targets from samples.

Affinity chromatography is usually carried out by applying a sample to the column under conditions that allow strong and specific binding of the affinity ligand to its target compound.[55] This is done in the presence of a mobile phase solution, known as the application buffer, which mimics the pH and natural or preferred conditions for binding between the affinity ligand and the target. As the target binds to the column under these conditions, most other sample components are washed away as a result of the selective nature of this interaction. The retained target is then later released from the column by using an elution buffer that either contains a competing agent that will displace the target from the affinity ligand (a method known as biospecific elution) or uses a change in conditions such as pH, ionic strength, or polarity of the mobile phase to decrease the strength of binding between the target and affinity ligand (a technique called nonspecific elution). After the target has been eluted for detection or further use, the application buffer can be passed again through the column and the system is allowed to regenerate before application of the next sample. For systems with weak to moderate binding strengths, it is also possible to use isocratic elution for both sample application and target elution. This second type of elution is usually employed in chiral separations and in a method known as weak affinity chromatography.[55,58]

Various supports can be used in affinity chromatography. For preparative work, it is most common to use carbohydrate-based materials such as cross-linked agarose and various modified forms of cellulose. Silica and glass also have been used as supports for both preparative and analytical applications of affinity chromatography. This first requires that these supports be modified to give them low nonspecific binding for biological molecules and to provide groups that can be used for the immobilization of affinity ligands. In addition, several polymeric materials have been used in this method, ranging from hydroxylated polystyrene to azlactone beads, agarose-acrylamide or dextran-acrylamide copolymers, and derivatives of polyacrylamide or polymethacrylate.[55,59]

The power of affinity chromatography lies in its selectivity and in the wide range of binding agents that can be used in this method. This has led to many applications for this type of LC in work with biological compounds. Bioaffinity chromatography (or biospecific adsorption) is the most common type of affinity chromatography and involves the use of a biological binding agent as the affinity ligand.[55,59] The purification of an enzyme by using an immobilized inhibitor, coenzyme, substrate, or cofactor is an example of this approach.[55,60,61] Another example is the use of affinity ligands that are lectins or nonimmune system proteins that can bind certain types of carbohydrate residues.[55] Lectins such as concanavalin A (Con A, which binds to α-D-mannose and α-D-glucose residues) and wheat germ agglutinin (which binds to D-N-acetylglucosamines) have been popular in recent years for the isolation of carbohydrate-containing compounds, such as glycopeptides, glycoproteins, and glycolipids. This method is sometimes referred to as lectin affinity chromatography.[55] Another set of binding agents that are used in bioaffinity chromatography are immunoglobulin-binding proteins, such as protein A (from *Staphylococcus aureus*) and protein G (from group G streptococci), which have been used for antibody purification and as secondary ligands for the biospecific adsorption of antibodies.[55]

Immunoaffinity chromatography (IAC) is a subset of bioaffinity chromatography that uses an antibody or antibody-related agent as the affinity ligand. The selectivity of this method has made it popular for the isolation of targets that have ranged from antibodies, hormones, and recombinant proteins to receptors, viruses, and cellular components. This method also can be used to detect specific target compounds directly or indirectly in a set of techniques known as chromatographic immunoassays (or flow-injection immunoanalysis).[55,62] Another important application of IAC is as a tool for target isolation and sample pretreatment before analysis by other methods. The use of IAC to isolate a specific target from a sample is known as immunoextraction, which can be combined either off-line or online with other analytical methods.[55,63] A related technique is immunodepletion, which is used to remove certain compounds from a sample before analysis of the remaining sample components. This last approach has been used in proteomics to remove high-abundance proteins from biological samples before the measurement and detection of lower abundance proteins in the same samples.[63]

Another group of methods in affinity chromatography are those that use nonbiological binding agents. Dye-ligand affinity chromatography uses an affinity ligand that is a synthetic dye such as Cibacron Blue 3GA, Procion Red HE-3B, or Procion Yellow H-A. This method is commonly used for the large- and small-scale purification of enzymes and proteins, including many protein-based biopharmaceuticals.[55,64] Dye-ligand affinity chromatography is a type of biomimetic affinity chromatography that uses an affinity ligand that is a mimic of a natural compound. Besides synthetic dyes, other binding agents that are used in this method are those generated by combinatorial chemistry and computer modeling or that are derived from peptide libraries, phage display libraries, aptamer libraries, and ribosome display libraries.[55]

Two other types of affinity chromatography that use nonbiological binding agents are immobilized metal-ion affinity chromatography (IMAC) and boronate affinity chromatography. In IMAC, the affinity ligand is a metal ion that is complexed with an immobilized chelating agent, such as Ni^{2+} complexed to a support containing iminodiacetic acid.[55] This technique is frequently used to isolate recombinant histidine-tagged proteins and has been used for the isolation or analysis of phosphorylated proteins in proteomics.[55] Boronate affinity chromatography uses an affinity ligand that is boronic acid or a related derivative. These affinity ligands are able to form covalent bonds with compounds that contain *cis*-diol groups; this feature makes these binding agents useful in the purification and analysis of polysaccharides, glycoproteins, ribonucleic acids, and catecholamines.[55,65] An important clinical application of boronate affinity chromatography is its use in the analysis and isolation of glycosylated hemoglobin in blood samples from diabetic patients.[66,67]

Many biological molecules occur as specific stereoisomers. Examples are amino acids, peptides, and proteins. As a result, it is not unusual for the different chiral forms of a drug to have some variations in their interactions with these biological agents. This, in turn, can lead to differences in the activity and toxicity of these drugs in the body.[68-70] In some cases only a particular stereoisomer of a drug may be active. In the absence of any chiral binding agent, the two mirror-image forms of a drug (or enantiomers) will have identical physical and chemical properties. These forms will not be separated in most types of chromatography, which generally use nonstereoselective (or achiral) stationary phases. However, it may be possible to separate the enantiomers and stereoisomers of a drug or target compound if the stationary phase is also chiral and can interact with these compounds in a stereospecific manner. This type of medium is known as a chiral stationary phase (CSP).[55,68,69]

The use of a CSP can be viewed as a subset of affinity chromatography, in that the resulting separation makes use of a biologically related binding agent or a mimic of such an agent.[55,71-73] For instance, carbohydrates, peptides, and proteins (including some enzymes and serum transport proteins) have all been used as CSPs because they are composed of chiral amino acids and sugars.[55,68] Cyclodextrins, which are cyclic polymers of glucose, are an important set of carbohydrates that have been used for separating many types of chiral compounds in both LC and GC.[68,69] CSPs also can be based on synthetic binding agents or molecularly imprinted polymers.[12,55,68-70] It is further possible in LC to carry out a chiral separation with an achiral column, such as a reversed-phase support, by placing a chiral binding agent such as a cyclodextrin in the mobile phase. In this last case, the separate forms of a chiral drug or compound may have different interactions with this mobile phase additive, which then leads to differences in their observed retention on the column.[68,69]

Hydrophilic Interaction Liquid Chromatography and Mixed-Mode Methods

Besides the traditional categories of LC, there are other methods that combine several separation modes. One example is hydrophilic interaction liquid chromatography (HILIC). HILIC is a type of partition chromatography that uses a polar stationary phase and in which chemicals partition between an organic-rich region in the mobile phase and a more polar water-enriched layer that is at or near the surface of a polar support. The surface of the support, which can often undergo hydrogen bonding or dipole-related interactions with the applied solutes, also may have charged groups that can take part in ionic interactions with these compounds while they are in the water-enriched layer.[74-76] These features make HILIC a variation of normal-phase chromatography that is combined with some of the retention characteristics of reversed-phase chromatography and IEC.

Several types of supports can be used in HILIC. The conventional form of HILIC uses a polar but noncharged surface, such as is present on unmodified silica. Other neutral groups that may be present on the supports for HILIC are amide, diol, or cyano groups. One variation of this method is the technique of electrostatic repulsion hydrophilic interaction liquid chromatography (ERLIC, or eHILIC), in which charged groups such as protonated amines or deprotonated carboxylic acids are present on the support and used to repel injected compounds with the same charge. Another form of HILIC is zwitterionic hydrophilic interaction liquid chromatography (ZIC-HILIC), in which zwitterionic groups are present on the support; these groups can interact with analytes that have a positive charge or negative charge or that are also zwitterions.[71,76]

HILIC and related methods have become popular in areas such as proteomics and glycomics. Advantages of these methods include (1) their ability to give better separations for polar compounds than can be obtained by reversed-phase chromatography, (2) the greater ease with which they can be used with aqueous samples and in solubilizing polar compounds compared to normal-phase chromatography, and (3) the ability to couple these methods with mass spectrometry. A possible limitation of HILIC for use in clinical laboratories is that biological fluids are highly polar and include a substantial quantity of salts that also can interact with polar stationary phases. Therefore, these specimens may need to be extracted or modified by the addition of a less polar solvent such as acetonitrile to promote compound interactions in HILIC.

Ion-pair chromatography (IPC) is another example of a mixed-mode LC method.[12] This technique combines columns that are used in reversed-phase chromatography with the ability to separate ionic compounds based on their charges, as is done in IEC. This method is carried out by adding an ion-pairing agent to the mobile phase for a reversed-phase column. The ion-pairing agent is usually a surfactant that has a charged group at one end and a nonpolar tail or group at the other end. Examples of ion-pairing agents are sodium dodecyl sulfate and perchlorate, for binding to positively charged ions, and t-butyl ammonium, for binding to negatively charged ions.

The purpose of the ion-pairing agent is to combine with ions of the opposite charge in the sample. This may involve the sample ions interacting with the charged end of the ion-pairing agent while the nonpolar tail of the same agent partitions into the nonpolar stationary phase of the reversed-phase column. Alternatively, the sample ion and ion-pairing agent may interact in the mobile phase and form a neutral complex that then interacts with the nonpolar stationary phase. The result in either case is the retention of charged analytes based on their ability to interact with the ion-pairing agent.

IPC is useful in the separation of charged compounds that are poorly resolved by IEC. This method not only combines the better efficiencies that are normally produced by reversed-phase columns, but it also has several parameters that can be varied to control and adjust its separations. These parameters include (1) the strength and type of solvent used in the mobile phase for the reversed-phase column, (2) the concentration and type of ion-pairing agent placed into this mobile phase, and (3) the ion content and pH of the mobile phase. Applications in which IPC has been employed include the separation and analysis of catecholamines, drugs, and nucleic acids.[12]

Restricted access media are a set of supports that combine SEC with another type of LC, which is usually reversed-phase chromatography.[12] This type of material is prepared in a manner so that the exterior of the support is inert or protected by a hydrophilic network with low nonspecific binding for proteins and biological compounds. The interior contains a stationary phase such as a nonpolar bonded phase. If the sizes of the pores or the size-exclusion properties are chosen properly, small solutes such as drugs can pass into the interior and be retained by the stationary phase that is located there. Larger compounds, such as proteins, will not be able to access this inner region and pass nonretained through the column. Columns that contain a restricted access support can be used for the direct injection of biological samples that may contain high concentrations of proteins (ie, which will elute nonretained) and are being used for measurement of small analytes. This approach can greatly simplify the process of sample preparation for such analytes.

Liquid Chromatography Instrumentation

The major components of an LC system used in HPLC are shown in Fig. 1.15.[21,77] First, there is a source for the mobile phase, or a solvent reservoir, which supplies a solvent or solution that goes into a pump for delivery to the rest of the LC system. This is followed by an injection valve or injection system, which allows samples to be placed into the mobile phase stream. Next, the mobile phase and sample enter and pass through the column, which contains the support and stationary phase. A control system also may be present to maintain a constant or well-defined temperature within the column. The column is followed by a detector to observe and measure the components of the sample as they exit the system. Modern systems also have a computer or control system to operate the liquid chromatograph and to gather data from the LC detector.

Mobile Phase Reservoirs and Delivery Systems

Solvents and solutions that are used as mobile phases in LC are contained in solvent reservoirs. In their simplest form, these reservoirs are glass bottles or flasks into which feed lines to the pump are inserted. Filters are often placed at the inlets of the feed lines to prevent any particles in the mobile phase from moving on to the rest of the LC system. Most mobile phase reservoirs also have a means of "sparging" the mobile phase by bubbling through a gas such as helium or nitrogen to remove dissolved air or oxygen that may interfere with the response of some detectors. The removal of air and oxygen, or degassing, also can be achieved by applying a vacuum to the reservoir by placing gas exchange devices or gas filters in the flow path leading from the mobile phase reservoir.[12]

The composition and strength of the mobile phase are factors that can be used to adjust and control a separation in LC. If the same mobile phase is used throughout the separation, this approach is known as isocratic elution.[7,10] If the composition of the mobile phase is varied over time, the method is called solvent programming.[7,12] Solvent programming begins with a weak mobile phase to allow chemicals with weak retention to have their strongest possible interactions with the column. A change is then made over time to a stronger mobile phase to also elute chemicals with moderate or high retention. This change in mobile phase composition can be made in one or more steps and may involve the use of a linear change or a nonlinear change over time.

A variety of techniques have been used to vary the composition of the mobile phase over time.[12,21] For instance, this might be done by using valves that alternate which solvents or solutions are being passed into the LC system at a given time. Solvent gradients may be generated by using the same type of valve linked to two or more solvent reservoirs and that passes these mobile phases into a mixing chamber and onto the inlet of a single pump. This method is known as

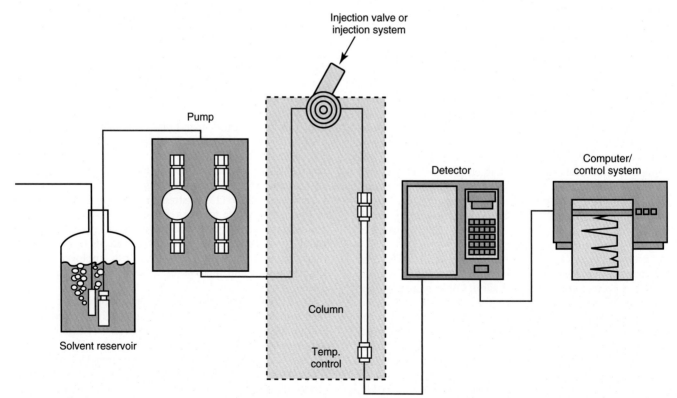

FIGURE 1.15 General design of a liquid chromatograph, as used in high-performance liquid chromatography. (Modified from a figure courtesy Restek Corporation, Bellefonte, Pennsylvania.)

low-pressure mixing. A second approach, known as high-pressure mixing, uses two or more pumps that are each linked to a different solvent or solution; the flow rates of the mobile phases that are being passed through these pumps are then varied to control the mixing ratio of these solvents or solutions. This combined solution is then passed through a mixing chamber and onto the column. These solvents and solutions can be mixed by using static mixers, which rely on flow-generated turbulence, or dynamic mixers, which use magnetic stirrers. Solvent miscibility and viscosity are two factors to consider when choosing which solvents or solutions are to be used in a solvent program. Both of these factors can affect the mixing characteristics of the two liquids, where inadequate mixing may result in poor chromatographic performance and inadequate separations.[21]

Several types of pumps have been used in LC.[12,21] Peristaltic and diaphragm-type pumps can be used with columns that can be operated at low pressures, as are encountered in classic and low- to medium-performance LC; however, these pumps are not usually suitable for HPLC. Reciprocating pumps and syringe pumps are instead used to achieve the higher pressures needed to deliver the mobile phase through HPLC columns. Reciprocating pumps are commonly used in HPLC for work at flow rates in the milliliter-per-minute range. In these pumps, a piston moves in and out of the solvent chamber, with check valves being used to keep the flow of the mobile phase moving from the pump inlet to the outlet. The reciprocating action of the piston in this type of pump does generate some pulsation in the pressure and mobile phase flow, which can increase the baseline noise seen with many LC detectors. These pulsations can be minimized by electronic control of the pump and by placing pulse dampers in the flow path. Syringe pumps make use of the continuous application of a syringe to the solvent chamber to deliver the mobile phase to the rest of the system. These pumps can deliver essentially pulse-free flow and can be used at much lower flow rates than reciprocal pumps (eg, flow rates in the microliter-to-minute range). However, syringe pumps are not as convenient to use as reciprocating pumps when carrying out solvent programming or during the application of even modest volumes of the mobile phase to a column.[12,21]

Until recently, the upper pressure limit of most HPLC applications has been approximately 6000 psi (41 MPa or 414 bar). In recent years, commercial instrumentation for LC has been developed that can operate up to 15,000 psi (103 MPa or 1034 bar).[33,36,78-80] These higher pressures are needed for work with small-diameter supports, which offer the potential for more efficient separations but also produce higher column back pressures. The use of these smaller support particles and these higher pressures has resulted in a method that is often called ultra-high performance liquid chromatography (UPLC or UHPLC).[33-36] Work at these higher pressures not only requires special pumps that are designed to operate under these conditions, but also requires tubing, connections, and columns that can be used under the same conditions.

Systems for HPLC have pressure sensors to detect any obstruction to flow. These sensors can shut down the entire system once a defined pressure limit has been reached, which is done to prevent damage to the components of the LC system. At very high pressures, some solvents become slightly compressible and a compensation for this solvent compression needs to be made to achieve constant flow rates.[33]

Another extreme condition that may be encountered during the operation of an LC system is when work is to be carried out at quite low flow rates, as might be needed for small-bore microfluidic columns or capillary columns. Work at flow rates below 10 μL/min may require specially designed pumping systems or flow splitting of the output from a standard HPLC pump. The use of low rates in the nanoliter-per-minute range is sometimes called nanoflow chromatography and has been combined with mass spectrometry through the use of nanospray interfaces, which can provide high ionization efficiencies.

Injection Systems and Sample Derivatization

Various approaches can be used to introduce a sample into an LC system.[12,21] The most widely used approach in HPLC is a fixed-loop injector that is switched into or out of the flow path by manual control or through the use of an autoinjector. When this valve is in the inject mode, the sample loop is switched into the flow path and the sample is carried downstream and into the column. The loop continues to be part of this flow path until it is switched back into the load or fill position.

Some important characteristics to consider when selecting an injection system are its (1) reproducibility, (2) the amount of sample carryover from one injection to the next, and (3) the range of volumes that can be injected. Some automated injection systems have the capability of injecting multiple aliquots of the same sample or of mixing a sample and a reagent for derivatization before injection. Some of these systems also are able to control the temperature of the samples before their injection. For instance, the refrigeration of samples before injection may be important during the analysis of specimens or analytes that have limited stability or when large batches of samples are to be analyzed.

Derivatization is sometimes used in LC to improve the response of a given compound or group of compounds to a particular detector (eg, an absorbance, fluorescence, or electrochemical detector, as will be discussed later). It is also possible to use derivatization in LC to alter the separation of a compound from other chemicals by changing the structure and retention of this compound on the column. The two main ways of carrying out derivatization in LC are (1) precolumn derivatization and (2) postcolumn derivatization. Precolumn derivatization is done before the sample is injected and can be used to alter a compound's retention or to increase its response to a particular type of detector. Postcolumn derivatization is carried out online as compounds elute from a column and is used only to improve the response of one or more of these compounds on the LC detector.[44,81]

Columns and Supports

A wide selection of columns are available for LC. These columns can have various combinations of packing materials and diameters or lengths. Columns for LC, and especially those used in HPLC, often include an inlet filter to remove particulate matter. In the use of LC and HPLC for chemical analysis, a short guard column that contains the same packing material also may be placed before a longer analytical column to protect and extend the usable life of the more expensive analytical column.[12]

The size used for a column in LC will depend on the desired application for this column. The column size used for off-line sample pretreatment or the low-performance isolation of compounds is often determined by the sample capacity that is needed for the separation. Examples of these columns include those used for applications such as desalting, purification of compounds based on IEC, and many types of affinity-based separations for sample pretreatment. Size-exclusion columns, such as small centrifugation columns that are used for desalting, can accommodate specimens with sizes up to about 10% of the column volume. The size of ion-exchange or affinity columns that are used for sample pretreatment and compound isolation will depend on the amount of compound that needs to be separated and the binding capacity of the packing material. This principle also applies to the use of other types of LC for sample pretreatment or compound isolation.

Modern column technology for HPLC has produced columns having various dimensions, with a trend toward smaller internal volumes.[12] These small volume columns are useful in combining LC with other methods, such as mass spectrometry, to produce hyphenated techniques (see Chapter 2). In the clinical laboratory, most conventional packed HPLC columns consist of tubes that are made of 316 stainless steel; however, polymers that are suitable for work at high liquid pressures also can be employed. These columns have typical internal diameters that range from 4 to 5 mm and lengths ranging from 5 to 30 cm (Table 1.3). Column end fittings, which ideally have a zero dead volume and frits to hold the support particles in the column, are used to connect the column to the injector and to a detector or other postcolumn devices.

In general, better efficiencies and lower detection limits are achieved with HPLC columns that have longer lengths and smaller inner diameters. These smaller inner diameter columns include narrow-bore columns, with approximate inner diameters of 2 to 3 mm, and microbore columns, with approximate inner diameters of 1 to 2 mm. In addition to providing improved efficiencies, these columns with small inner diameters also can require lower flow rates and smaller volumes of the mobile phase for their operation than conventional packed columns.

Capillary columns are sometimes used in LC. For instance, packed capillary columns can be used that have inner diameters of 0.1 to 0.5 mm and lengths of 20 to 200 cm. Open tubular capillary columns for LC also can be constructed by placing a thin film or coating of the stationary phase onto the inner wall of a fused silica tube. These open tubular columns have typical inner diameters of 0.01 to 0.075 mm and lengths of 1 to 100 cm. Both types of capillary columns are used with flow rates in the mid-to-low microliter-per-minute range.

Many types of particles and support formats have been developed for LC.[30,82] The most common type of support in LC is a packed bed of small particles.[82] The supports in modern HPLC columns may have particle diameters for porous supports that are in the range of 1.8 to 10 μm, with a typical value of 5 μm. The lower end of this diameter range is representative of the supports that are used in the UPLC.[36] A smaller diameter for these supports provides better efficiency for the chromatographic system, but it also leads to an increase in back pressure across the column. As mentioned previously, the back pressure generated by a packed bed that contains such a support will vary inversely with the square of the particle diameter. Thus a twofold reduction in the particle size will result in approximately a fourfold increase in back pressure. Low-to-medium performance separations, which have much lower operating pressures than HPLC, typically use packing materials such as cross-linked dextran or agarose that have support particles with diameters of 50 to 200 μm.

In the porous support particles that are usually employed in LC, the mobile phase flows around the support but not through the particle. However, this means compounds must travel within the particle by means of diffusion, which is a relatively slow process that can be a major source of band-broadening. The distance that these compounds must diffuse can be reduced by using a nonporous support or a pellicular support, in which the latter has a thin porous layer or porous shell.[10,30] The use of these supports results in a more efficient separation and less band-broadening because of diffusion-based processes. Another approach for minimizing this band-broadening is to use perfusion particles. This type of support has small pores that contain most of the stationary phase and larger pores that allow the mobile phase to pass both through and around the support particles. The presence of these large flow-through pores decreases the distance compounds must diffuse to reach the stationary phase and helps decrease band-broadening.[10,30]

Another alternative support that can be used to improve efficiency is a monolithic support.[83-85] This type of support consists of a continuous porous bed that is prepared from an inorganic or organic polymer. Monoliths may be made from silica or various polymers. Monolith columns have bimodal pore structures with large pores (approximately a few microns in diameter) that allow the mobile phase to flow through the support and smaller pores (with typical pore sizes of 10 to 20 nm) that provide a large internal surface

TABLE 1.3 Typical Column Sizes Used in Analytical High-Performance Liquid Chromatography

Type of Column	Typical Inner Diameter (ID) and Lengths	Typical Flow Rate Range
Conventional packed column	4–5 mm ID × 5–30 cm	1–3 mL/min
Narrow-bore column	2–3 mm ID × 5–15 cm	0.2–0.6 mL/min
Microbore column	1–2 mm ID × 10–100 cm	0.05–0.2 mL/min
Packed capillary	0.1–0.5 mm ID × 20–200 cm	0.1–20 μL/min
Open tubular column	0.01–0.075 mm ID × 1–100 cm	0.05–2 μL/min

Portions of the data in this table are based on Poole, CF, Poole, SK. *Chromatography today*. New York: Elsevier, 1991.

area to contain the stationary phase. These supports can provide efficient and fast separations, while also providing lower back pressures than particle-based columns at high flow rates. The low back pressure of a monolithic column makes it possible to use this type of column with a flow gradient (eg, increasing the flow rate at the end of a separation) and allow several such columns to be coupled in series to improve the efficiency and resolution of an LC separation. These columns also can have reasonably high sample capacities. Commercial monolithic rods are encased in inert polytetrafluoroethylene tubing and housed in stainless steel tubes. The inert tubing eliminates voids that may occur at the interface between the stainless steel and the monolith, thus improving the resolution of the column. Capillary monolithic columns also are available. One area of clinical interest in which monolithic columns have been used is in reversed-phase separations of peptides and proteins.[46-48,86]

Temperature Control

The control of column temperature can be an important factor in determining the reproducibility and efficiency of an LC separation.[21] Unlike in GC, in which temperature gradients are often employed, in LC a constant column temperature is usually maintained. Temperature control of an LC column can be achieved by a variety of techniques. These techniques include the use of temperature-controlled (1) column chambers, (2) water jackets, (3) blankets, and (4) heating/cooling blocks. In addition, operation at high flow rates might require a heater/exchanger, which is usually a coil of tubing with good heat exchange properties that is placed before the column inlet.

During the operation of an LC separation, a stable column temperature is required to generate reproducible retention times. In addition, an increase in the column temperature will (1) lower the mobile phase viscosity, (2) increase the rates of mass transfer between mobile phase and stationary phase, and (3) allow the use of higher flow rates, which in turn will lead to a shorter analysis time. The degree to which the temperature can be increased is determined by the boiling point and vapor pressure of the mobile phase, as well as the thermal stability of the analytes in the injected samples. In some instances, the stability of the samples and analytes may require separations to be carried out at reduced temperatures. One common example of this occurs in the use of LC for the isolation and preparation of proteins, which is often performed in cold rooms or in refrigerated cabinets to decrease the rates of protein denaturation and proteolytic degradation. Some systems for temperature control have the ability to operate below room temperature through the action of Peltier coolers or other types of refrigeration. Features to consider in selecting a system for temperature control in LC include (1) the usable temperature range of the system, (2) the constancy of the temperatures it can provide, and (3) the number and sizes of columns that the system can accommodate.

Liquid Chromatography Detectors

Many types of detectors can be used in LC (Table 1.4).[12,21] Some common LC detectors are (1) absorbance detectors, (2) fluorescence detectors, (3) electrochemical detectors, (4) refractive index detectors, and (5) mass spectrometric detectors. A key component for most of these detectors is the flow cell through which the mobile phase and eluting compounds from the column must pass. As these components travel through the flow cell or into the detector, a signal is generated that can be used to monitor the eluting chemicals and

TABLE 1.4 Examples of Detectors Used in Liquid Chromatography

Type of Detector	Principle of Operation	Range of Application	Detection Limit
Absorbance detector	Measures absorbance of light at a given wavelength or set of wavelengths	Compounds with chromophores that can absorb ultraviolet or visible light	10^{-10}–10^{-9} g
Fluorescence detector	Measures ability of chemicals to absorb and reemit light through fluorescence	Compounds with fluorophores	10^{-12}–10^{-9} g
Electrochemical detector	Measures current or charge as a result of chemical oxidation or reduction	Electrochemically active compounds	10^{-11}–10^{-9} g
Conductivity detector	Measures change in conductivity of the mobile phase as ions elute from the column	General for ionic solutes	10^{-9} g
Refractive index detector	Measures change in refractive index of the mobile phase as compounds elute the column	Universal	10^{-7}–10^{-6} g
Mass spectrometry	Production of gas phase ions, followed by separation/analysis of these ions based on their mass-to-charge ratios	Universal: Full-scan mode Selective: Selected ion monitoring mode	10^{-10}–10^{-9} g (full-scan mode) $\leq 10^{-12}$ g (SIM mode)
Evaporative light scattering detector	Light scattering by chemicals after solvent evaporation	Nonvolatile compounds	10^{-9} g
Charged aerosol detector	Measurement of ions produced from chemicals by using a corona discharge	Nonvolatile compounds	$<10^{-9}$ g

Portions of the data in this table are based on Poole, CF, Poole, SK. *Chromatography today*. New York: Elsevier, 1991.

measure the amount of these chemicals that are present. Many LC detectors are nondestructive and can be used individually or linked together in series. In addition, a postcolumn reactor may be present between the column and detector to derivatize some of the eluting compounds and generate products that have a stronger and more specific signal on the detector.

Absorbance Detectors. The absorption of UV or visible light is often used to detect compounds as they elute from a liquid chromatographic column.[12,21] Many of the absorbance detectors (also referred to as photometers or spectrophotometers) used in LC can measure the absorption of UV light with wavelengths in the range 190 to 400 nm or of visible light with wavelengths in the range of 400 to 700 nm. Many organic compounds with aromatic groups or double or triple bonds absorb UV light between 250 and 300 nm. Many other organic compounds can absorb in the range of 190 to 220 nm, at which amide bonds, carboxylic acids, and many other groups can have substantial absorption of light. In addition, some ions, inorganic compounds and metal complexes can be detected by their absorption of light in the UV or visible range.

There are several types of absorbance detectors that can be used in LC.[2,12,21] Fixed-wavelength absorbance detectors have the simplest design and are used to monitor absorbance at a particular wavelength or wavelength band. For instance, detection is often done with a UV absorbance detector at 254 nm, which is a wavelength absorbed by many unsaturated organic compounds and corresponds to an intense emission line that is produced by a mercury arc lamp. A fixed-wavelength absorbance detector can be extremely sensitive and is capable of operating with detection at 0.005 absorbance unit full scale. Fixed-wavelength absorbance detectors that have greater flexibility in their design can be obtained by using other, less intense emission lines of a mercury arc lamp. In addition, a phosphor can be placed between the light source and the flow cell, with the light that is emitted by this agent then being passed through the flow cell. This approach is used in dual-wavelength detectors that operate at two fixed wavelengths (eg, 254 and 280 nm). The intense emission lines at 214 or 229 nm that are produced by a zinc or cadmium arc lamp, respectively, may be used for detection at lower wavelengths, where many organic compounds have strong absorption of light.

A second type of detector in this category is a variable-wavelength absorbance detector.[2,12,21] This detector operates at a wavelength that is selected from a given wavelength range. The ability to have a detector that operates at the absorption maximum for a given chemical or set of chemicals can greatly enhance the applicability and selectivity of such a device. Another advantage of this detector is its ability to operate at low-UV wavelengths (eg, 190 nm), at which a number of clinically important compounds absorb light (eg, cholesterol). However, at these lower wavelengths many solvents and mobile phases also absorb light. Important exceptions are water, acetonitrile, and methanol, which are frequently used in reversed-phase chromatography.

A photodiode array detector also can be used in LC.[2,12,21] This is an absorbance detector that uses an array of small detector cells to measure the change in absorbance at many wavelengths simultaneously. This array makes it possible to record an entire spectrum for a compound because it elutes from a column, which can be valuable in identifying overlapping peaks.[12,32,44] This type of detector can yield spectral data over a wide wavelength range (eg, 190 to 600 nm) in approximately 10 ms. During operation, the photodiode array detector passes polychromatic light through the flow cell. The transmitted light is then dispersed by a diffraction grating and directed to a photodiode array, at which the intensity of transmitted light is measured at multiple wavelengths across the spectrum. Such detectors have been helpful in the identification of drugs in samples such as urine and serum.[87]

During the use of an absorbance detector, it is necessary to use solvents, ion-pairing agents, and buffers that have little or no absorption of light at the wavelengths of interest; this is needed to maintain a low background signal. Water, acetonitrile, methanol, isopropanol, and hexane are solvents that allow UV detection down to wavelengths of 200 nm. Phosphate buffers also can be used under these detection conditions. Many other solvents and buffers have substantial absorbance in the UV, which may limit their use over this wavelength range.

There are a number of other factors to consider in the use of absorbance detectors. For instance, flow cells with small volumes should be used in absorbance detectors for HPLC to avoid the introduction of significant extracolumn band-broadening. Another issue with the operation of these detectors is the outgassing and bubble formation that can occur as the mobile phase exits the high-pressure region within the column and enters the lower pressure region in the flow cell. Because these detectors can be quite sensitive, these bubbles can lead to noise in their response and degrade their signal-to-noise ratio. Effective degassing of the mobile phase and the use of some back pressure across the detector can help minimize this bubble formation. However, care must also be taken in this last approach to avoid exceeding the usable pressure range of the detector.

Fluorescence Detectors. Fluorescence occurs when a chemical absorbs light at one wavelength and reemits light at a different, longer wavelength.[12,21] Fluorescence detectors with flow cells are used in LC to detect fluorescent compounds as they elute from the column. These detectors are generally much more selective and have better limits of detection than absorbance detectors for chemicals that are naturally fluorescent or can be converted into a fluorescent derivative. Both precolumn and postcolumn derivatization have been used to modify chemicals for use with this type of detector.[81] For example, amino acids and other primary amines are often labeled with a dansyl or fluorescamine tag, followed by their HPLC separation and detection through fluorescence. Some fluorescence detectors for LC use fixed wavelengths for both the excitation and emission wavelengths that are employed for monitoring compounds. However, variable-wavelength fluorescence detectors are also available. Deuterium lamps, xenon arc lamps, and lasers have all been used as light sources in such detectors.

Electrochemical and Conductivity Detectors. Various types of electrochemical detectors can be used in LC. This combination is sometimes known as liquid chromatography/electrochemical detection (LC-EC).[2,12] In an amperometric electrochemical detector, an electroactive chemical enters the flow cell, where it may be oxidized or reduced at an electrode that is held at a constant potential; the current needed for or generated by this process is then detected.[88] The use of multiple electrodes and cyclic changes in the applied voltage can allow the detection of multiple components at different potentials and provides for regular cleaning of the electrode. Electroactive compounds that are of clinical interest and that can be readily examined by HPLC with electrochemical detection include urinary catecholamines, ascorbic acid, and thiol-containing compounds such as homocysteine. In addition, electrochemically active tags (eg, bromine) can be added to compounds such as unsaturated fatty acids or prostaglandins for use with this type of detector.

Coulometric detectors are also used in LC. This type of detector measures the amount of charge that is required for a given electrochemical reaction. When placed in series, such detectors can be used to detect and measure coeluting compounds that differ in their half-wave potentials (ie, the potential at half of the maximum signal) by 60 mV or more. These detectors are selective, sensitive, and have reasonably wide linear ranges. Coulometric detectors are used in clinical laboratories during the analysis of metanephrines, vanillylmandelic acid, homovanillic acid, and 5-hydroxyindole acetic acid in human urine.

A conductivity detector in LC measures the ability of the mobile phase and its contents to conduct a current when they are placed in an electrical field.[2,12] This type of detector is often used in combination with IEC. For instance, conductivity detectors with relatively low sensitivities have been used to monitor salt gradients during IEC. Conductivity detectors are also used to monitor the elution of charged analytes in ion chromatography.[2,89] The signal resulting from the conductivity of a specific ion will be related to its concentration, charge, and mobility. This means such a detector is best suited for work with small inorganic and organic ions, which have high mobilities. Conductivity detectors have been used to measure compounds such as sulfate in biological fluids.

Refractive Index Detectors. A refractive index detector in LC measures the change in the refraction of light as chemicals pass with the mobile phase through a flow cell.[12,21] An important advantage for this type of detector is that it can monitor substances such as alcohols, polyethylene glycol, salts, and sugars that do not give a usable response on absorbance or fluorescence detectors.[90] An RI detector also can be valuable in work in which the nature or spectroscopic properties of an analyte have not yet been determined. One disadvantage of this type of detector is it does not have limits of detection as low as absorbance or fluorescence detectors, and it has a response that can be sensitive to changes in the mobile phase composition and temperature.

Mass Spectrometry. LC can be combined with mass spectrometry, giving a combined technique known as LC-MS.[91-94] This is a sensitive and specific technique that has seen increasing applications in clinical and research laboratories and in fields such as proteomics, metabolomics, and small molecule analysis (see Chapters 2, 4 and 6).[60,95-99] This method is similar to GC-MS in that the combined use of LC with mass spectrometry makes it possible to both measure chemicals and identify them based on the masses of their molecular ions or fragment ions. When used in the full-scan mode to look at all or most ions, the mass spectrometer in LC-MS acts as a general detector. If the mass spectrometer is instead used for looking at particular ions, this device then acts as a selective detector.

Several types of ionization methods and mass analyzers can be employed in LC-MS (see Chapter 2). A common combination is the use of electrospray ionization with a quadrupole mass analyzer.[2] Other possible ionization methods that can be used in LC-MS are chemical ionization or photoionization (see Chapter 2). For many applications, the specificity of tandem mass spectrometers allows short HPLC separations to be used because most compounds do not need to be completely separated for them to be detected by the mass spectrometer.

A critical element in linking HPLC to a mass spectrometer is their interface. For example, the interface between an LC and a mass spectrometer has the challenging task of removing solvent from the mobile phase and placing the remaining sample components in a charged form and in the gas phase that can be analyzed by the mass spectrometer. This process requires that the buffers used in LC-MS be sufficiently volatile to avoid overloading and contaminating the interface. For the same reason, a switching valve is often used to divert salts and other nonretained components that elute early in the LC separation to a waste container. The same switching valve can then direct later eluting components to the mass spectrometer for analysis (see Chapter 2 for an extensive discussion of mass spectrometry).

Other Liquid Chromatography Detectors. Several detectors have been developed for LC to detect nonvolatile compounds.[90,100] An example is an evaporative light-scattering detector (ELSD). In an ELSD, the solvent is evaporated by nebulizing it with a stream of gas as the mobile phase and its contents exit the column. Nonvolatile chemicals that were present in the mobile phase will remain as particles in the gas phase, and these particles can be detected by measuring their ability to scatter light. The degree of this light scattering will be proportional to the mass of the nonvolatile substances. Potential applications for an ELSD include its use in the analysis of lipids, sugars, and other compounds that are difficult to monitor by absorbance or fluorescence detectors.

Another type of evaporative detector is a charged aerosol detector.[100] This detector ionizes chemicals by using a corona discharge and measures the ion current that is produced. This type of detector has a good response for many compounds. A disadvantage of both the ELSD and the charged aerosol detector is that they are destructive. This means sample components cannot be collected for further analysis after passing through these detectors and that these detectors cannot be followed directly online by another detector.

A number of other detectors have been used in LC, although many of these have been used primarily for research applications. Dynamic light-scattering detectors measure the scattering of light by chemicals that are eluting from a column, which provides a signal that is related to the size of these chemicals. This type of detector has been useful in characterizing the size of large molecules and complexes.[101] Nuclear magnetic resonance (NMR) spectroscopy has been combined with LC[12] for applications such as lipoprotein analysis, metabolomics, and the characterization of drug metabolites.

Data Acquisition and System Control

When using simple LC systems it is possible to perform injections and make pump adjustments manually, with the results being recorded by a computer or comparable data acquisition device. For automated or high-volume applications, there usually is a need to automate both the injection of samples and the chromatographic system. The system controller for an HPLC will often manage (1) sample injections, (2) solvent delivery, (3) system flow rate and temperature, (4) control the detectors, and (5) acquisition of data from these detectors. Modern control systems also usually provide an auditable record of the analyst, LC method, calibration conditions, control samples, and specimens that were analyzed.

Data acquisition systems can collect thousands of data points from an individual run. These data can then be used to identify and characterize a set of peaks based on parameters such as the retention times, areas, heights, and widths of these peaks. Comparison of these parameters with those that have been generated by reference materials and standards makes it possible to identify and measure the compounds in these peaks. The hardware and software that are used for data analysis in LC become more critical as the amount of collected data becomes large, as can often occur during the use of photodiode array detectors or LC-MS. Libraries of spectra or other databases can be searched as part of this process to aid in the identification of chemicals (eg, peptides and nucleic acid sequences) based on the chromatograms and signals that are generated during the separation.

Laboratory Safety in Liquid Chromatography

Standard laboratory procedures for the storage, handling, and disposal of chemicals and solvents should be followed when using LC and HPLC. For instance, many of the organic solvents that are used as mobile phases or solvents in LC are flammable and should be treated with appropriate precautions for such chemicals. The waste solvents, samples, and column effluent should be collected in a suitable container and stored appropriately before disposal. The release of pressure in a traditional LC system or HPLC system is not usually a major hazard, because liquids compress only slightly and therefore accumulate little energy; however, work at the higher pressures of UPLC may require some additional precautions.

OTHER CHROMATOGRAPHIC METHODS

In addition to LC and GC, and the use of columns or open tubular supports, there are a variety of other chromatographic methods that can be used for chemical separation and analysis. Some important examples are supercritical fluid chromatography (SFC) and planar chromatography. The use of multidimensional separations based on chromatography is another area of continued interest.

Supercritical Fluid Chromatography

SFC is a type of chromatography in which the mobile phase is a supercritical fluid.[12,102,103] A supercritical fluid is a state of matter that has properties between those of a gas and a liquid and that is formed when the temperature and pressure exceed a particular critical point in a chemical's phase diagram. Carbon dioxide is one chemical that can be easily converted into a supercritical fluid for use in SFC. The formation of supercritical fluid carbon dioxide occurs at or above a temperature of 31.1° C and at or above a pressure of 73.9 bar (72.9 atm). Under these conditions, carbon dioxide has a density that approaches that of a liquid, so it can interact with and solvate chemicals; this feature allows supercritical fluid carbon dioxide to be used in dissolving many hydrophobic compounds.[104,105] However, a supercritical fluid also has a lower viscosity and a higher diffusion coefficient than liquid, which allows it to provide efficiencies that are closer to those seen when using a gas as the mobile phase. As a result, SFC has performance characteristics that are between those of LC and GC.[102,103]

SFC can be used with many columns that are available for either LC or GC and can be carried out on systems that are modified versions of LC or GC instruments. It is necessary for these systems to have both pressure and temperature control to keep the mobile phase in the state of a supercritical fluid as it passes through the column. A variety of organic modifiers have been mixed with carbon dioxide to serve as the mobile phase, and solvent programming or temperature programming can be used for elution in this method. It is also common for pressure programming (or density programming) to be used in SFC, as the mobile phase strength of a supercritical fluid will change with its density.[21,102] This technique has been used for the analysis of lipids and other hydrophobic compounds. In addition, SFC has been applied to pharmaceutical research and to the analysis of natural products. However, because of its need for specialized equipment, SFC has found limited use in clinical laboratories.

Planar Chromatography

Another alternative type of chromatographic method is planar chromatography. In planar chromatography, the stationary phase is coated or placed onto a flat surface, or plane.[12] The sample is added as a small spot or band on this surface. This support is then placed into an enclosed container with the bottom edge in contact with the mobile phase and the sample band located above this point of contact (Fig. 1.16). The mobile phase is usually allowed to travel

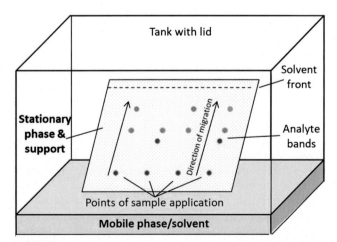

FIGURE 1.16 General operation and system components of thin-layer chromatography. In this example, the mobile phase moves up a glass plate containing a thin layer of a support by means of capillary action.

across the plane by means of capillary action. After this movement has occurred for a given period, the support is removed from the mobile phase and dried before the analysis or measurement of the separated sample components.

The planar surface that is used in this method may be a sheet of paper, giving a method known as paper chromatography, or some other type of surface, resulting in a method known as thin-layer chromatography (TLC).[12,106] In paper chromatography, the stationary phase is a layer of water or a polar solvent that is coated onto paper. In TLC, a thin layer of particles (made from a material such as silica, microparticulate cellulose, or alumina) is usually spread uniformly on a glass plate, plastic sheet, or aluminum sheet. When this layer of particles is made up of a material with a small diameter (eg, silica with diameter of around 4.5 μm), the resulting technique is known as high-performance thin-layer chromatography.[12]

In planar chromatography, retention is described as a function of the distance that compounds have traveled in a given amount of time (Fig. 1.17). This differs from column chromatography or open tubular chromatography, in which retention is instead described by using the time or volume of mobile phase that is needed for compounds to travel a given distance (eg, the length of the column). The retention of chemicals in methods such as paper chromatography and TLC can be described either in terms of the migration distances that these chemicals have traveled from their point of application in a set amount of time, or by comparing these distances to the distance that has been traveled by the mobile phase in the same amount of time. This second approach can be used to calculate a measure of retention that is known as the retardation factor (R_f), which is defined as follows,

$$R_f = D_s/D_f$$

where D_s is the distance traveled by a chemical from its point of application, and D_f is the distance traveled by the mobile phase, or solvent front, in the same amount of time. The value for R_f will always be between 0 and 1. Chemicals that have high retention with the stationary phase will have low values R_f, and chemicals that have low retention will have R_f values that approach 1.[7,12]

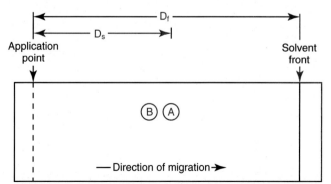

FIGURE 1.17 General example of a separation obtained by planar chromatography. In this example, compound B is more strongly retained and migrates a shorter distance than compound A. D_f, distance traveled by the mobile phase, or solvent front, from the point of sample application in the same amount of time as allowed for sample migration; D_s, distance traveled by an analyte (A, in this example) from the point of sample application.

Chemicals can be identified in planar chromatography based on the position of their bands and by comparing their retention to reference compounds that have been examined on the same plate or surface as the unknown samples. The detection characteristics of the reference compounds also can be compared with the chemicals in the unknown sample. If the R_f value for an unknown substance and the R_f value for a reference compound do not match within the allowed tolerance, the compounds can be said to be different. If the R_f values do match, then confirmation can be made by comparing the detection properties of these compounds (eg, the color of their bands or their response to a color-forming reagent). Software and databases are also available for compound identification in planar chromatography that allow for searching libraries of both absorption spectra and R_f values. Additional confirmation can be obtained by comparing the unknown compound and the reference compound under a different set of separation conditions.

The separated components in planar chromatography often can be detected by their natural color, by their response to UV light (eg, through their fluorescence), or through their visualization with chemical reagents that form colored products.[12] In some cases, these chemicals may be allowed to react with labeled antibodies for their detection or they may be detected by using radiolabels and autoradiography. Their bands also may be removed from the planar surface for analysis by a method such as mass spectrometry or NMR spectroscopy.

Paper chromatography and TLC tend to be used primarily for qualitative analysis. In addition, they can be used in multidimensional separations (see next section). One application of these techniques in clinical laboratories is their use in the analysis of amniotic fluid to determine lecithin-to-sphingomyelin ratios. Another application is their use in the screening of urine for drugs or metabolites such as amino acids that accumulate during hereditary disorders. Planar chromatography is relatively simple, inexpensive to conduct, and can be used for the simultaneous analysis of multiple samples. However, the application of these methods in clinical laboratories has been decreasing in recent years because of the lack of automation in traditional planar methods and their general lack of ability to perform precise quantitative measurements for chemicals.

Multidimensional Separations

Another area of growing interest in chromatography is in the area of multidimensional separations. These separations involve the use of two or more separation methods on a sample, in which each of these methods ideally uses a different mechanism for resolving components of the sample. A multidimensional separation can allow a large increase in peak capacity by combining chromatographic methods in which each separation step (or dimension) is performed sequentially.

Multidimensional separations can be carried out in various ways. For instance, this might be done by collecting fractions from one method and then analyzing these fractions by a second method. It is also possible in some cases to couple two chromatographic methods together. This can be done if the second method is faster than the first and if the mobile phase used for elution in the first method is compatible with the conditions needed for sample application in the second method.[14,107]

Planar chromatography is one approach that can be used for two-dimensional separations, but HPLC methods also can be employed. Peak capacities greater than 10,000 have been achieved for two-dimensional HPLC separations; however, this can require prolonged analysis times for the sample and usually means that multiple runs per sample must be conducted in the second dimension.

It is possible to link chromatographic methods with other analytical methods or detectors to create multidimensional methods. One common example is liquid chromatography-tandem mass spectrometry (LC-MS/MS), in which LC is used to separate chemicals based on their interactions with a given mobile phase and stationary phase, while a mass spectrometer is used to separate and analyze the gas phase ions that are generated at any given point in the chromatogram.[91] The addition of another dimension based on mass spectrometry, as occurs in LC-MS/MS to look at fragment ions that are produced from a given parent ion,[92,93,94,108] can further increase the ability to resolve or detect multiple components without extending the chromatographic component of the analysis time. This combined approach can enable the practical analysis of hundreds or thousands of components in a single specimen, as occurs during the analysis of samples in metabolomics[109] and proteomics.[110]

REFERENCES

1. Ettre LS. Nomenclature for chromatography. *Pure Appl Chem* 1993;**65**:819—72.
2. Hage DS, Carr JD. *Analytical chemistry and quantitative analysis*. New York: Pearson; 2011.
3. Miller JM. *Chromatography: concepts and contrasts*. 2nd ed. Malden: Mass: Wiley-InterScience; 2009.
4. Ullman MD, Burtis CA. Chromatography. In: Burtis CA, Ashwood ER, Bruns DE, editors. *Tietz textbook of clinical chemistry*. 4th ed. Philadelphia: WB Saunders; 2006. p. 141—63.
5. Ettre LSMS. Tswett and the invention of chromatography. *LC GC* 2003;**21**:458—67.
6. Cserháti T. Carbon-based sorbents in chromatography: new achievements. *Biomed Chromatogr* 2009;**23**:111—8.
7. Inczedy J, Lengyel T, Ure AM. *International Union of Pure and Applied Chemistry: compendium of analytical nomenclature: definitive rules 1997*. Malden, MA: Blackwell Science; 1997 [Chapter 9].
8. McNair HM, Miller JM. *Basic gas chromatography*. 2nd ed. Malden: Mass: Wiley-InterScience; 2009.
9. Snyder LR, Kirkland JJ, Dolan JW. *Introduction to modern liquid chromatography*. 3rd ed. New York: Wiley; 2009.
10. Majors RE, Carr PW. Glossary of liquid-phase separation terms. *LC GC* 2001;**19**:124—62.
11. van Deemter JJ, Zuiderweg FJ, Klinkenberg A. Longitudinal diffusion and resistance to mass transfer as causes of nonideality in chromatography. *Chem Eng Sci* 1956;**5**:271—89.
12. Poole CF, Poole SK. *Chromatography today*. New York: Elsevier; 1991.
13. Said AS. Comparison between different resolution equations. *J High Resolut Chromatogr* 2005;**2**:193—4.
14. Gilar M, Daly AE, Kele M, et al. Implications of column peak capacity on the separation of complex peptide mixtures in single- and two-dimensional high-performance chromatography. *J Chromatogr A* 2004;**1061**:183—92.
15. Neue UD. Theory of peak capacity in gradient elution. *J Chromatogr A* 2005;**1079**:153—61.
16. Petersson P, Heaton FA, Euerby MR. Maximizing peak capacity and separation speed in liquid chromatography. *J Sep Sci* 2008;**31**:2346—57.
17. Krull I, Swartz M. Quantitation in method validation. *LC GC* 1998;**16**:1084—90.
18. Grinstein LS, Rose RK, Rafailovich MH. *Women in chemistry and physics: a biobibliographic sourcebook*. Westport: Greenwood Press; 1993. p. 129—35.
19. Laitinen HA, Ewing GW. *A history of analytical chemistry*. New York: Maple Press; 1977 [Chapter 5].
20. James AT, Martin AJP. Separation and identification of methyl esters of saturated and unsaturated fatty acids from n-pentanoic to n-octadecanoic acids. *Analyst* 1952;**77**:915.
21. Ewing GW, editor. *Analytical instrumentation handbook*. 2nd ed. New York: Marcel Dekker; 1997.
22. Hinshaw JV. A compendium of GC terms and techniques. *LC GC* 1992;**10**:516—22.
23. Eiceman GA, Gardea-Torresdey J, Overton E, et al. Gas chromatography. *Anal Chem* 2002;**74**:22771—80.
24. Drozd J. *Chemical derivatization in gas chromatography*. Amsterdam: Elsevier; 1981.
25. Ji Z, Majors RE. Porous-layer open-tubular capillary GC columns and their applications. *LC GC* 1998;**16**:620—32.
26. Breadmore MC, Dawod M, Quirino JP. Recent advances in enhancing the sensitivity of electrophoresis and electrochromatography in capillaries and microchips (2008—2010). *Electrophoresis* 2011;**32**:127—48.
27. Eiceman GA. Instrumentation of gas chromatography in clinical chemistry. In: Meyers RA, editor. *Encyclopedia of analytical chemistry*. Chichester: John Wiley & Sons; 2000. p. 10671—9.
28. Jain V, Phillips JB. Fast temperature programming on fused-silica open-tubular capillary columns by direct resistive heating. *J Chromatogr Sci* 1995;**33**:55—9.
29. Etxebarria N, Zuloaga O, Olivares M, et al. Retention-time locked methods in gas chromatography. *J Chromatogr A* 2009;**1216**:1624—9.
30. Majors RE. A review of HPLC column packing technology. *Am Lab* 2003;**10**:46—54.
31. Katz E, Eksteen R, Schoenmakers P, et al., editors. *Handbook of HPLC*. New York: Marcel Dekker; 1998 [Chapter 10].
32. Lough WJ, Wainer IW. *High performance liquid chromatography: fundamentals principles and practice*. New York: Blackie Academic; 1995.
33. Gilpin RK, Zhou W. Ultrahigh-pressure liquid chromatography: fundamental aspects of compression and decompression heating. *J Chromatogr Sci* 2008;**46**:248—53.
34. McNair JE, Lewis KC, Jorgenson JW. Ultrahigh-pressure reversed-phase liquid chromatography in packed capillary columns. *Anal Chem* 1997;**69**:983—9.
35. Thompson JW, Mellors JS, Eschelbach JW, et al. Recent advances in ultrahigh-pressure liquid chromatography. *LC GC* 2006;**24**:16—20.
36. Wu N, Clausen AM. Fundamental and practical aspects of ultrahigh pressure liquid chromatography for fast separations. *J Sep Sci* 2007;**30**:1167—82.
37. Fornstedt T. Characterization of adsorption processes in analytical liquid-solid chromatography. *J Chromatogr A* 2010;**1217**:792—812.

38. Hurtubise RJ. Adsorption chromatography. In: Cazes J, editor. *Encyclopedia of chromatography*. 2nd ed, vol. 2. Boca Raton: Taylor & Francis; 2005. p. 21–4.
39. Karger BL, Snyder LR, Horvath C. *An introduction to separation science*. New York: Wiley; 1973.
40. Martin AJP, Synge RLM. A new form of chromatography employing two liquid phases. I. A theory of chromatography. II. Applications to the microdetermination of the higher monoamino acids in proteins. *Biochem J* 1941;**35**:1358–68.
41. Poole CF. New trends in solid-phase extraction. *Trends Anal Chem* 2003;**22**:362–73.
42. Poole CF, Poole SK. Foundations of retention in partition chromatography. *J Chromatogr A* 2009;**1216**:1530–50.
43. Cooper WT. Normal-phase liquid chromatography. In: Myers RA, editor. *Encyclopedia of analytical chemistry*. New York: John Wiley & Sons; 2000. p. 11428–42.
44. Ravindranath B. *Principles and practice of chromatography*. New York: Wiley; 1989.
45. Majors RE. Current trends in HPLC column usage. *LC GC* 1997;**15**:1008–15.
46. Causon TJ, Nordborg A, Shellie RA, et al. High temperature liquid chromatography of intact proteins using organic polymer monoliths and alternative solvent systems. *J Chromatogr A* 2010;**1217**:3519–24.
47. Jandera P, Urban J, Skeríková V, et al. Polymethacrylate monolithic and hybrid particle-monolithic columns for reversed-phase and hydrophilic interaction capillary liquid chromatography. *J Chromatogr A* 2010;**1217**:22–33.
48. Tang J, Gao M, Deng C, et al. Recent development of multi-dimensional chromatography strategies in proteome research. *J Chromatogr B Analyt Technol Biomed Life Sci* 2008;**866**:123–32.
49. Claessens HA, van Straten MA. Review on the chemical and thermal stability of stationary phases for reversed-phase liquid chromatography. *J Chromatogr A* 2004;**1060**:23–41.
50. Forgacs E. Retention characteristics and practical applications of carbon sorbents. *J Chromatogr A* 2002;**975**:229–43.
51. Lesellier E, West C. Description and comparison of chromatographic tests and chemometric methods for packed column classification. *J Chromatogr A* 2007;**1158**:329–60.
52. Dolan JW, Snyder LR. Gradient elution chromatography. In: Meyers RA, editor. *Encyclopedia of analytical chemistry*. Chichester: John Wiley & Sons; 2000. p. 11342–60.
53. Tarvers RC, Church FC. Use of high-performance size-exclusion chromatography to measure protein molecular weight and hydrodynamic radius. *Int J Peptide Protein Res* 1985;**26**:539–49.
54. Hage DS. Affinity chromatography: a review of clinical applications. *Clin Chem* 1999;**45**:593–615.
55. Hage DS, editor. *Handbook of affinity chromatography*. 2nd ed. Boca Raton: CRC Press; 2005.
56. Roque AC, Lowe CR. Affinity chromatography: history, perspectives, limitations and prospects. *Methods Mol Biol* 2008;**421**:1–21.
57. Lee WC, Cheng CH, Pan HH, et al. Chromatographic characterization of molecularly imprinted polymers. *Anal Bioanal Chem* 2008;**390**:1101–9.
58. Wikstroem M, Ohlson S. Computer simulation of weak affinity chromatography. *J Chromatogr* 1992;**597**:83–92.
59. Hermanson GT, Mallia AK, Smith PK. *Immobilized affinity ligand techniques*. New York: Academic Press; 1992.
60. Chen G, Pramanik BN. Application of LC/MS to proteomics studies: current status and future prospects. *Drug Discov Today* 2009;**14**:465–71.
61. Wilchek M, Miron T, Kohn J. Affinity chromatography. *Methods Enzymol* 1984;**104**:3–55.
62. Hage DS, Nelson MA. Chromatographic immunoassays. *Anal Chem* 2001;**73**:198A–205A.
63. Moser AC, Hage DS. Immunoaffinity chromatography: an introduction to applications and recent developments. *Bioanalysis* 2010;**2**:769–90.
64. Janson JC, editor. *Protein purification: principles, high resolution methods, and applications*. 3rd ed. Hoboken: Wiley; 2011.
65. Scouten WH, editor. *Solid phase biochemistry*. New York: Wiley; 1983. p. 149–87.
66. Bouriotis V, Stott J, Galloway A, et al. Measurement of glycosylated haemoglobins using affinity chromatography. *Diabetologia* 1981;**21**:579–80.
67. Mallia AK, Hermanson GT, Krohn RI, et al. Preparation and use of a boronic acid affinity support for the separation and quantitation of glycosylated hemoglobins. *Anal Lett* 1981;**14**:649–61.
68. Allenmark S. *Chromatographic enantioseparations: methods and applications*. 2nd ed. New York: Ellis Horwood; 1991.
69. Armstrong DW. Direct enantiomeric separations in liquid chromatography and gas chromatography. In: Issaq HJ, editor. *A century of separation science*. New York: Marcel Dekker; 2002 [Chapter 33].
70. Guebitz G, Schmid MG. Chiral separation principles: an introduction. *Methods Mol Biol* 2004;**243**:1–28.
71. Okamoto Y, Ikai T. Chiral HPLC for efficient resolution of enantiomers. *Chem Soc Rev* 2008;**37**:2593–608.
72. Stalcup AM. Chiral separations. *Annu Rev Anal Chem (Palo Alto Calif)* 2010;**3**:341–63.
73. Dejaegher B, Mangelings D, Vander Heyden Y. Method development for HILIC assays. *J Sep Sci* 2008;**31**:1438–48.
74. Ward TJ. Chiral separations. *Anal Chem* 2008;**80**:4363–72.
75. Hao Z, Xiao B, Weng N. Impact of column temperature and mobile phase components on selectivity of hydrophilic interaction chromatography (HILIC). *J Sep Sci* 2008;**31**:1449–64.
76. Lienqueo ME, Mahn A, Salgado JC, et al. Current insights on protein behavior in hydrophobic interaction chromatography. *J Chromatogr B* 2007;**849**:53–68.
77. LaCourse WR. Column liquid chromatography: equipment and instrumentation. *Anal Chem* 2002;**74**:2813–32.
78. Fallas MM, Neue UD, Hadley MR, et al. Investigation of the effect of pressure on retention of small molecules using reversed-phase ultra-high pressure liquid chromatography. *J Chromatogr A* 2008;**1209**:195–205.
79. Jorgenson JW. Capillary liquid chromatography at ultrahigh pressures. *Annu Rev Anal Chem (Palo Alto Calif)* 2010;**3**:129–50.
80. Varma D, Jansen SA, Ganti S. Chromatography with higher pressure, smaller particles and higher temperature: a bioanalytical perspective. *Bioanalysis* 2010;**2**:2019–34.
81. Lunn G, Hellwig GC. *Handbook of derivatization reactions for HPLC*. New York: Wiley-InterScience; 1998.
82. Unger KK, Skudas R, Schulte MM. Particle packed columns and monolithic columns in high-performance liquid chromatography: comparison and critical appraisal. *J Chromatogr A* 2008;**1184**:393–415.

83. Kobayashi H, Ikegami T, Kimura H, et al. Properties of monolithic silica columns for HPLC. *Anal Sci* 2006;**22**:491−501.
84. Vlakh EG, Tennikova TB. Applications of polymethacrylate-based monoliths in high-performance liquid chromatography. *J Chromatogr A* 2009;**1216**:2637−50.
85. Svec F, Huber CG. Monolithic materials: promises, challenges, achievements. *Anal Chem* 2006;**78**:2100−8.
86. Zacharis CK. Accelerating the quality control of pharmaceuticals using monolithic stationary phases: a review of recent HPLC applications. *J Chromatogr Sci* 2009;**47**:443−51.
87. Lambert WE, Van Bocxlaer JF, De Leenheer AP. Potential of high-performance liquid chromatography with photodiode array detection in forensic toxicology. *J Chromatogr B Biomed Sci Appl* 1997;**689**:45−53.
88. Wang C, Xu J, Zhou G, et al. Electrochemical detection coupled with high-performance liquid chromatography in pharmaceutical and biomedical analysis: a mini review. *Comb Chem High Throughput Screen* 2007;**10**:547−54.
89. Haddad PR, Nesterenko PN, Buchberger W. Recent developments and emerging directions in ion chromatography. *J Chromatogr A* 2008;**1184**:456−73.
90. Kou D, Manius G, Zhan S, et al. Size exclusion chromatography with corona charged aerosol detector for the analysis of polyethylene glycol polymer. *J Chromatogr A* 2009;**1216**:5424−8.
91. Gross ML, Caprioli RM, Niessen W. *The encyclopedia of mass spectrometry. hyphenated methods*, vol. 8. Amsterdam: Elsevier; 2006.
92. Shushan B. A review of clinical diagnostic applications of liquid chromatography-tandem mass spectrometry. *Mass Spectrom Rev* 2010;**29**:930−44.
93. Vogeser M, Seger C. A decade of HPLC-MS/MS in the routine clinical laboratory: goals for further developments. *Clin Biochem* 2008;**41**:649−62.
94. Vogeser M, Seger C. Pitfalls associated with the use of liquid chromatography-tandem mass spectrometry in the clinical laboratory. *Clin Chem* 2010;**56**:1234−44.
95. Gergov M, Nokua P, Vuori E, et al. Simultaneous screening and quantification of 25 opioid drugs in post-mortem blood and urine by liquid chromatography-tandem mass spectrometry. *Forensic Sci Int* 2009;**186**:36−43.
96. Hoofnagle AN. Quantitative clinical proteomics by liquid chromatography: tandem mass spectrometry—assessing the platform. *Clin Chem* 2010;**56**:161−4.
97. Makawita S, Diamandis EP. The bottleneck in the cancer biomarker pipeline and protein quantification through mass spectrometry—based approaches: current strategies for candidate verification. *Clin Chem* 2010;**56**:212−22.
98. Maurer HH. Current role of liquid chromatography—mass spectrometry in clinical and forensic toxicology. *Anal Bioanal Chem* 2007;**388**:1315−25.
99. Maurer HH. Perspectives of liquid chromatography coupled to low- and high-resolution mass spectrometry for screening, identification, and quantification of drugs in clinical and forensic toxicology. *Ther Drug Monit* 2010;**32**:324−7.
100. Sinclair I, Charles I. Applications of the charged aerosol detector in compound management. *J Biomol Screen* 2009;**14**:531−7.
101. Mogridge J. Using light scattering to determine the stoichiometry of protein complexes. *Methods Mol Biol* 2004;**261**:113−8.
102. Chester TL, Pinkston JD, Raynie DE. Supercritical fluid chromatography and extraction. *Anal Chem* 1992;**64**:153R−70R.
103. Taylor LT. Supercritical fluid chromatography. *Anal Chem* 2008;**80**:4285−94.
104. Bamba T. Application of supercritical fluid chromatography to the analysis of hydrophobic metabolites. *J Sep Sci* 2008;**31**:1274−8.
105. Li F, Hsieh Y. Supercritical fluid chromatography—mass spectrometry for chemical analysis. *J Sep Sci* 2008;**31**:1231−7.
106. Sherma J, Fried B, editors. *Planar chromatography*. New York: Taylor & Francis; 2003.
107. Shalliker RA, Gray MJ. Concepts and practice of multidimensional high-performance liquid chromatography. *Adv Chromatogr* 2006;**44**:177−236.
108. Wang S, Miller A. A rapid liquid chromatography—tandem mass spectrometry analysis of whole blood sirolimus using turbulent flow technology for online extraction. *Clin Chem Lab Med* 2008;**46**:1631−4.
109. Gowda GA, Zhang S, Gu H, et al. Metabolomics-based methods for early disease diagnostics. *Expert Rev Mol Diagn* 2008;**8**:617−33.
110. Lu B, Xu T, Park SK, et al. Shotgun protein identification and quantification by mass spectrometry. *Methods Mol Biol* 2009;**564**:261−88.

Mass Spectrometry

*Alan L. Rockwood, Mark M. Kushnir, and Nigel J. Clarke**

ABSTRACT

Background
Mass spectrometry is a powerful analytical technique used to identify and quantify analytes using the mass-to-charge ratio *(m/z)* of ions generated from a sample. It is useful for the analysis of a wide range of clinically relevant analytes, including small molecules, proteins, and peptides. When mass spectrometry is coupled with either gas or liquid chromatographs, the resultant analyzers have expanded analytical capabilities with widespread clinical applications, including quantitation of analytes from myriad body tissues and fluids. In addition, because of its ability to identify and quantify proteins, mass spectrometry is widely used in the field of proteomics.

Content
This chapter describes the basic concepts and definitions of mass spectrometry. Techniques based on mass spectrometry require an ionization step wherein an ion is produced from neutral atoms or molecules. Electron impact and chemical ionization (CI) are often used in gas chromatography—mass spectrometry. In liquid chromatography—mass spectrometry, electrospray ionization (ESI) and atmospheric pressure CI are the most commonly used techniques. In microbiology, a desorption/ionization technique termed MALDI (matrix-assisted laser desorption ionization) is employed. Each of these ionization techniques is described in detail, and advantages of the techniques are highlighted. Once molecules are ionized, resultant ions are analyzed using either beam type analyzers (eg, quadrupole, or time-of-flight [TOF]) or trapping mass analyzers (eg, ion trap). Mass analyzers also can be combined to form tandem mass spectrometers, which allow further expending capabilities of the technique. Clinical applications of mass spectrometry are provided to illustrate the role of this technique in the analysis of clinically relevant analytes.

Mass spectrometry (MS) is a powerful qualitative and quantitative analytical technique that is used to identify and quantify a wide range of clinically relevant analytes. When coupled with gas or liquid chromatographs, mass spectrometers allow expansion of analytical capabilities to a variety of clinical applications. In addition, because of its ability to identify and quantify proteins, MS is a key analytical tool in the field of proteomics.

We begin this chapter with a discussion of the basic concepts and definitions of MS, followed by discussions of MS instrumentation and clinical applications, and we end the chapter with a discussion of logistic, operational, and quality issues. In this chapter it is impossible to cover all concepts in a field as vast as MS, even if focus is limited to clinical applications. The Clinical and Laboratory Standards Institute (CLSI) has published recommendations on clinical MS that can serve as a good next step to study this topic and another gateway into the extensive literature on this topic.[1,2]

BASIC CONCEPTS AND DEFINITIONS

MS is the branch of physical chemistry (often also considered a branch of analytical chemistry) that deals with all aspects of instrumentation and the applications of this technique. *Molecular mass* (sometimes referred to as *molecular weight*) is measured in *unified atomic mass units* (u), also known as the *dalton* (Da), equal to 1/12 of the atomic mass of the most abundant isotope of a carbon atom in its lowest energy state, defined as 12 Da. Although the term *atomic mass unit* (amu) has been regarded as equivalent to the Da, it is only approximately equal to the dalton and now is considered an obsolete unit; its use to refer to the dalton is strongly discouraged.

Most MS data are presented in units of mass-to-charge ratio, or *m/z*, where *m* is the molecular weight of the ion (in daltons) and *z* is the number of charges present on the measured molecule. For small molecules (<1000 Da) there is typically only a single charge and therefore the *m/z* value is the same as the mass of the molecular ion. However, when larger molecules such as proteins or peptides are

---

*The authors gratefully acknowledge the original contributions by Thomas M. Annesley, Nicholas E. Sherman, and Larry D. Bowers on which portions of this chapter are based. We also wish to acknowledge technical assistance by Leita Rogers, Jacquelyn McCowen-Rose, and Martha Fowles and helpful suggestions from N. Leigh Anderson, Julianne C. Botelho, Pierre Chaurand, David K. Crockett, Ulrich Eigner, Steven A. Hofstadler, Andrew N. Hoofnagle, Gary H. Kruppa, Donald Mason, Michael Morris, Maria M. Ospina, and Hubert W. Vesper.

measured, they typically carry multiple ionic charges and therefore the z value is an integer greater than 1. In these cases the m/z value will be a fraction of the mass of the ion.

All MS techniques require an initial *ionization* step in which an ion is produced from a neutral atom or molecule. Ions are formed in the ion source of the mass spectrometer. The development of versatile ionization techniques has allowed MS to become the excellent broad-spectrum analytical technique it is today; this was highlighted when, in 2002, John Fenn and Koichi Tanaka shared the Nobel Prize for their development of electrospray[3,4] and laser desorption[5-7] ionization, respectively. In the ion sources most commonly used with MS instruments in clinical chemistry, ionization in positive ion mode results from the addition of one (or more) protons to the basic sites on the molecule. This is referred to as protonation and leads to formation of a positively charged ion. The mass of the ion is greater than the mass of the uncharged neutral molecule by the added mass of one proton, approximately 1 Da, or multiples of a single proton mass (in case of multiply charged ions). Negatively charged ions (negative ion mode of MS operation) can be generated by the loss of a proton or addition of a negatively charged moiety (such as a hydroxyl group).

Ions may also be produced by removal of one or more electrons from a molecule using electron ionization (EI). This ionization method is historically the dominant ionization method used in MS (most commonly in gas chromatography—mass spectrometry [GC-MS] instruments) and is still used in some applications, but other ionization methods are now more frequently used in clinical laboratories. The removal of one electron produces a positively charged ion and reduces the mass by approximately 5×10^{-4} Da relative to the neutral molecule from which the ion is produced. This relatively small mass shift is often considered to be negligible and therefore ignored. More rarely, ions may be produced by addition of one electron to a molecule, producing a negative ion with mass approximately 5×10^{-4} Da greater than the neutral molecule from which the ion is produced. This small mass increment is also often considered to be negligible relative to the neutral molecule.

Ions formed in the ion source are separated according to m/z values in a mass analyzer. The term *mass analyzer* is in common use, although more correctly it would be termed an *m/z analyzer* given the fact that mass spectrometers separate ions according to m/z, not mass. This chapter will use the terms *mass analyzer* and *m/z analyzer* interchangeably.

While in a mass analyzer, ions may undergo *fragmentation*, whereby energy is imparted into the ionized analyte, causing internal bonds to break and resulting in the production of multiple independent unconnected chemical species. *Fragmentation can occur within different regions of the mass spectrometer and can occur due to the deliberate action of the operator or excessive energy imparted into the parent molecule as it is being ionized or passes through the vacuum region of the* mass analyzer. An unfragmented ion of the intact molecule is referred to as the *molecular ion*, whereas the species that occur on fragmentation of the molecular ion are called the *fragment ions*. There is a certain ambiguity in the term molecular ion because in many cases the structure of the ion is not identical to the structure of the original neutral molecule (eg, differing by the addition or removal of one or more protons, so the term *molecular ion* must be thought of in terms of an unfragmented ion whose structure is closely related to, but not necessarily identical to, the original uncharged molecule.

If the ionization of the analyte in the source produces little fragmentation, it is referred to as being soft, and the most abundant peak in the mass spectrum (the *base* peak) is often the molecular ion. If the ion source produces extensive fragmentation it is referred to as hard ionization, and the base peak in the resulting spectra may be one of the fragment ions. By convention, the base peak in a mass spectrum is assigned a relative abundance value of 100%.

Fragment ions that are formed in a separate dissociation cell (also known as the collision cell) inside a *tandem mass spectrometer* are known as *product ions, and the technique is called tandem mass spectrometry (MS/MS)*. Ions that give rise to product ions are known as precursor ions. A tandem mass spectrometer consists of two mass spectrometers operated in sequence (MS/MS in space) or a single mass spectrometer capable of sequential fragmentation and measurement of ions within a single region of space (MS/MS in time). Most commonly in the clinical diagnostic methods, precursor ions are dissociated into product ions between the two stages of m/z analysis (MS/MS in space).

A *mass spectrum* is represented by the relative abundance of each ion plotted as a function of m/z (Fig. 2.1). As mentioned earlier, for small molecules, usually each ion has

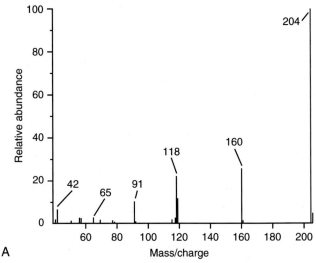

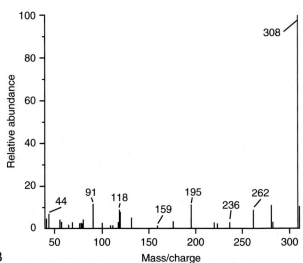

FIGURE 2.1 Mass spectrum of the pentafluoropropionyl (A) and carbethoxyhexafluorobutyryl (B) derivatives of D-methamphetamine.

a single charge ($z = 1$); thus the *m/z* ratio is equal to the mass of the ion and is approximately 1 Da greater than the neutral molecule from which the ion is formed. However, in some cases, the charge may be represented by an integer number greater than 1, in which case the *m/z* ratio is not equal to the mass of the ion, but rather is some fraction of the mass of the ion.

An ion may be positively charged, in which case the number of electrons in the ion is less than the sum of the number of protons in all nuclei of the ion, or negatively charged, in which case the number of electrons is greater than the number of protons. By convention, in MS z is taken as an absolute value. For example, $z = 1$ for both Na^+ and Cl^-.

Chemical interferences as well as higher background noise are more common for analytes with *m/z* 200 to 500 than for *m/z* less than 200 and *m/z* greater than 500. Monitoring ions with higher *m/z* often results in lower limits of detection because of the lower background noise and lower occurrence of isomers and isobars of the targeted molecules (ie, superior signal to noise).

A peak in a mass spectrum can be characterized by its *resolution* $[(m/z)/(\Delta m/z)]$, where $\Delta m/z$ is the width of the mass spectral peak. This parameter characterizes the ability of a mass spectrometer to separate nearby masses from each other. Typically the width of the peak is measured at 50% of the height of the peak and is referred to as the full width half height (FWHH) or full width half maximum (FWHM) resolution. A second frequently encountered definition for resolution is 10% valley. It defines $\Delta m/z$ as the distance between two peaks of equal intensity, spaced so that the valley between the peaks is 10% of the peak height (Fig. 2.2). *This is a more conservative definition than FWHM because for a given quoted resolution (eg, 2000) the peaks are narrower under the 10% valley definition, hence better separated. High resolution is a desirable property in MS because it can help reduce interferences from nearby peaks in the mass spectrum, thereby allowing it to achieve a higher specificity.*

By setting the relative abundance of the base peak to 100% and therefore using the relative, rather than absolute, abundance of each ion fragment, instrument-dependent variability is minimized and the mass spectrum can be compared with mass spectra obtained on other instruments. Because fragmentation at specific bonds depends on their chemical nature and strength of the bonds, the mass spectrum can be interpreted in terms of the molecular structure of the analyte. In some cases, the chemical structure of the analyte can be deduced or at least reconciled with features found in the mass spectrum. Computer-based libraries of mass spectra are also available to assist in identification of the analyte(s) based on fragmentation pattern. In some applications, the mass spectrum of an analyte may be matched against mass spectra in a database, thereby identifying the analyte by its mass spectral *fingerprint*. In general, an unknown is considered to be identified if the relative abundances of three or four ion fragments agree within ±20% of those from a reference compound and the relative abundances of the fragments, monoisotopic and isotopic ions of the molecular ion are in agreement with the relative abundances of the reference mass spectrum.

When interfaced to a liquid or gas chromatograph, the mass spectrometer functions as a powerful detector, providing structural information in real time on individual analytes as they elute from a chromatographic column. Depending on the operating characteristics of the mass spectrometer and the chromatographic peak width, multiple mass spectral scans can be acquired across the peak. The data also can be displayed as a function of time to yield a *total ion chromatogram in which at each time point in the chromatogram the abundances of all ions in a mass spectrum are summed to constitute a single point in the ion chromatogram, regardless of m/z.*

The mass spectrometer can be considered to be close to a universal detector because molecules of many identities may be ionized and then detected in a mass spectrometer. Furthermore there are different MS operation modes and different types of fragmentation that can be applied to provide different types of data, giving more information about the measured compound(s). Finally, the instrument data system can analyze and display the collected data in various manners, allowing the operator to selectively extract information from the acquired data.

For example, it is possible to display chromatograms of only preselected ions acquired during data acquisition—that is, representing data from only part of the mass spectrum. The resultant display of data is called an *extracted ion chromatogram*, displaying signal intensity plotted as a function of time; peak heights or peak areas can be integrated for use in quantitative analysis. Use of the extracted ion chromatogram allows selecting data corresponding to the analyte of interest, as identified by its *m/z*, while disregarding data corresponding to different *m/z*. With high-resolution instruments, specificity of analysis can be improved by use of narrow *m/z* windows for plotting extracted ion chromatograms. Such data processing results in a reduced number of overlapping chromatographic peaks from ions of nearby *m/z* thus improving the quantitative accuracy and the specificity (Fig. 2.3).

Sample preparation is critical to successful MS, particularly when dealing with complex matrices, such as are commonly encountered in clinical chemistry. This typically involves one or more of the following steps: (1) protein precipitation followed by centrifugation or filtration, (2) solid-phase extraction, (3) liquid-liquid extraction, (4) affinity enrichment, or (5) *derivatization* (see Chapter 3).

Derivatization is the process of chemically modifying the target compound(s) to be more favorably analyzed by MS. Derivatization usually involves the addition of some well-defined functional group. The goals of derivatization vary, depending on the application, but typically include (1) increased volatility, (2) greater thermal stability, (3) modified chromatographic properties, (4) greater ionization efficiency, (5) favorable fragmentation properties, or a combination of these.

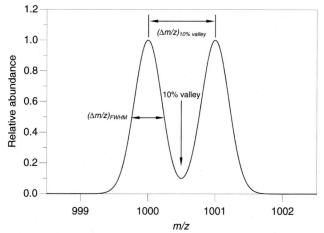

FIGURE 2.2 Parameters used to define resolution in mass spectrometry. *FWHM*, Full width half maximum.

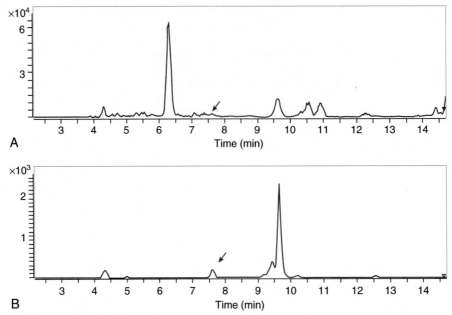

FIGURE 2.3 Extracted ion chromatograms for a peptide ion of m/z 761.3718 using a window of 1 Da (A) and 0.0076 Da (B). *Arrows* point on peak that is completely hidden in the chemical background noise while resolved from the noise using mass extraction window of 0.0076 Da.

Analysis by MS can be used to target specific known compounds (targeted analysis) or seek to identify one or more unknown compounds in a sample (screening). When only one or a few targeted analytes are of interest for quantitative analysis and their mass spectrum is known, the mass spectrometer is set to monitor only those ions of interest. This selective detection technique is known as *selected ion monitoring* (SIM). Because SIM focuses on a limited number of ions, more data points are collected for the selected m/z, which *results in better, more precise measurements*. The SIM data acquisition increases the signal-to-noise ratio for the analyte of interest, improves the lower limit of detection, and enables more accurate quantitation. One drawback of SIM is that it is based around measurement of a nominal analyte mass. Most biological samples are highly complex, and thus it is not unexpected to find multiple compounds with very close or identical masses in the matrix. In those cases, chromatography can aid in separation of these isobars; however, they still can affect a SIM result should the isobar not be separated completely from the analyte to be measured. By using a triple-quadrupole mass analyzer, a method known as selected reaction monitoring (SRM) (or generalized for the analysis of many ions at the same time, multiple reaction monitoring [MRM]) can be used to help alleviate such potential issues. This is where the first quadrupole instrument is set to transmit the m/z of the molecular ion, the analyte is caused to fragment in the second quadrupole and the third quadrupole is set to transmit the m/z of one or more known fragment ions from the analyte. In this manner data similar to those gathered by SIM can be produced but with added specificity from the structural information gathered by the use of the fragment ion as a gatekeeper. A more detailed description of MRM is given in the section of this chapter on tandem mass spectrometers.

Screening methods (used here in the analytical chemistry sense, not to be confused with screening in a clinical or medical sense) are less common in clinical chemistry than the analysis of target compounds; the main task for screening methods is qualitative identification of unknowns in a sample. In most cases this reduces to the problem of matching chromatographic retention time and fragment ion patterns, that is, mass and abundance patterns of either fragment ions generated in the source of a single-stage mass spectrometer or product ions from a collision cell in a tandem mass spectrometer.

A chemical element may be composed of a single or multiple isotopes. Each isotope of an element has the same number of protons in its nucleus but different numbers of neutrons. For example, naturally occurring carbon is composed primarily of two isotopes: ^{12}C, whose nuclei contain six protons and six neutrons, and ^{13}C, whose nuclei contain six protons and seven neutrons. (Here we ignore ^{14}C, which is generally of negligible abundance compared to the other two isotopes.) The natural abundance of ^{12}C is approximately 98.9%, and the natural abundance of ^{13}C is approximately 1.1%. Some elements, such as arsenic, have only a single isotope in the naturally occurring state, whereas other elements, such as tin, may have as many as 10 naturally occurring isotopes.

When a compound is made of multiple atoms the isotope pattern is a convolution of the isotope patterns of the individual atoms. To illustrate with a simple example, carbon monoxide (CO) has the following combinations of isotopes $^{12}C^{16}O$ (molecular weight 28), $^{13}C^{16}O$ (molecular weight 29), $^{12}C^{17}O$ (molecular weight 29), $^{12}C^{18}O$ (molecular weight 30), and $^{13}C^{18}O$ (molecular weight 31).

Nitrogen (N_2) is isobaric with CO; that is, it has nearly the same mass. However, the accurate masses of the isotope peaks of isobars may differ. For example, the monoisotopic mass of $^{12}C^{16}O^+$ (the isotopic peak composed of the most abundant atomic isotopes) has an accurate mass of 27.9944 Da, whereas the isobaric mass 28 Da peak of N_2^+ has an accurate molecular mass of 28.0056 Da. The small difference in masses of isobars can be used to infer the chemical formula of a compound or to confirm the identity of a target compound. This technique requires a mass analyzer capable of mass accuracy of a few parts per million, is limited to compounds of a few hundred Da or less, and it is not capable of discriminating between isobars of the same chemical formula.

Compounds that have the same chemical formulas but different chemical structures are also isobars and might therefore be referred to as strict isobars because they have exactly the same mass. Succinic acid and methylmalonic acid provide an example of a pair of compounds that are strict isobars because they have the same chemical formula ($C_4H_6O_4$). Unlike isobars that have nearly the same masses but different accurate masses, strict isobars cannot be separated or distinguished by MS alone, although they often can be separated if MS is combined with a separation method or if tandem MS is applied.

Isotopic information also can be used to identify a compound in a different way. Using CO^+ and N_2^+ as examples, the first three isotopic peaks of CO^+ have a relative abundance pattern of 0.986, 0.011, and 0.002, whereas the first three isotope peaks of N_2^+ has a pattern of 0.993, 0.007, and 0.000. Differences in the isotopic pattern can be used to infer the chemical formula of an unknown or to confirm chemical identity of a target compound. This technique requires accurate measurement of relative isotopic peak abundances and is sometimes used in conjunction with accurate mass measurements, particularly when using TOF mass spectrometers.

A distinct advantage of the mass spectrometer is that it can distinguish between ions of the same chemical formula that have different masses because of the different isotopic composition. To illustrate with a trivial and simple example, $^{12}C^{16}O^+$ has a different mass than $^{12}C^{18}O^+$, and these two forms can be separated and detected in a mass spectrometer. One can take advantage of this fact by using artificially labeled forms of a target analyte. The labeling consists of substitution of a less common isotope for one or more of the atoms in the analyte, for example, substituting ^2H for ^1H, ^{13}C for ^{12}C, or ^{15}N for ^{14}N. The substituted molecule is prepared artificially and added to the sample as an internal standard, which behaves nearly identically to the native compounds during sample preparation and chromatographic separation. In this respect, ^{13}C or ^{15}N is generally preferred over ^2H labeling because ^2H-labeled compounds sometimes exhibit chromatographic shifts compared to unlabeled compounds, whereas ^{13}C- or ^{15}N-labeled compounds generally do not. A quantitative analysis can then be carried out by comparison of the signal from the native compound versus the artificially added labeled version of the compound spiked into the sample.

An internal standard should be selected to have a sufficient number of isotopic ions so that no naturally occurring isotopes (such as ^2H or ^{13}C) of the analyte of interest would significantly contribute to the signal of the internal standard. For the methamphetamine derivatives shown in Figure 2.4, A, an internal standard with at least three ^2H or ^{13}C atoms is preferred, because contribution of the natural abundance of these isotopes to the molecular ion $[(M + 3)^+]$ would be $\approx 0.1\%$. The position of the stable isotope atoms within the molecule and the number of isotopic ions within the structure is also important for adequate performance of the methods.[8] For example, the m/z 204 ion for methamphetamine represents the aliphatic portion of the molecule (loss of the aromatic ring). If three deuterium atoms were located on the aromatic ring of the pentafluoropropionyl derivative of methamphetamine, the native and the isotope-labeled molecules would both yield the m/z 204 ion. This m/z 204 ion would therefore fail to distinguish the native compound from the isotope-labeled compound and would therefore not be useful as an internal standard. On the other hand, if ^2H labeling were to occur in the aliphatic portion of the molecule, the fragment ion analogous to the m/z 204 fragment ion would be labeled, producing a higher m/z than 204 (eg, 207), and the ion could be useful as an internal standard. The same comments apply to the compound illustrated in Figure 2.4, B.

These concepts of internal standard selection must be modified slightly when applied to tandem MS. For example, it is possible for a native compound and the internal standard to have product ions of the same m/z, because they can be distinguished by their differing precursor ion masses.

When using hydrogen (^2H) labeling the stable isotope must be located on atoms from which it will not be exchangeable with hydrogen atoms in solution or in gas phase (in the ion source). For example, deuterium labeling of an acidic hydrogen position would be useless because the ^2H would easily exchange with protons in the matrix, making the original labeling moot. Certain other labeling positions also must be avoided. The hydrogen atoms in the constituent

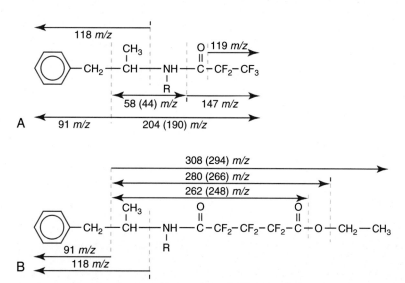

FIGURE 2.4 Fragmentation patterns for the pentafluoropropionyl (A) and carbethoxyhexafluorobutyryl (B) derivatives of methamphetamine (R = CH_3) and amphetamine (R = H; *masses in parentheses*). Compare the predicted masses with the spectrum shown in Figure 2.1. Note that for the pentafluoropropionyl derivative, only one ion [204 (190) m/z] is characteristic of the aliphatic portion of the molecule.

groups of alcohols, amines, amides, and thiols all may readily exchange with hydrogen ions in an aqueous matrix.

A technique of quantitative analysis of compounds relative to their isotopic analogs added to the samples at known or fixed concentration is called *isotope dilution analysis* or isotope dilution mass spectrometry (IDMS). The IDMS technique has been used to develop definitive methods for a number of clinically relevant analytes, including drugs of abuse and disease markers.

MS is often referred to as a highly sensitive technique. *Sensitivity* is a somewhat problematic term because it is used in two different ways. In an official definition[9,10] it means the slope of a calibration curve (or more generally, a change in signal vs the change in concentration), but far more commonly it is used to signify the ability to detect or quantify an analyte at very low concentration; that is, a highly sensitive technique would be able to detect or quantify a very low concentration of the target analyte.

INSTRUMENTATION

A mass spectrometer consists of (1) an ion source, (2) a vacuum system, (3) a mass analyzer, (4) a detector, and (5) a computer (Fig. 2.5).

Ion Source

Many approaches have been used to form ions in both high-vacuum and near-atmospheric pressure conditions. EI and CI are ionization techniques used when gas-phase molecules are introduced directly into an ion source operated at very low pressure, often from a gas chromatograph. In other analyses, such as high-performance liquid chromatography–mass spectrometry (HPLC-MS), ESI, *atmospheric pressure chemical ionization* (APCI), and *atmospheric pressure photoionization* (APPI) ion sources are often used.[11-14] Ionization in these three ion sources takes place at atmospheric pressure. Other ionization techniques include (1) *inductively coupled plasma* (ICP), (2) MALDI (see Chapter 4), (3) *atmospheric pressure matrix-assisted laser desorption ionization*, and others. This chapter will limit its discussion primarily to ion sources of interest to clinical applications of MS. The CLSI documents C-50A and C-62A contain recommendations for matching the capabilities of different ion source technologies to various application classes.[1,2]

Electron Ionization

In EI, gas-phase molecules are bombarded by electrons emitted from a heated filament and attracted to a collector electrode (Fig. 2.6). To prevent filament oxidation, as well as to minimize scattering of the electron beam, this process must occur in a vacuum. EI is typically performed using electrons with a kinetic energy of 70 eV; collision of electrons having such energy with most organic molecules results in formation of *radical* cations, that is, a structure that is both a positively charged ion and a radical.[15] A radical is a molecule or ion with an unpaired electron. The radical ion then often undergoes unimolecular rearrangement and dissociation to produce a cation and an uncharged radical:

$$AB^{+-} \rightarrow A^+ + B^-$$

Positive ions are drawn out of the ionization chamber by an electrical field. The cations are then electrostatically focused and introduced into the mass analyzer. EI is primarily used as an ion source in GC-MS. Because the same ion energy (70 eV) is used in commercial instruments using EI, and because the fragmentation pattern is only weakly dependent on small deviations from 70 eV, fragmentation patterns observed using an EI source are reasonably reproducible among the GC-MS instruments. The fragmentation pattern

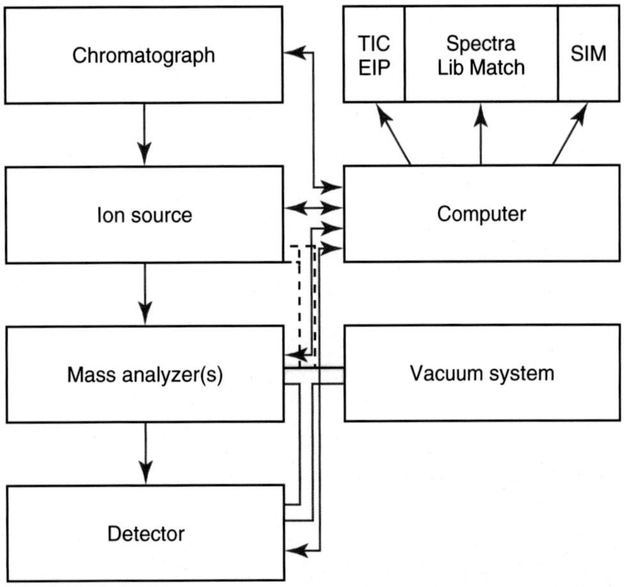

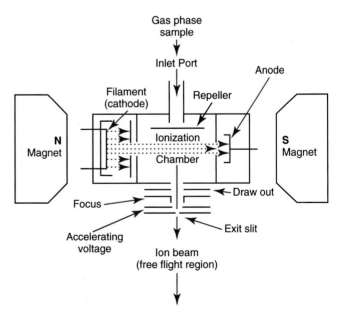

FIGURE 2.5 Block diagram of the components of a chromatograph–mass spectrometer system. The mass analyzer and the detector are always under vacuum. The ion source may be under vacuum or under near-atmospheric pressure conditions, depending on the ionization mode. The computer system is an integral part of data acquisition and output. *EIP,* Extracted ion profile; *SIM,* selected ion monitoring; *TIC,* total ion current.

FIGURE 2.6 Electron impact ion source. The small magnets are used to collimate a dense electron beam, which is drawn from a heated filament placed at a negative potential. The electron beam is positioned in front of a repeller, which is at a slightly positive potential compared with the ion source. The repeller sends any positively charged fragment ions toward the opening at the front of the ion source. The accelerating plates strongly attract the positively charged fragment ions.

is therefore often used as a fingerprint to identify compounds by matching mass spectra of unknowns to the entries in the mass spectral libraries.

Chemical Ionization

CI is a soft ionization technique in which a proton is transferred to, or abstracted from, a gas-phase molecule by a reagent gas molecule such as methane, ammonia (NH3), isobutane, or water. The reagent gas is supplied into a CI ion source at a pressure of about 0.1 torr. (Note: For virtually all practical purposes, torr is equivalent to millimeters of mercury and is the more customary term used in the field of MS). An electron beam produces reactive species through a series of ion-molecule reactions, intermediate species (such as methonium [CH_5^+] if methane is the CI reagent gas); further ion-molecule reactions can cause analyte ions to become charged, usually via attachment of a proton. In most cases, relatively little fragmentation occurs, and for the majority of the molecules, only molecular ions (in the form of a protonated version of the neutral molecule) are observed in the mass spectra; the lack of fragmentation enhances sensitivity of detection. Negative ion electron capture CI has become popular for quantification of drugs such as benzodiazepines. Negative ions are formed when thermalized electrons are captured by electronegative functional groups, such as chlorine or fluorine atoms within structure of the molecule. Negative ion CI often leads to very low limits of detection.

Electrospray Ionization

ESI is a soft ionization technique in which a sample is ionized at atmospheric pressure before introduction into the mass analyzer.[16,17] An effluent from a separation device, typically an HPLC, is passed through a narrow metal or fused silica capillary to which a 1- to 5-kV voltage has been applied (Fig. 2.7, A). The partial charge separation between the liquid and the capillary results in instability in the liquid that in turn results in expulsion of charged droplets from a Taylor cone, which forms at the tip of the capillary (Fig. 2.8). In many variations of ESI a coaxial nebulizing gas aids in nebulization and helps direct the charged droplets toward a counter-electrode as well as speeding up the evaporative process. As droplets evaporate while migrating through the atmospheric pressure region, they expel smaller droplets as the charge-to-volume ratio exceeds the Raleigh instability limit. The adducts of the molecules (with solvent molecules, NH_3, etc.) are desolvated to form bare ions that are typically formed in the ionization process. However, other ionization products are sometimes observed, such as metal ion adducts, or ions formed by redox processes. Ions then pass through a sampling cone and one or more extraction cones (skimmers) before entering the high-vacuum region of the mass analyzer.

One feature of ESI is the production of multiple charged ions, particularly from peptides and proteins. It is common to observe approximately one charge for every 10 to 15 amino acid residues in a protein. For example, for a molecule of mass 20,000, 20 charges supplied by the addition of 20 protons would be detected at m/z approximately 1000 [or more correctly, m/z 1001 = (20,000 + 20)/20]. This phenomenon greatly extends the accessible mass range of an instrument. It is frequently observed that a distribution of charges occurs; in cases of multiple charged molecules, one usually observes a series of peaks, with each peak corresponding to a different number of charges. Multiply charged ions are also observed for nucleic acid polymers, particularly when ESI mass spectra are acquired in negative ion mode.

It should be noted that Figure 2.7 is a simplified illustration of the probe being directed toward the sampling cone of the mass detector. To enhance performance and minimize contamination of the mass analyzer, modern hardware configurations have offset the probe and/or the

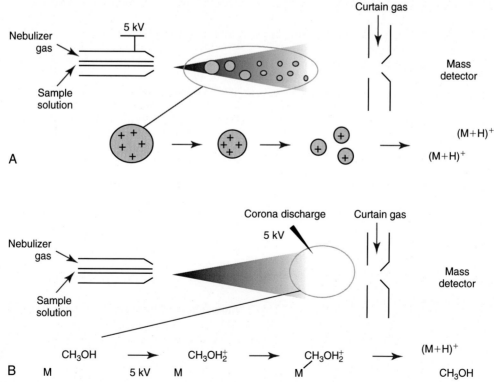

FIGURE 2.7 Schematics of (A) electrospray and (B) atmospheric pressure chemical ionization sources. Note the different points where ionization occurs, as described in the text.

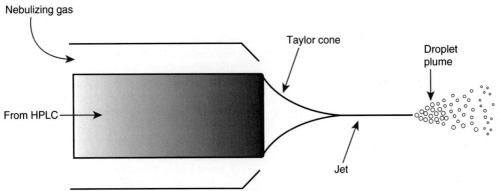

FIGURE 2.8 Simplified conceptual schematic of electrospray ion source showing Taylor cone. *HPLC*, High-performance liquid chromatography. (Published with permission from Eclipse Business Media Ltd. Reprinted from MS Solutions, Issue 7 [2010]. http://www.sepscience.com/Information/Archive/MS-Solutions/235-/MS-Solutions-7-Adjusting-Electrospray-Voltage-for-Optimum-Results.)

mass detector relative to the sampling cone; in some instruments the spray is orthogonal to the sampling cone.

ESI tends to be an efficient ion source for polar compounds or for molecules that are present as ions in solution, which includes a majority of biomolecules. ESI, along with APCI, allows an effective interface between a liquid chromatograph and a mass spectrometer. ESI and APCI have become the most widely used ion sources in mass spectrometers used in clinical laboratories.

As already mentioned, electrospray is considered a soft ionization source. However, it is also possible to generate fragment ions before mass analysis using a technique known as nozzle-skimmer dissociation,[18] infrared multiphoton dissociation,[19] and thermally induced dissociation.[20] In these methods, ions are heated before entering the mass analyzer, and this causes ions to dissociate into fragment ions. In nozzle-skimmer dissociation, a higher than normal voltage gradient is applied in the first low-pressure region of the electrospray interface, resulting in collisional heating of the ions. In infrared multiphoton dissociation, ions are subjected to an intense bombardment of infrared photons, resulting in heating of the ions, followed by dissociation. In thermally induced dissociation, ions are activated and dissociated by excess heating of the gas used to transport ions from atmospheric pressure to the vacuum system of the mass spectrometer.

Atmospheric Pressure Chemical Ionization

APCI is similar to ESI in the sense that ionization takes place at atmospheric pressure, involves nebulization and desolvation, and uses the same design of the ion extraction cone as ESI. The major difference lies in the mode of ionization. In APCI, no high voltage is applied to the inlet capillary. Instead, the mobile phase from the separation device gets evaporated and the vapor passes by a needle with applied current.[21] This process generates a corona discharge. Somewhat analogously to the processes occurring in a CI source, ions generated by the corona discharge undergo variety of ion-molecule reactions such as the following:

$$CH_3OH + H^+ \rightarrow CH_3OH_2^+$$

$$A(analyte) + CH_3OH_2^+ \rightarrow AH^+ + CH_3OH$$

or

$$H_2O + H^+ \rightarrow H_3O^+$$

$$A(analyte) + H_3O^+ \rightarrow AH^+ + H_2O$$

Because solvent molecules from the evaporated mobile phase (eg, water, methanol, acetonitrile) are present in the vapor in excess relative to the sample constituents, they are predominantly ionized early in the ion molecule cascade of reactions and then act as a reagent gas that reacts secondarily to ionize analyte molecules (see Fig. 2.7, *B*). The products of these secondary reactions may contain clusters of solvent and analyte molecules, and thus a heated transfer tube or a countercurrent flow of a curtain gas, such as nitrogen, is used to decluster the ions. As with ESI, APCI is a soft ionization technique that produces relatively little fragmentation. However, unlike ESI, APCI uses much higher heat and this can cause pyrolysis of the compound leading to loss of metabolically induced modifications to targeted compounds (glucuronidation, glutathionylation, etc.) and therefore can cause issues with analyses (in the methods that do not chromatographically resolve peaks of unconjugated and conjugated targeted molecules). On the other hand, in methods using ESI at lower temperature, those modifications tend to remain intact. When compared with EI, the mass spectra produced by APCI, ESI, and other soft ionization techniques typically have fewer fragments and are less useful for analyte identification by mass spectral fingerprinting. However, because the ion current is concentrated into a single mass spectral peak (or relatively few mass spectral peaks), APCI and other soft ionization sources are well matched to the requirements of tandem MS (discussed later) and are well suited for quantitative analysis. APCI and ESI are the most commonly used ion sources in quantitative analysis. However, in the case of nonpolar compounds such as steroids and some drug molecules, APCI is often a more efficient ion source than ESI.

Atmospheric Pressure Photoionization

APPI is a relatively new and less frequently used ion source in clinical chemistry that provides a complementary ionization approach to ESI or APCI. The physical configuration of an APPI source is similar to that for APCI, but an ultraviolet photon flux (typically Krypton lamp that emits photons at 10 Ev) is used instead of a corona discharge needle to generate ions in the gas phase.[22-24] In APPI, an ionizable dopant, such as toluene or acetone, is often infused coaxially to the nebulizer to provide a source of ions that participate in charge or proton transfer to analyte molecules, thus increasing the efficiency of analyte ionization. APPI has a similar range of application to APCI and could be more useful than APCI for compounds of very low polarity, such as some steroids.

Inductively Coupled Plasma

ICP, as ESI, APCI, and APPI, is an atmospheric pressure ionization method. However, unlike most atmospheric pressure ionization methods, which are soft (ie, producing little fragmentation), ICP is the ultimate in hard ionization, typically leading to complete atomization of the sample during ionization. Consequently, its primary use is for elemental analysis. In the clinical laboratory, it is particularly useful for trace element analysis in tissues or body fluids. ICP is extremely sensitive (eg, parts per trillion limits of detection) and is capable of extremely wide dynamic ranges.

After sample preparation, which generally includes the addition of an internal standard such as yttrium and sometimes includes an acid digestion step, the sample is introduced into the ion source, usually via a nebulizer fed by a peristaltic pump. The nebulized sample is transmitted into hot plasma generated at atmospheric pressure by inductively coupling power into the plasma using a high-powered, radiofrequency (RF) generator (Fig. 2.9).[25] The temperature of the plasma is typically 6000 to 10,000 K (comparable to the temperature of the surface of the sun). Sample is introduced in the plasma, and ions are transmitted to the mass analyzer through a series of differential pumping stages. The atmospheric sampling apparatus is conceptually similar to that of other atmospheric pressure ion sources, such as electrospray, except that the device must withstand the extremely high temperatures generated by the plasma.

ICP-MS is comparatively free from most interference. However, some interfering species can be extremely troublesome. Most interfering species are small polyatomic ions formed in the torch via ion-molecule reactions. For example, argon oxide (ArO^+) interferes with iron at m/z 56. One solution to this problem is to use a reaction cell, which consists of a moderate-pressure gas region in front of the m/z analyzer,[26] with a reactant gas, such as NH_3, bled into the reaction cell. The reactant gas reacts with polyatomic interferences and removes them before introduction into the m/z analyzer. A related technique uses a nonreactive collision gas, which removes interferences using collisions, relying on differences in collision cross-sections between polyatomic ions and atomic ions. Another approach to removing interferences of the same nominal mass is to use a high-resolution mass spectrometer, which is capable of resolving species with similar nominal mass.[26] For example, the masses of ArO^+ and $^{56}Fe^+$ differ by 0.022 Da—a difference that may be resolved using a high-resolution mass spectrometer.

Matrix-Assisted Laser Desorption Ionization

MALDI is another type of soft ionization method that typically produces singly charged ions. MALDI and related techniques rely on energy transfer processes from a pulsed laser beam to the sample for ion generation. In most cases, the analyte is dissolved in a solution containing a solid phase matrix, a small molecular weight UV-absorbing compound, and this solution is placed on a target and dried. A pulsed laser irradiates the dried spots, triggering ablation, and desorption of the sample and matrix material; ions produced in the process are accelerated and introduced into the mass analyzer (Fig. 2.10). In other cases a layer of the solid matrix is deposited on the target and allowed to crystalize, and then the sample applied on top. The sample causes the top portion of the matrix to solubilize and mix with the sample before recrystallizing. In this way the sample is maintained in the very outer layer of the matrix, and this can help with enhancing the sensitivity and reducing the background noise. A related technique includes atmospheric pressure matrix-assisted laser desorption/ionization, in which the MALDI process occurs at atmospheric pressure rather than reduced pressure (see Chapter 4).

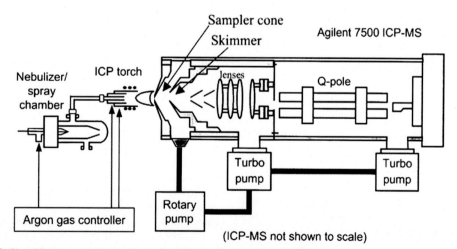

FIGURE 2.9 Simplified conceptual schematic of inductively coupled plasma—mass spectrometer *(ICP-MS)*. Q-pole, Quadrupole. (From Kannamkumarath S, Wrobel K, Wrobel K, et al. Capillary electrophoresis—inductively coupled plasma-mass spectrometry: an attractive complementary technique for elemental speciation analysis. *J Chromatogr A* 2002;975:245—266.)

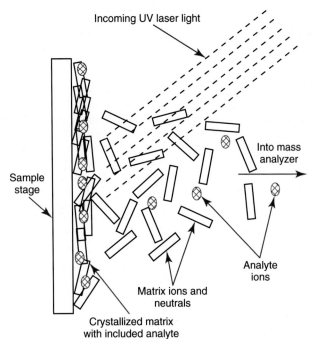

FIGURE 2.10 A generic view of the process of matrix-assisted laser desorption ionization. Co-crystallized matrix and analyte molecules are irradiated with an ultraviolet *(UV)* laser. The laser vaporizes the matrix, producing a plume of matrix ions, analyte ions, and neutrals. Gas-phase ions are directed into a mass analyzer.

Ionization Methods of Potential Interest

Desorption electrospray ionization (DESI)[27] and direct analysis in real time (DART)[28] are two relatively new ionization methods that generate ions from surfaces at atmospheric pressure. Most applications of DESI and DART to date have been directed toward minimal sample preparation. Paper spray MS is another emerging ionization method of potential interest. It is essentially a version of electrospray MS, but the spray is generated at the point of a triangular cut in a solvent-wetted piece of paper rather than from a capillary. It has the feature of easily integrating paper chromatography with MS.[29] It also has been used for rapid measurement of therapeutic drugs from dried blood spots without the need for complex sample preparation and separation.[30]

Ionization Methods of Historical Interest

The older literature includes several ionization methods or sample introduction methods that, although promising, or even widely used at one time, hold little interest for current practice in clinical chemistry. Nevertheless, it is useful to be aware of them because they may still be referred to in the literature from time to time. Some of these include (1) *fast atom bombardment*, (2) *thermospray*, (3) *direct liquid introduction*, (4) *plasma desorption*, (5) *field ionization*, (6) *field desorption*, (7) *secondary ion mass spectrometry*, and (8) *laser desorption*. ESI and APCI ion sources have largely rendered these techniques obsolete.

Vacuum System

With the exception of certain ion trap mass spectrometers, ion separation in any of the mass analyzers requires that the ions do not collide with other molecules during their interaction with magnetic or electric fields. This requires the use of a vacuum from 10^{-3} to 10^{-9} torr, depending on mass analyzer type. The length of the ion path in the analyzer must be less than the mean free path length, unless collisions play a role in mass analysis.

Fourier transform ion cyclotron resonance (FT-ICR) requires the lowest pressure (10^{-9} torr). The quadrupole ion trap (QIT) tolerates the highest pressure (10^{-3} to 10^{-5} torr), a pressure range in which some collisions occur between ions and background gas. Routine quality assurance checks for vacuum leaks should include evaluation of the presence of air and water in the mass spectra.

Efficient high-vacuum pumps generally do not operate well near atmospheric pressure. Thus the vacuum system must have a positive displacement (mechanical) vacuum pump to evacuate the system to a pressure at which the high-vacuum pumps are effective. Mechanical pumps require routine maintenance, such as ballasting and replacing the pump oil.

Although diffusion pumps are the least expensive and most reliable high-vacuum pump, they are rarely used outside of some very specialized mass spectrometers that typically are not used in the clinical laboratory. Cryopumps are another class of pumps that are sometimes used in specialized mass spectrometers but not in the clinical laboratory. In modern instruments, the most common high-vacuum pumps are turbomolecular (often referred to as "turbo") pumps; they have largely replaced diffusion pumps and cryopumps because they are more convenient to use. A key consideration in the design of the vacuum system is pumping speed. The ability of the pump to maintain the vacuum by removing any gas (or solvent vapor) that enters the system determines the maximum flow rate of gas introduced into the mass spectrometer. In general, higher pump capacities are associated with lower detection limits because noise arising from the gas background is reduced.

Mass Analyzers, Tandem Mass Spectrometers, and Ion Detectors

The term *mass spectrometry* is somewhat a misnomer because mass spectrometers do not measure molecular mass, but rather they measure the mass-to-charge ratio. This fact is fundamental to the physical operating principles of mass spectrometers and consequently affects all aspects of instrumentation design, instrument operation, and interpretation of results. The symbol *m/z* is used to denote mass-to-charge ratio and conventionally has been defined as a dimensionless quantity[31] (see also http://goldbook.iupac.org/M03752.html).

However, a "dimensionless" mass-to-charge ratio is not consistent with equations of ion motion in the presence of electric and magnetic fields, which require units of mass divided by charge. Furthermore, the *m/z* scale sometimes is loosely discussed in terms of daltons (also known as unified atomic mass units [u]), although strictly speaking, Da is a unit of mass, not mass-to-charge ratio. Despite these somewhat confusing nomenclature issues, the present chapter generally follows convention by discussing *m/z* in terms of mass and Da.

To help avoid some of the confusion surrounding the use of *m/z*, it has been proposed that it should be defined explicitly as quantity having units of mass-to-charge ratio, with mass specified in daltons and charge specified in elementary charges, this proposed unit would be called the *Thomson* (Th) in honor of one of the pioneers of MS.[32-34] This terminology is sometimes seen in the literature but has not been widely adopted.

General Classes of Mass Spectrometers

Mass spectrometers are broadly classified into two groups: beam-type instruments and trapping-type instruments. In a beam-type instrument, the ions make one pass through the instrument and then strike the detector, where they are destructively detected. The entire process, from the time an ion enters the analyzer until the time it is detected, generally takes microseconds to milliseconds.

In a trapping-type analyzer, ions are held in a spatially confined region of space through a combination of magnetic and/or electrostatic and/or RF electrical fields. The trapping fields or supplemental fields are applied and manipulated in ways that allow m/z measurements to be performed. Trapping times may range from a fraction of a second to minutes, although most clinical applications are at the low end of this range.

Examples of trapping-type instruments include QITs, linear ion traps (which, along with QITs, also depend on RF electric fields), ICR mass spectrometers, electrostatic ion traps, and orbitraps (which are a type of electrostatic ion traps).

Detection of the ions in a trapping-type instrument may be destructive or nondestructive, depending on the specific type of mass spectrometer used. In this context, *destructive* means that ions are destroyed in the detection process. Additional discussions of mass analyzers, tandem mass spectrometers, and ion detectors can be found in the literature,[33,35] and the CLSI documents C-50A and C-62A contain recommendations for matching the capabilities of different m/z analyzers to various application classes.[1,2]

Beam-Type Designs. The main beam-type mass spectrometer designs are (1) quadrupole, (2) magnetic sector, and (3) TOF. It is convenient to categorize beam-type instruments into two broad categories, those that produce a mass spectrum by scanning the m/z range over a period (quadrupole and magnetic sectors) and those that acquire instantaneous snapshots of the mass spectrum (TOF). This categorization is not hard and fast. Certain instrument designs have been adapted to scanning or nonscanning operations. Nevertheless, the categorization is a useful one because it covers the majority of instruments currently available and because scanning and nonscanning instruments are adapted for different optimal usages.

Quadrupole. Quadrupole mass spectrometers, sometimes known as quadrupole mass filters (QMFs), are currently the most widely used mass spectrometers, having displaced magnetic sector mass spectrometers as the standard instrument. Although these instruments lag behind magnetic sector instruments in terms of (1) sensitivity, (2) upper mass range, (3) resolution, and (4) mass accuracy, they offer an attractive and practical set of features that account for their popularity, including (1) ease of use, (2) flexibility, (3) adequate performance for most applications, (4) relatively low cost, (5) small size, (6) noncritical site requirements, and (7) highly developed data collection software systems.

A quadrupole mass spectrometer consists of four parallel electrically conductive rods arranged in a square array (Fig. 2.11). The four rods form a long channel through which the ion beam passes. The beam enters near the axis at one end of the array, passes through the array in a direction generally parallel to the axis, and exits at the far end of the array. The ion beam entering the quadrupole array may contain a mixture of ions of various m/z values, and in different modes of operation, different mass ranges can be selected. If a very narrow m/z range is selected (eg, $\Delta m/z < 1$) only ions of the specified m/z will be transported through the device to reach the detector. Ions outside this narrow range are ejected radially. The $\Delta m/z$ range represents a pass band, analogous to the pass band of an interference filter in optics. This is why quadrupole mass spectrometers are often referred to as *mass filters* rather than *mass spectrometers*.

Separation of ions in QMS is based on a superposition of RF and constant direct current, or DC potentials applied to the quadrupole rods. DC voltages are applied to the electrodes in a quadrupolar pattern. For example, a positive DC potential is applied to electrodes 1 and 3, as indicated in Figure 2.12, and an equivalent negative DC potential is applied to electrodes 2 and 4. The DC potentials are relatively small, on the order of a few volts. Superimposed on the DC potentials are RF potentials, also applied in a quadrupolar fashion. RF potentials range up to the kilovolt range, and frequency is on the order of 1 MHz. In the most frequently used mode of operation the frequency is fixed and highly stable, derived from a crystal controlled oscillator.

The physical principles underlying the operation of a quadrupole mass spectrometer are rigorously described by solutions of a complicated differential equation, the Mathieu equation.[36] When an ion is subjected to a quadrupolar RF field, its trajectory is described qualitatively as a combination of fast and slow oscillatory motions. For descriptive purposes, the fast component will be ignored here. The slow

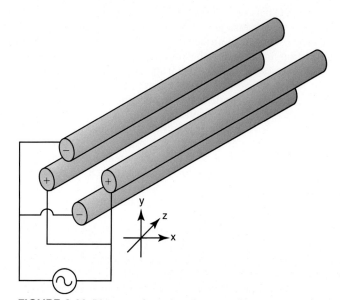

FIGURE 2.11 Diagram of quadrupole mass filter, including the radiofrequency part of voltages applied to the quadrupole rods.

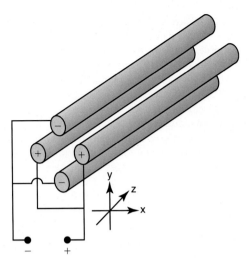

FIGURE 2.12 Direct current voltages applied to quadrupole rod assembly.

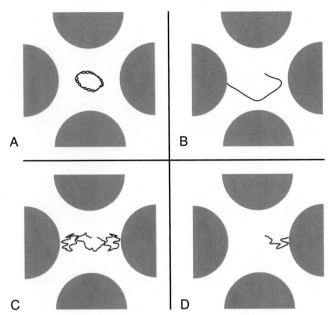

FIGURE 2.13 Ion trajectories showing confinement and ejection in quadrupole mass filters. A, Ion confinement by radiofrequency (RF)-only field. B, Ion ejection by RF-only field. C, Ion confinement with a combination of RF and direct current (DC) fields. D, Ion ejection with a combination of RF and DC fields. All trajectories were simulated using Simion software. (Courtesy Scientific Instrument Services, Ringoes, New Jersey.)

component oscillates about the quadrupolar axis; this resembles the motion of a particle in a fictitious harmonic *pseudopotential*. The frequency of this oscillation is sometimes called the *secular frequency*.

Effective force associated with the pseudopotential is directed inward toward the quadrupolar axis and is proportional to the distance from the axis. It therefore acts as a confining force, preventing ions from being ejected radially from the quadrupolar assembly. Figure 2.13, *A*, shows an example of an ion confined by an RF-only quadrupole. Below a certain m/z cutoff frequency (which depends on the frequency and amplitude of the RF field), ions are ejected rather than confined. Figure 2.13, *B*, shows an example of an ion ejected by an RF-only quadrupolar field. This establishes the low mass cutoff for the m/z pass band. The effective confining force is strongest just above the low m/z cutoff and then decreases asymptotically toward zero at high m/z.

The DC part of the quadrupolar potential is independent of m/z. Positive ions are attracted toward the negative poles. Negative ions are attracted toward the positive poles. Attraction increases as the distance from the quadrupolar axis increases. Because a quadrupolar DC potential always has both negative and positive poles, the quadrupolar DC potential always contributes to ejection in at least one direction. Whether ejection of an ion of a particular m/z actually occurs depends on whether the ejecting force caused by the quadrupolar DC potential overcomes the effective confining force caused by the pseudopotential generated by the RF field. Above a certain m/z value, the DC part dominates and ions are ejected radially from the device. This establishes an upper m/z limit for ion transmission. Figures 2.13, *C* and *D*, show examples of ion trajectories under the influence of combined RF-DC fields, one being confined and the other being ejected. Trajectories in Figure 2.13 were calculated using the Simion ion optics computer program.[37]

A rigorous description of low- and high-mass cutoffs is found in so-called stability diagrams, which graphically describe the lower and upper m/z cutoffs of a quadrupole mass spectrometer in terms of parameters related to voltages, frequencies, and m/z. However, a full discussion of the stability diagram is outside of the scope of this chapter.

The combination of lower and upper m/z limits establishes a pass band ($\Delta m/z$) and ultimately a resolution [$(m/z)/(\Delta m/z)$]. With relatively few exceptions, quadrupole instruments are limited to a resolution of a few hundred to several thousand, which is sufficient to achieve isotopic resolution for singly charged ions of m/z as high as several thousand.

A quadrupole MS may be operated in SIM mode or scanning mode. In SIM mode, both DC and RF voltages are fixed. Consequently, both the center of the pass band and the width of the pass band are fixed. For example, the mass spectrometer may be set to pass ions of m/z 363 ± 0.5. Both the center m/z and the $\Delta m/z$ are adjusted by the appropriate choice of DC and RF.

In the scanning mode of operation, the RF and/or DC voltages are continuously varied to scan a range of the specified m/z values. As with the SIM mode, the $\Delta m/z$ is determined by the RF and DC voltages. Usually the scan function is designed to maintain a constant $\Delta m/z$ across the full m/z range. Thus the resolution increases as m/z increases. The value of $\Delta m/z$ is frequently chosen in the range 0.5 to 0.7 to resolve isotopic peaks of singly charged species across the full m/z range.

Magnetic Sector. Because magnetic sector mass spectrometers are rarely used in clinical laboratories, they will not be described in detail here. It should be noted, however, that these classic mass spectrometers are easy to understand (given a basic understanding of physics); are versatile, reliable, and highly sensitive; and in their "double focusing" design are capable of very high m/z resolution and mass accuracy. However, they are typically very large, expensive, and have the reputation of being difficult to use.

Consequently, other instruments have largely displaced magnetic sector mass spectrometers.

Time-of-Flight. TOF mass spectrometry (TOF-MS) is a nonscanning technique whereby a full mass spectrum is acquired as a snapshot rather than by sweeping through a sequential series of *m/z* values while acquiring the data. It is described here as a snapshot because, although ions of different *m/z* arrive at the detector sequentially (low *m/z* first), the samples are loaded into the ion source with little or no *m/z* discrimination with regard to time, and the duration of the acquisition of a single mass spectrum is measured in microseconds. One implication of this is that if the composition of the sample stream being presented to the mass spectrometer changes with time, there is essentially no distortion of the mass spectrum resulting from this time dependence, whereas with scanning-type mass spectrometers the mass spectrum may be distorted because of the interaction between scan time of the mass spectrometer and the changing concentration of the sample stream. This is particularly significant when dealing with fast chromatography coupled to MS.

TOF mass spectrometers have several advantages, including (1) a nearly unlimited *m/z* range, (2) high acquisition speed, (3) high mass accuracy, (4) moderate to high resolution, (5) moderate to high sensitivity, (6) absence of spectral distortions when used in conjunction with fast separations and narrow chromatographic peaks, and (7) reasonable cost. TOF-MS is also well adapted to pulsed ionization sources, which is an advantage in some applications, particularly with MALDI and related techniques.[38]

A major advantage of modern TOF mass spectrometers is that they are capable of acquiring accurate mass measurements, sometimes loosely referred to as *exact mass*, which is typically accurate to a few parts per million (ppm). This allows TOF measurements to confirm the molecular formula of a compound and assist with identification of unknowns in the mass spectra. TOF mass spectrometers are conceptually simple to understand because they are based on the fact that in vacuum a lighter ion travels faster than a heavier ion, provided that both have the same kinetic energy.

Figure 2.14 presents a simplified conceptual diagram of a TOF mass spectrometer. It resembles a long pipe wherein ions are created or injected at the source end of the device and are then accelerated by the applied potential of several kilovolts. The ions travel down the flight tube and strike the detector at the far end. The time it takes to traverse the tube is known as the flight time; this is related to the mass-to-charge ratio of the ion.

The flight time for an ion of mass m and kinetic energy E to travel a distance L in a region free of electric fields is given by:

$$t = L\left(\frac{m}{2E}\right)^{1/2}$$

A sample calculation for an ion of molecular weight 200 Da (3.32×10^{-25} kg) with a kinetic energy of 10 keV (1.60×10^{-15} J), traveling through a distance of 1 m, yields a flight time of 10.18 µs, and an ion of molecular weight 201 takes just 25 ns longer. To accurately capture such fleeting signals, the data recording system must operate on an

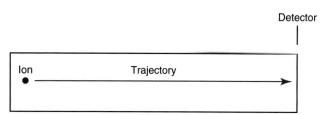

FIGURE 2.14 Diagram of simplified time-of-flight mass spectrometer.

approximately 1 ns or shorter time-scale. Advances in signal processing electronics have made these mass analyzers possible at a modest cost, and this has been a major factor in the rise in popularity of TOF-MS.

TOF is inherently a pulsed technique; it couples readily to pulsed ionization methods, with MALDI being the most common example, although TOF is also coupled with continuous ion sources such as EI, ESI, and APCI. However, the continuous nature of these sources causes a mismatch between continuous introduction of ions from the ion source and a pulsed detection with TOF-MS. This mismatch is overcome by using a technique known as orthogonal acceleration TOF-MS (OA-TOF-MS), in which the ion beam is injected orthogonal to the axis of the TOF-MS.[39-41]

During the injection period, the acceleration voltage is turned off. Once the injection region is filled with the traversing beam, the acceleration voltage is quickly turned on and the TOF timing cycle starts. The process is cycled repeatedly. The overall duty cycle for this method can be more than 10%; this represents a vast improvement over the traditional method of gating the ion beam for TOF analysis. For full spectrum capability with continuous ion sources, orthogonal injection TOF mass spectrometers are generally considered to have the lowest detection limits of all mass spectrometers. However, for the monitoring of a single *m/z* rather than a full mass spectrum, the use of SIM mode with a quadrupole MS provides superior sensitivity.

Improved resolution is an additional benefit conferred by OA-TOF-MS. Although a complete explanation is beyond the scope of this chapter, in brief, orthogonal acceleration reduces the resolution-degrading effects that would normally accompany the kinetic energy variations of individual ions in the ion beam.

Use of an ion mirror is another technique often employed in TOF-MS design to improve resolution by compensating for kinetic energy variations.[42,43] Such instruments are known as *reflectrons*. To date, TOF-MS has had a limited impact in clinical chemistry with only a few commercially offered TOF-MS assays, such as insulin-like growth factor I[44,45] and drug screening,[46] but it could potentially play a greater role in the future. For example, full-spectrum capability, high resolution (up to 40,000 in some current instruments), high speed (10 to 100 stored spectra per second), and high mass accuracy of TOF-MS seem ideally suited to applications such as high-speed drug screens in toxicology when combined with fast chromatographic sample introduction.

Another area in which TOF-MS provides an advantage is high-mass analysis, where its mass range is nearly unlimited. In MALDI-TOF, for example, it is not unusual to detect proteins with molecular weights exceeding 100,000 (see

Chapter 4). The ability for high-mass analysis is expected to increase in importance as clinical laboratories embrace proteomic-based diagnostic methods.

Trapping Mass Spectrometers. In contrast to beam-type designs, these mass spectrometers are based on the trapping of ions to capture and hold ions for an extended length of time in a small region of space. Trapping times vary from a fraction of a second to minutes. Compared with beam-type instruments, the division between scanning and nonscanning instruments has less meaning for ion-trapping instruments. The main practical difference between scanning and nonscanning instruments is related to distortions in chromatographic peak shape (or peak skewing). These arise from the finite scan time of a mass spectrometer relative to the time-scale of the width of a chromatographic peak. The result is that the abundances of the peaks in mass spectra collected during the rising or falling portions of a chromatographic peak are distorted relative to the true mass spectrum. In other words, as an instrument collects a mass spectrum, scanning from low m/z to high m/z, the peak intensities observed for low m/z will reflect the concentration of analyte that elutes earlier than the concentration of analyte that elutes when detecting high m/z. As a result, mass spectra collected at the beginning of a chromatographic peak may have different relative peak intensities compared with those of spectra collected at the end of a chromatographic peak. In terms of producing skewed spectra, trapping devices are more similar to nonscanning instruments, such as TOF (does not cause skewing), than to scanning instruments. This is because the sample is captured in an instant and then analyzed at leisure. Because the sample is captured in an instant, no skewing of the spectra occurs, regardless of whether the *mass* analysis is performed by a scanning or a nonscanning technique.

Traditionally, ion traps have been classified as (1) a QIT, which relies on RF fields to provide ion trapping; (2) a linear ion trap, which is closely related to the QIT in its operating principles; (3) an ion cyclotron resonance (ICR) mass spectrometer, which relies on a combination of magnetic fields and electrostatic fields for trapping, and (4) an orbitrap, a more recent introduction into the field of ion trap MS.[47]

Quadrupole Ion Trap. QITs are relatively compact, inexpensive, and versatile instruments that are excellent for (1) exploratory studies, (2) structural characterization, and (3) qualitative identification. They are also used for quantitative analysis, although precision of measurements is inferior when compared with quadrupole-based instruments.

Operation of the QIT is based on the same physical principle as the quadrupole mass spectrometer described earlier. Both devices make use of the ability of RF fields to confine ions. However, the RF field of an ion trap is designed to trap ions in three dimensions rather than to allow the ions to pass through as in a QMF, which confines ions in two dimensions. This difference has a large impact on the operation and limitations of the QIT.

The physical arrangement of a QIT is different from that of a QMF. If an imaginary axis is drawn through the *y*-axis of the quadrupole rods, and the rods are rotated around the axis, a solid ring with a hyperbolic inner surface results from the *x*-axis pair of rods. The two *y*-axis rods form two solid end caps. A diagram of an ion trap is given in Figure 2.15. The description of the fields within the electrodes must now include a radial component and an axial (between the end caps) component. These design features have an effect on the conditions required for ion confinement when compared with a QMF, although the qualitative description of ion confinement discussed previously is valid.

A discussion of the several types of scanning experiments in QIT is beyond the scope of this chapter, except to mention that ions may be ejected from the trap in an m/z-dependent fashion for detection using an external electron multiplier.

Some advantages of QITs are an ability to perform multiple stages of tandem MS (MS^n), high sensitivity, and decoupling of the mass analysis from scanning, so no mass spectral peak skewing is seen in GC-MS and HPLC-MS. However, ion-ion repulsion effects (caused by large numbers of similarly charged species in a small space within the trap) limit the number of ions that can be trapped, simultaneously reducing dynamic range and producing mass misassignments at high signal levels. The previously mentioned features make QITs not well suited for quantitative analyses, which are typically required for majority of applications in clinical laboratories.

Linear Ion Trap. The linear trap is an RF ion trap that is based on a modified linear QMF. Rather than being a pass-through device, as in a traditional linear QMF, electrostatic fields are applied to the ends to prevent ions from exiting out the ends of the device. When trapped in this manner, ions can be manipulated in many of the same ways as in a QIT. An advantage of the linear quadrupole ion trap is that the trapping field can be turned off at will and the device operated as a normal QMF. Furthermore the trapping volume available within the QMF is much greater than the traditional QIT, allowing greater capacity of the ions to be trapped before ion-ion repulsion becomes an issue. Thus a single device combines most of the features of a QIT and QMF and is extremely versatile. Commercial triple quadrupole mass spectrometers are being offered in

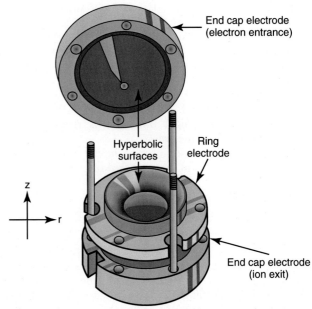

FIGURE 2.15 Diagram of quadrupole ion trap. *r*, Radial direction; *z*, axial direction.

which the third quadrupole is modified to function either as a linear trap or as a conventional third quadrupole mass spectrometer as selected by the user.

Ion Cyclotron Resonance. The ICR-MS excels in high-resolution and high mass accuracy measurements.[48] Measurements at resolution exceeding 1 million are not unusual. ICR is a trapping technique that shares many of the advantages of RF ion traps (QIT or linear ion traps). However, there are even more ways to manipulate ions in an ICR-MS than in RF ion traps, and MS^n (multiple stages of MS/MS) measurements are easily done with an ICR-MS. Sensitivity of an ICR-MS is generally high. Furthermore, sampling is decoupled from spectral acquisition, so no peak skewing is seen in chromatographic experiments—a feature that ICR shares with TOF and QIT, and the signal acquisition times are typically longer than for other types of mass analyzers.

Fourier transform ion cyclotron resonance–mass spectrometry (ICR-MS) is based on the principle that ions immersed in a magnetic field undergo circular motion (cyclotron motion). A typical ICR-MS uses a high-field (3 to 12 tesla) superconducting magnet. Within this field and within a high vacuum is mounted a cell typically composed of six metal electrodes, arranged as the faces of a cube. Ions are suspended inside the cell and undergo cyclotron motion, which keeps ions from being lost radially (the radial direction being defined as perpendicular to the magnetic field lines). A low (~ 1 V) potential is applied to the end caps to keep ions from leaving the trap axially. Thus the combination of electric and magnetic fields keeps ions confined within the cell.

Ions circulating in the ICR cell induce an electrical current in two parallel detection electrodes. The detection electrodes are on opposite sides of the ICR cell and are arranged parallel to the magnetic field. After certain mathematical operations are performed on the signal (principally a Fourier transform [FT]), a mass spectrum is recovered. Each m/z is associated with a specific cyclotron frequency, and each m/z value that is present in the sample produces a peak in the transformed signal. Because of the frequent use of FT in ICR, the technique is often referred to as FT-ICR or FTMS.

Although this technique has many advantages, including (1) high mass accuracy, (2) ultra-high resolution, and (3) the ability to perform MS^n, ICR-MS has several disadvantages, including (1) high instrument costs; (2) very demanding site requirements, in terms of both space and access restrictions; (3) requirement for a high-field superconducting magnet; (4) relatively long signal acquisition time, which limits the number of scans that can be acquired during the elution of a chromatographic peak; (5) safety concerns related to high magnetic fields; (6) demagnetization of credit cards and other magnetically encoded strips; (7) high costs of operation and maintenance because the instruments consume liquid helium and must never be allowed to run out of helium; and (8) necessity of a highly skilled individual to operate the instrument.

Orbitrap. The suitability of a new type of mass analyzer, the orbitrap, for clinical analysis has yet to be proved. However, the high resolution and mass accuracy of the orbitrap suggest that it has potential for use in clinical laboratories. The orbitrap mass analyzer has resolution and mass accuracy approaching that of an ICR mass spectrometer but does not require a magnetic field. This innovation minimizes many of the ICR disadvantages listed previously.

The principles of mass analysis in an orbitrap are based on an early ion storage device—the Kingdon trap.[49] After many variations over the years, Makarov and associates[47,50-52] developed a modified version that was commercialized in 2006. The commercial instrument can easily achieve resolutions up to 100,000 and parts per million or even sub–parts per million mass accuracy, has four orders of magnitude dynamic range, and has sampling decoupled from spectral acquisition (as in the ICR). The resolution and mass accuracy are typically approximately 2 orders of magnitude greater than with a quadrupole mass spectrometer.

Orbitrap-MS is based on trapping within electrostatic fields.[53] The actual device is a spindle-like central electrode surrounded by a barrel-like outer electrode.[54] When ions are introduced perpendicular to the central electrode and a radial potential is applied between electrodes, the ions spiral (orbit) around the central electrode and are effectively trapped in a radial direction. Trapping in the axial direction is assisted by the shape of the electrodes, together with the potentials that are applied to the electrodes. Ion trapping therefore involves both orbital motion around the central electrode and axial oscillations.

The trapping potential in the axial direction is of the form of a harmonic oscillator, and because the frequency of a harmonic oscillator is independent of oscillation amplitude, this frequency is very stable and well behaved. The m/z can be calculated from the frequency of axial oscillation:

$$\omega = 2\pi f = (km/z)^{-1/2}$$

where ω is angular velocity, f is frequency, m/z is the mass-to-charge ratio, and k is a constant determined by the trap geometry, dimensions, and applied potential. (To be dimensionally correct, m/z in this equation must have units of mass divided by charge, which differs from the currently accepted definition of m/z as a unitless number[31] (see also http://goldbook.iupac.org/M03752.html).

The image current (current induced by a motion of ions passing near a conductor) made in the outer electrode induced by the ion motion is acquired in the time domain and can be Fourier-transformed to produce a frequency spectrum that is then converted to m/z using the previous equation.

With the ability to perform accurate mass measurements, especially when combined with a linear ion trap or quadrupole to form a hybrid tandem mass spectrometer, orbitrap mass analyzers have excellent capabilities for proteomics research. One recent publication noted anomalous isotope ratios observed under high-resolution operating conditions.[55] It is a curious point that the anomalies are compound-dependent, and generally increase with increasing resolution. A theoretical explanation for these anomalies has been given.[56]

Tandem Mass Spectrometers

Tandem MS, or MS/MS, has become the dominant MS-based technique used in clinical laboratories, where it has found extensive application in the quantitative analysis of routine samples.[57] However, it is also a useful technique for structural characterization and compound identification and therefore often used for exploratory work. The most important features of this technique are its very high selectivity, ability to measure very low concentrations of analyte(s), and ability to multiplex the measurement of multiple analytes in a single method. Susceptibility of MS/MS to interferences is typically very low, especially if MS/MS is combined with chromatographic separation. The reason is that a detected compound is separated and characterized by three physical properties: chromatographic retention time, precursor ion mass, and product ion mass. Because of its high specificity, low consumable cost, and a potentially high sample throughput, increasingly more clinical laboratories are using tandem mass spectrometers for the routine analysis of samples.

The physical principle of MS/MS is based on the use of two mass spectrometers (or mass filters) arranged sequentially in tandem, with a collision cell placed between the two mass filters. The first filter is used to select a *precursor ion* of a particular m/z. The precursor ion is directed into the collision cell, where ions collide with background gas molecules and are broken into smaller product ions. The second mass filter acquires the mass spectrum of the product ions.

A variety of scan functions are possible with MS/MS. A product ion scan involves setting the first mass spectrometer (also called mass filter 1, MF1, MF1, MS1, or Q1) to select a given m/z, and scanning through the full mass spectrum of product ions using the second mass spectrometer or mass filter, MF2. This scan function is often used for structural characterization.

A precursor ion scan reverses this relationship, with the second mass filter, MF2, set to select a specific product ion, and MF1 is scanned through the spectrum of precursor ions. The scan tells which precursor ions produce a specific product ion—a capability that is often used to analyze for specific classes of compounds. For example, acylcarnitines are often analyzed using precursor ion scan mode by acquiring signal from all the precursors of the m/z 85.

In a constant neutral loss scan, the two mass filters are scanned synchronously, with a constant m/z offset between precursor and product ion. This scan indicates which ions lose a particular neutral fragment. For example, an offset of 176 m/z units would select for ions losing a glucuronide moiety in the dissociation process.

The most commonly used scan function in MS/MS is MRM (also referred as SRM). In this type of acquisition a series of precursor/product ion pairs are monitored, with the mass spectrometer set to step through the table of parent/product ion pairs in a cyclic fashion. MRM acquisition is primarily used for quantitative analysis of target compounds and is an analog to the SIM type of acquisition used in GC-MS.

As with single-stage mass spectrometers, MS/MS are roughly categorized as beam-type instruments and trapping instruments. The most popular beam-type instrument is the triple quadrupole. In this instrument, the first quadrupole (Q1) functions as MF1 and the third quadrupole (Q3) functions as MF2. Between these two quadrupoles is another quadrupole, Q2, which functions as the collision cell. The pressure is raised in Q2 (eg, $>10^{-3}$ torr) by the addition of a nonreactive gas (nitrogen, helium, argon, etc.) to the point that ions traversing Q2 undergo multiple collisions, leading to deposition of energy onto the analyte and subsequent fragmentation of the precursor ion(s) into smaller fragments, followed by separation and detection of the product ions in a subsequent stage of mass analysis. The Q2 is operated as an RF-only quadrupole, ideally passing all ions regardless of m/z. The technique is typically used to fragment molecular or pseudomolecular ions in order to obtain analyte-specific fragments, which can be used for elucidating structure of the molecules of interest or in selective analysis of targeted molecules. In cases in which the collision-induced dissociation spectrum contains a large number of ions, the experienced investigator often can deduce the structure of a molecule from the mass spectrum of the product ion.

Two magnetic sector instruments also have been operated in tandem, with a collision cell placed between the two mass analyzers. These instruments permit high-resolution selection of both precursor and product ions. However, they are now rarely used because of the high cost and cumbersome operation. A single magnetic sector mass spectrometer (in the form of a double focusing mass spectrometer) has also been used as an MS/MS by a technique known as *linked scanning*. A product ion scan by linked scanning involves low resolution for the first m/z selection and high resolution for the second m/z selection.

Hybrid mass spectrometers include a combination of two different types of mass spectrometers in a tandem arrangement. The combination of a magnetic sector mass spectrometer with a quadrupole mass spectrometer was an early instrument of this type. More popular today is the combination of a quadrupole for the first stage of m/z selection and a TOF for the second m/z analyzer. Subsequently, linear ion trap and quadrupole mass analyzers have been combined with an orbitrap. Hybrid instruments are presently used mainly for proteomics research. These instruments cannot perform the true precursor ion scans or constant neutral loss scans, though it is possible to mimic these functions by postprocessing data, provided the full precursor-product ion map was generated in the experiment.

QIT, linear ion trap, and ICR mass spectrometers also can be used as MS/MS. Unlike beam-type instruments, which are referred to as "tandem in space," trapping mass spectrometers are "tandem in time," meaning that ions are held in one region of space while the parent ion is selected and dissociated and the daughter ion is analyzed sequentially in time in the same region of space. The ability to perform MS/MS is inherent in the design of most trapping mass spectrometers. Generally, little or no additional hardware is required, and tandem capability is supplied via software. An exception is the orbitrap, which is not amenable to MS/MS when used alone. However, when incorporated into a hybrid instrument, with a different type of mass spectrometer supplying the first stage of MS (such as a linear ion trap or a quadrupole), and with the orbitrap providing the final stage of MS, MS/MS is possible in an orbitrap-based instrument.

Most trap-based instruments are capable of multiple stages of MS. Thus, product ions may be further dissociated to produce another generation of product ions (MS/MS/MS, or MS^3). In principle, any number of dissociation stages may be performed (MS^n). This capability finds its greatest use in structural characterization, such as in the sequencing of peptides, and is less useful for quantitative analysis.

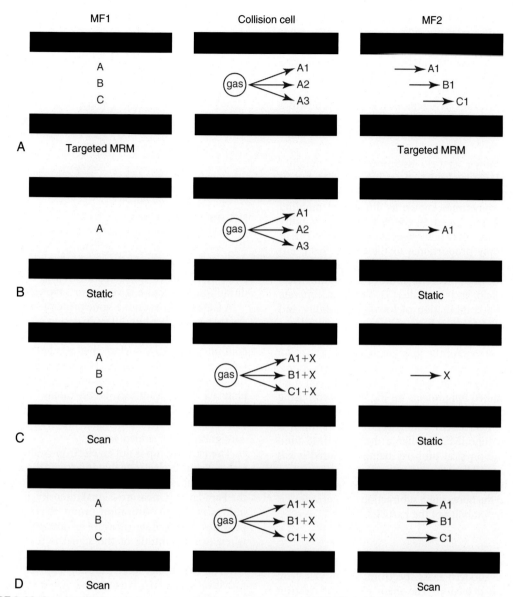

FIGURE 2.16 Scan modes in mass spectrometry/mass spectrometry (MS/MS). A, Multiple reaction monitoring (MRM), in which A, B, C, A1, B1, and C1 are ions. Monitoring of MS/MS transitions $A \rightarrow A1$, $B \rightarrow B1$, and $B \rightarrow C1$ is multiplexed. For simplicity, only the dissociation of A is shown in a collision cell in the figure. B, MRM of a single compound, where only one MS/MS transition is monitored. C, Precursor ion scan, in which A, B, C, and X are all ions. The second mass filter (MF2) is fixed to monitor the mass-to-charge ratio (m/z) corresponding to ionic species X, and the first mass filter (MF1) is scanned through a range of m/z values. D, Constant neutral loss scan, in which X is uncharged and A, B, C, A1, B1, and C1 are ions. The two mass filters are scanned with a constant m/z offset between the two corresponding to the mass of X.

Although trapping designs are extremely versatile (such as allowing multiple stages of fragmentation), these instruments are unable to perform true precursor ion scans or constant neutral loss scans, as illustrated in Figure 2.16, C and D, respectively. However, it is possible to simulate the effect of precursor ion scans or constant neutral loss scans by taking a series of product ion scans, one for each possible parent ion m/z. This will generate a complete MS/MS map. From this complete map data can be selected to simulate these two scan modes because a precursor ion scan is just a subset of the complete MS/MS map, as is a constant neutral loss scan. However, this procedure can be quite time-consuming, which would make it impractical in some applications.

Ion Mobility

Although strictly speaking, ion mobility spectrometers (IMS) are not mass analyzers, they are nevertheless often included as part of the field of MS, either as part of a hyphenated technique (eg, IMS-MS) or as a substitute for an MS analyzer.[58-60] Ion mobility spectrometers are like mass spectrometers in the sense that they require the analyte to be ionized, but the separation mechanism is different. Rather than separating ions by their mass-to-charge ratio, ions are separated according to their mobility in an electric field. Thus ion mobility can be regarded in some respects as a form of gas phase electrophoresis.

The simplified schematic of a conventional ion mobility spectrometer strongly resembles a TOF-MS, but rather than following a collisionless trajectory, ions undergo many collisions as they drift under the influence of an electric field. Other configurations for measuring gas phase mobility are also possible,[58-60] but these will not be reviewed in detail here.

An IMS may operate at atmospheric pressure or at reduced pressure but not under a high vacuum because collisions are necessary for its operation. When used in conjunction with a mass spectrometer it is possible to place the mobility device before the first mass analyzer or following one or more stage of mass analysis.

A technique known as field asymmetric ion mobility spectrometry (FAIMS) is also based on ion mobility, but in this case ions are not separated strictly according to their mobility. FAIMS, sometimes known as differential mobility spectrometry (DMS), is based on the fact that the mobility of a gas phase ion is not strictly constant; that is, the drift velocity is not simply proportional to the electric field, but rather at high field there is a deviation from the proportional relationship. FAIMS uses a combination of an asymmetrical high-voltage RF field and a smaller DC field to separate ions according to a combination of low-field mobility and high-field mobility. FAIMS is beginning to find applications in clinical MS when used as a filtering device positioned between the ion source and the mass analyzer.[61]

IMS also has been used alone for clinical applications, without being combined with MS. Notably, it has been used to separate unmodified lipoproteins on the basis of size using a differential mobility analyzer (DMA, not to be confused with DMS). The instrument configuration differs from a conventional drift tube IMS, but like a drift tube IMS (and unlike FAIMS), the physical property being measured is gas phase ion mobility. After the separation, each lipoprotein particle is directly detected and counted as it exits the separation chamber, and the lipoprotein subfraction categorization is made based on the mobility of the particles.[62-64]

Detectors

With the exception of ICR-MS, orbitrap, and some ICP-MS instruments, most modern mass spectrometers use electron multipliers for ion detection. The main classes of electron multipliers used as MS detectors include the (1) discrete dynode multipliers; (2) continuous dynode electron multipliers (CDEMs), also known as channel electron multipliers; and (3) microchannel plate electron multipliers, also known as multichannel plate electron multipliers. Although different in design, all three work on the same physical principle. Additional types of detectors used in mass spectrometers are the Faraday cup, image current detection, and photomultipliers.

Figure 2.17 presents a conceptual diagram of the operation of a discrete dynode electron multiplier. When an ion strikes the first dynode, it causes the ejection of one or more electrons (secondary electrons) from the dynode surface. The electron is accelerated toward the second dynode by a voltage difference of approximately 100 V. On striking the second dynode, this electron causes the ejection of additional electrons, typically 2 or 3. The second group of electrons is then accelerated toward the third dynode and, on striking the third dynode, causes the ejection of several more electrons. This process is repeated through a chain of dynodes, numbering between 12 and 24 for most designs. The cascade process typically produces a gain of 10^4 to 10^8,

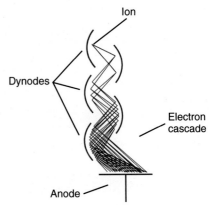

FIGURE 2.17 Discrete dynode electron multiplier showing dynode structure and generation of electron cascade.

meaning that one ion striking the first electrode produces a pulse of 10^4 to 10^8 electrons at the end of the cascade. The duration of the pulse is very short, typically less than 10 ns.

A CDEM works on the same principle as a discrete dynode electron multiplier but differs in design. The set of dynodes of a discrete dynode electron multiplier is replaced by a single continuous resistive surface that acts both as a (continuous) voltage divider to establish the potential gradient and as the secondary electron-generating surface. A microchannel plate electron multiplier is essentially a monolithic array of miniaturized CDEMs fabricated in a single wafer or disk of glass. Sometimes these are stacked into a chevron configuration for added gain.

The Faraday cup is not an electron-multiplying device, but rather a simple electrode that intercepts the ion beam directly. This current is amplified using electronic amplifiers. Because the Faraday cup measures signal intensity directly, rather than indirectly (as in saturation-prone electron multipliers), it provides an absolute measure of ion current and is useful when the magnitude of the signal is too high for electron multiplier–based detection. Some instruments use both electron multiplier and Faraday cup–based detection to provide extended dynamic range—a capability that is especially useful for elemental analysis of trace and toxic elements by ICP-MS.

Detection in ICR occurs via image current detection. This is closely related to the Faraday detection cup in the sense that the ion current is detected directly. However, ions are not destroyed in the process of image current detection and are available for remeasurement. This feature is one of the keys to the versatility of ICR mass spectrometers. Image current detection is also used in the orbitrap.

Closely linked to the detection system is the electronic and signal processing system. In instruments that use electron multiplier detection (the vast majority of mass spectrometers), the raw signal from the detector is processed in one of two ways: (1) individual pulses (corresponding to individual ions) may be counted, as in ion counting systems, or (2) the signal may be converted to a digital representation of the analog signal using an analog-to-digital converter, as in analog detection.

Computer and Software

Because of their (1) mass resolution capabilities, (2) scanning functions, (3) ability to automatically switch between positive to negative ionization modes, and (4) speed with which multiple *m/z* signals are acquired, modern MS instruments generate enormous quantities of raw data. In addition, the use of MS in such areas as (1) proteomics, (2) biomarker discovery, (3) synthetic combinatorial chemistry, (4) high-throughput drug discovery, (5) pharmacogenomics, (6) toxicology, and (7) therapeutic drug monitoring requires that MS manufacturers provide powerful computers and software.

In toxicology laboratories, one important function of the data system is library searching to assist in compound identification. Several commercial libraries, including the Wiley Registry of Mass Spectral Data; the NIST Mass Spectral Database; and the Pfleger, Maurer, and Weber drug libraries, are available. In addition, many laboratories generate their own libraries. The quality and number of available spectra, the search algorithm, and whether condensed or full spectra are searched are all important factors in spectral matching.

In proteomics and biomarker discovery, complex mass spectra from single proteins, protein mixtures, or protein digests corresponding to complex samples are obtained. Data systems aid in characterization of spectral data to identify such properties as intact protein mass, amino acid subsequences, and posttranslational modifications. Fragmentation information also can be compared with peptide databases to identify structural mutations that may be present.

The most important function of software in MS systems is data collection and processing. Chromatographic peaks are integrated using data analysis software, and integrated peak intensities or peak areas serve as the basis for quantitative analysis. Calibration curves are generated during data processing, and quantitative results from individual samples are generated using the calibration curves; the data systems also contain report generation capabilities.

Deconvolution protocols have been developed that identify and characterize the mass spectra corresponding to the coeluting peaks. In addition to proprietary deconvolution protocols embedded in the data systems of mass spectrometers supplied by some vendors, there is a freely available deconvolution software program known as AMDIS.[65]

CLINICAL APPLICATIONS

Mass spectrometers coupled with gas or liquid chromatographs (GC-MS or LC-MS) serve as versatile analytical instruments that combine the resolving power of a chromatograph with the specificity of a mass spectrometer.[66] Such instruments are powerful analytical tools that are used by clinical laboratories to identify and quantify biomolecules. The instruments are capable of providing structural and quantitative information in real time on individual analytes as they elute from a chromatographic column. Specific applications of these coupled instruments can be found in Chapters 5 and 6.

Gas Chromatography—Mass Spectrometry

GC-MS has been used for the analysis of biological samples for several decades. This technique is used by the US National Institute of Standards and Technology and other agencies for the development of definitive methods to qualify standard reference materials and assign accurate concentration to reference materials of many clinically relevant analytes, including cholesterol, glucose, steroid hormones, creatinine, and urea nitrogen.

One of the most common applications of GC-MS is drug testing for clinical or forensic purposes. Many drugs have relatively low molecular weight and nonpolar and/or volatile properties, making these compounds particularly suitable for analysis by GC. Electron impact ionization with full scan mass detection is the most widely used approach for comprehensive drug screening. Unknown compounds can be identified by matching full mass spectrum of unknown peaks with a mass spectral library or a database. In addition, vendors have recently introduced GC tandem quadrupole (GC-MS/MS) mass spectrometers, which should expand the capability of GC-MS to perform improved targeted and untargeted analysis, thus enhancing existing screening and mass spectral identification capabilities of GC-MS.

GC-MS has many applications beyond drug testing. Numerous xenobiotic compounds are readily analyzed by GC-MS. Applications for anabolic steroids, pesticides, pollutants, and inborn errors of metabolism have been described.[67-69]

One important limitation to GC-MS is the requirement that compounds be sufficiently volatile to allow transfer from the liquid phase to the mobile carrier gas and thus to elute from the analytical column to the detector. Although many biologic compounds are amenable to chromatographic separation with GC, numerous other compounds are too polar or too large to be analyzed with this technique. In many cases, chemical derivatization is necessary to create sufficiently volatile forms of compounds. Knapp's classic work on derivatization[70] may be consulted for more information.

Despite its limitations, GC-MS has several positive attributes. High-efficiency separations have been achieved with numerous commercial capillary columns. This technique allows achieving high-efficiency chromatographic separation and excellent limits of quantification, and it allows use of commercial mass spectral libraries for identification of sample constituents. For some of the analytes, such as organic acids, GC-MS has advantages of higher specificity compared to soft ionization techniques used in LC-MS.

Liquid Chromatography—Mass Spectrometry

As discussed earlier, several interface techniques have been developed for coupling a liquid chromatograph to a mass spectrometer, notably ESI and APCI, which have allowed LC-MS and LC-MS/MS to be successfully applied to analysis of a wide range of compounds. In theory, as long as a compound can be dissolved in a liquid, it can be introduced into an LC-MS system. Thus, in addition to low molecular weight polar and nonpolar analytes, large molecular weight compounds, such as proteins, can be analyzed using this technique (see Chapter 6).

LC-MS/MS has gained momentum in the arena of toxicology screening and confirmation.[71-73] A majority of the currently used methods for targeted analysis use MRM acquisition using mass transitions corresponding to drugs of interest (see Fig. 2.16, *A*). For example, within the chromatographic time window of 1.0 to 2.0 minutes, the MRM transitions for selected sympathomimetic amines might be monitored. During the next defined time window, a new set of MRM transitions are monitored and so on for the rest of the chromatographic run. A related approach is the use of targeted MRM, in which recognition of a chromatographic peak containing a preselected MRM transition triggers a product ion scan in a process called *information-dependent acquisition,* also known as *data-dependent acquisition.* One benefit of this approach is the ability to provide confirmation of the identity of the peaks identified during the analysis.

Coupling of TOF-MS to GC or LC provides a new approach to the identification of unknowns.[46,74] Because TOF is capable of achieving high mass resolution and high sensitivity, the need for compound fragmentation may be minimized, allowing compound identification based on retention time and accurate mass.[75]

The number of quantitative LC-MS/MS assays introduced for the measurement of clinically important compounds has markedly increased. For example, a few compounds that have been of special interest include (1) immunosuppressant drugs,[76] (2) biogenic amines,[77] (3) 25(OH)-vitamin D,[78] (4) antiretroviral drugs,[79,80] (5) psychoactive drugs,[81-83] (6) methylmalonic acid,[84,85] (7) thyroid hormones,[86] and (8) steroids.[87,88] When quantification of a specific compound is desired, the most effective approach is MRM analysis (see Fig. 2.16, *B*). With MRM acquisition, both mass filters MF1 and MF2 are set in a static mode, whereby only precursor ions specific for the compound and the internal standard being measured are passed through MF1. This preselected precursor ion is then fragmented in the collision cell, and molecule-specific fragment ions derived from the compound of interest are passed by MF2 to the detector. Because only one ion is monitored in MF1 and typically two molecule-specific fragment ions monitored in MF2, as opposed to scanning for multiple ions, the MRM approach allows much greater specificity as well as lower limits of quantification.

Another area in which MS/MS is used clinically is screening and confirmation of genetic disorders and inborn errors of metabolism.[89,90] The ability to analyze multiple compounds in a single analytical run makes this technique an efficient tool for screening purposes. In this application, in some cases MS/MS is of sufficient selectivity to eliminate the need to incorporate LC separation, a simplification that allows high-throughput analysis.

Electrospray-MS/MS is also used for carnitine and acylcarnitine analysis to detect organic acidemias and fatty acid oxidation defects.[91,92] In the methods for acylcarnitine and amino acid analysis, these compounds vary widely in their polarity, which creates problems with consistency of response factors. To address this issue, most methods use a butyl ester derivatization of the carboxyl group to force cationic character on the amino acids and thus enhance the ionization efficiency.[93]

Acylcarnitines can be analyzed without derivatization,[94] but most often are analyzed as butyl esters using a *precursor ion scan* mode of acquisition (see Fig. 2.16, *C*).[95,96] This type of acquisition makes use of the fact that acylcarnitines have a common collision-induced *m/z* 85 product ion (represented by *X* in Fig. 2.16, *C*) that is selectively monitored in MF2. MF1 is set to scan for precursors with *m/z* 85, thus detecting and identifying acylcarnitines present in the sample (see Fig. 2.16, *C*). By incorporating in the method stable isotope-labeled analogs of the targeted acetylcarnitines, it is possible in addition to identification, to establish concentration of the acylcarnitines of interest.

Analysis of amino acids by LC-MS/MS is typically performed using traditional MRM monitoring but also can be performed using a data acquisition mode known as *constant neutral loss* (see Fig. 2.16, *D*).[88] Butyl derivatives of α-amino acids share a common neutral product, butylformate, which has a mass of 102 Da (represented by *X* in Fig. 2.16, *D*). By scanning for both product (MF2) and precursor (MF1) ions, and by keeping a constant offset between the two mass *m/z* analyzers (eg, a difference of 102 *m/z* units), any *m/z* differences that equal 102 Da can be used to detect and identify amino acids present in the samples.

One advantage of LC-MS/MS relative to GC-MS is that in many cases it allows avoidance of derivatization of the target compounds, but in some cases derivatization is useful for LC-MS/MS as well. An example of butyl ester derivatization was discussed previously. In this example, the derivative has more favorable fragmentation properties than the underivatized compounds. Similarly, the dibutyl ester of methylmalonic acid (MMA), when run in positive ion mode ESI, has more favorable MS/MS spectra than the underivatized compound run in negative ion mode ESI; in addition, the dibutyl esters of dicarboxylic acids are selectively ionized in positive ion mode ESI, whereas monocarboxylic organic acids are not efficiently ionized and are therefore not detected by the mass spectrometer.[84] By using MMA extraction at conditions specific for acidic compounds and detection specific for polycarboxylic acids, it is possible to perform the LC-MS/MS analysis using isocratic chromatographic separation without the need for reconditioning and reequilibration of the chromatographic column between injections.[84]

The most frequent reason for using derivatization in LC-MS and LC-MS/MS is to achieve improved ionization efficiency. Gao and colleagues[97] have emphasized this issue in an extensive discussion of derivatization in ESI and APCI MS.

Product ion scan is another mode of acquisition using MS/MS; in this scan mode, the first mass filter is fixed to pass a specific *m/z* and the second mass filter is scanned over a specified range of *m/z* values. Product ion scan mode is very useful for structural elucidation, such as in peptide sequencing, but is less useful for routine quantitative analysis.

Matrix-Assisted Laser Desorption Ionization Mass Spectrometry

MALDI (typically coupled with a TOF analyzer) has been used to analyze many different classes of compounds. Notably, it has been widely applied in discovery applications for the detection and identification of proteins and peptides (see Chapter 6). Primary limitations include high background noise and a higher coefficient of variation that seems inherent in the MALDI ionization process. In addition, MALDI is essentially a batch-type process that does not interface naturally with online separation processes using chromatographic techniques (eg, HPLC, capillary electrophoresis).

MALDI-TOF is often used to determine the identity of proteins through peptide mass fingerprinting. This technique has been used to identify a large number of two-dimensional (2D) gel spots for the bacterial pathogen *Pseudomonas*

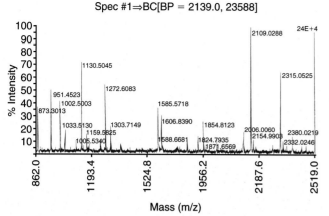

FIGURE 2.18 Example of a matrix-assisted laser desorption ionization–time-of-flight spectrum showing peptides generated in a tryptic digest of a spot cored from a two-dimensional sodium dodecyl sulfate polyacrylamide gel electrophoresis. The 16 most abundant *m/z* values were submitted to the MS-Fit database for searching against the nonredundant database. The results for this search are shown in Table 2.1.

aeruginosa.[98] The procedure generally involves in-gel tryptic digestion followed by accurate mass measurement of the peptides produced during the digestion. The generated mass list is then compared with theoretical tryptic masses for proteins in a database (Fig. 2.18 and Table 2.1). This procedure, which works best for organisms with complete and annotated genomes, is very rapid because 100 or more samples may be deposited on a single MALDI target plate and automatically processed. In the previous example,[98] the group rapidly identified a large number of proteins that were expressed differently among the studied bacteria. In addition, it was found that some proteins were listed as "hypothetical," meaning they were previously undescribed or confirmed to be expressed, and that the theoretical molecular weight and/or isoelectric point (pI) in some cases were different from those measured in the gel, indicating possible loss of terminal amino acids and/or posttranslational modifications.

One clinical application of MALDI-TOF that has proved its clinical utility is identification of microorganisms (discussed in more depth in Chapter 4). Identification of the bacteria is performed by fingerprinting proteins and peptides extracted from cultures using gentle conditions.[99] The basis of this technique is that different bacteria express unique mixtures of proteins and peptides; when samples are analyzed using MALDI-TOF, the bacteria-specific mass spectra are observed in the 2- to 20-kDa mass range, allowing database searching and classification based on the protein mass fingerprint. One of the disadvantages[99] of the technique is the lack of actual protein information and the relative lack of specificity to different strains of the same bacteria. The protein mass fingerprints must be catalogued (entered in mass spectral library) for each bacterium and validated to be specific and reproducible for a given extraction method.

TABLE 2.1 Example of Printout of Bacterial Identification Through Peptide Mass Fingerprinting Using Matrix-Assisted Laser Desorption Ionization MALDI-Time of Flight*

Rank	Mowse Score	# (%) Masses Matched	Protein Mw (Da)/pI	Species	NCBInr.81602 Accession #	Protein Name
1	1.07e+008	14/16 (87%)	101754.9/ 9.15	*Saccharomyces cerevisiae*	6321275	(Z72685) ORF YGL163c

1. 14/16 matches (87%). 101754.9 Da, pI = 9.15. Acc. #6321275. *Saccharomyces cerevisiae*. (Z72685) ORF YGL163c.

m/z Submitted	MH+ Matched	Delta ppm	Start	End	Peptide Sequence (Click for Fragment Ions)	Modifications
870.4746	870.4797	−5.8732	598	606	(K) GVGGSQPLR(A)	
873.3981	873.3929	5.9793	774	779	(K) DCFIYR(F)	$C^2H^2O^2$
951.4901	951.4900	0.1050	814	821	(R) LFSSDNLR(Q)	
1002.5385	1002.5373	1.2224	515	522	(K) NFENPILR(G)	
1033.5513	1033.5543	−2.8793	46	55	(K) NTHIPPAAGR(I)	
1130.6349	1130.6322	2.4037	120	128	(R) LSHIQYTLR(R)	
1130.6349	1130.6322	2.4037	514	522	(R) KNFENPILR(G)	
1159.6039	1159.6071	−2.7957	56	67	(R) IATGSDNIVGGR(S)	
1272.6508	1272.6483	1.9865	734	746	(K) AGGCGINLIGANR(L)	$C^2H^2O^2$
1303.7573	1303.7599	−1.9457	270	280	(K) ILRPHQVEGVR(F)	
1585.7190	1585.7215	−1.5602	446	459	(K) NCNVGLMLADEGHR(L)	$C^2H^2O^2$
1606.8861	1606.9029	−10.4650	22	35	(R) LVPRPINVQDSVNR(L)	
2138.0756	2138.0704	2.4250	747	765	(R) LILMDPDWNPAADQQALAR(V)	
2315.1093	2315.0951	6.1321	401	423	(K) SSMGGGNTTVSQAIHAWAQAQGR(N)	
2388.0671	2388.0731	−2.5004	293	313	(K) DYLEAEAFNTSSEDPLKSDEK(A)	

MH+, Ion formed by attachment of a proton to molecule M; MOWSE, MOlecular Weight SEarch method; MW, molecular weight; m/z, mass-to-charge ratio

*A generated mass list is compared with theoretical tryptic masses for proteins in a database. Match quality is used for pathogen identification.

Some of these drawbacks were addressed in a MALDI technique that targets ribosomal proteins.[100] This technique was evaluated using 1116 isolates collected in a routine clinical microbiology laboratory and was described as being fast, reliable, and easy to use. More than 95% of clinical isolates were correctly identified, and most of the previously incorrectly identified isolates were assigned to the correct genus or a closely related genus. Bacterial identification by MALDI MS is rapidly becoming a routine method in microbiology laboratories.

MALDI MS has the reputation of being a nonquantitative technique. However, some progress has been made toward its use as a quantitative technique.[101] If this application becomes routine, it could have major benefits for clinical MS because the time to acquire a mass spectrum by MALDI is only a few seconds. This could dramatically improve throughput. However, it seems likely that this application will require off-line separation (or sample fractionation) before loading on the MALDI target, to obtain sufficient selectivity necessary for clinical applications.

Inductively Coupled Plasma Mass Spectrometry

ICP-MS is used for the determination of trace and toxic elements in many types of samples (see references 10, 102, and 103). However, it is known that the toxicity of an element may depend on the organic or inorganic state in which the element is present. In these cases, it is more important to ascertain the concentrations of toxic species rather than the total concentration of the element. To extend the usefulness of this technique, GC and HPLC systems have been coupled to ICP-MS to separate different compounds containing the targeted element before ICP-MS analysis.[104]

Proteomics, Genomics, and Metabolomics

The past 20+ years have seen tremendous progress in genomics, with hundreds of genomes completed or near completion and many now parsed and annotated. This information is highly complex, mainly because of the myriad changes that occur to proteins produced from the genome throughout the life cycle of a cell, but potentially will provide a better understanding of the cellular functions and allow discovery of novel disease biomarkers.[105] In the mid-1990s, MS came to the forefront of analytical techniques used to study proteins, and the term *proteomics* was coined. Although the definition of proteomics is still debated, for the present discussion it is taken to encompass knowledge of the structure, function, and expression of all proteins in the biochemical or biological contexts of all organism.[106] In a more basic and practical sense, proteomics refers to the identification and quantification of proteins and their posttranslational modifications in a given system or systems. Proteome analysis is a powerful tool for investigating (1) biomarkers of disease, (2) antigens of pathogens, (3) drug target proteins, and (4) posttranslational modifications, as well as for other investigations. This is a challenging task in that a given gene may have many distinct chemical protein isoforms. In addition, many other molecules (metals, lipids, etc.) interact with proteins in a noncovalent fashion. Therefore in a genome, such as the human genome, a repertoire of more than a million proteins may require identification and quantification. Two foundations are necessary to begin this daunting challenge. The first is the basic sequence expected for each possible protein in a cell (ie, information from a completed genome). The second is instrumentation, which currently consists of advanced mass spectrometers that identify and quantify protein isoforms in an automated fashion at very low limits of detection. Both foundations are now essentially in place. However, the goals previously stated are far from being reached, and considerable advances will need to be made in the field of systems biology for better understanding of the biological systems.

Currently, MS is routinely used to accomplish many tasks in proteomics. The most basic task is protein identification. The typical approach is known as the *bottom-up* method, whereby proteins are separated—by gel electrophoresis or by solution-based methods—and then digested. The resulting enzymatic fragments are analyzed and used to identify the protein(s) present. This process is time-consuming and has many pitfalls. Increasingly, much research has been devoted to analysis of mixtures of proteins. These mixtures are derived from biological fluids, cellular compartments, tissue, or immunoprecipitation. Currently, both instrumentation and data analysis software are not sufficiently advanced to allow unambiguous identification of all the proteins in highly complex biological samples. As a result, much emphasis has been placed on separation methods and enrichment techniques for preparing samples for analysis of proteins and peptides.

Another approach that was shown to enable sequencing of intact proteins and posttranslational modifications is known as the top-down method. Top-down proteomics involves identification of proteins in complex mixtures without prior digestion of proteins into peptides. Approaches used for protein top-down characterization include extraction of the proteins from samples, fractionation and analysis of the samples using high-resolution accurate mass MS/MS with CID, higher energy collision dissociation, and electron-transfer dissociation fragmentation. Main benefits of the top-down analysis are in the ability to detect in the samples proteins containing posttranslational modifications and their sequence variants.

Many research groups have introduced methods allowing handling of highly complex biological samples. The most popular approaches include subcellular fractionation, multidimensional chromatography, affinity enrichment, and multiplexing. By combining these approaches, several thousand protein species can be identified routinely. Obviously these numbers are better than those obtained through bottom-up methods from gels, but they still fall far short of those necessary for complete understanding of biological systems.

The term *proteomics* is often used in the context of biomarker discovery. To date, very few markers have been discovered using proteomics methods that have migrated to the clinical laboratory. Some of the reasons for the dearth of new protein biomarkers have recently been discussed.[107] From a broader view, however, proteomics also may include the application of MS for the analysis of known protein and peptide biomarkers. For example, mass spectrometric methods for the analysis of carbohydrate-deficient transferrin have been developed, including a reference method[108] and a method for routine patient testing. Additional areas of application of MS include analysis of the proteome of the pathogenic mold *Aspergillus fumigatus* with the aim of identifying vaccine candidates and new allergens.[109]

Analysis of thyroglobulin, a widely used marker of the recurrence of thyroid cancer, has been described by several groups.[110-112] The initial digestion of the thyroglobulin by trypsin also digests and therefore removes autoantibodies which cause interference in immunoassays. Currently assays based on the above principle are offered in several commercial laboratories and represents substantial progress in the application of MS for routine analysis of proteins as well as providing important information to treating physicians.

Promising proof-of-principle research has been performed on the characterization of hemoglobinopathies by MS.[113,114] Most methods for hemoglobin analysis use MS to detect separate hemoglobin chains or peptide products from enzymatic digests of hemoglobin. However, as shown by Rockwood and coworkers,[115] and shortly thereafter by Ganem and associates,[116,117] the retention of higher order structure, such as noncovalent complexes, is possible when ions are transferred from solution to gas phase, and hemoglobin tetramers have been observed by MS.[118] Another hemoglobin application, a reference method for hemoglobin A_{1c} using MS, has been approved by the International Federation of Clinical Chemistry and Laboratory Medicine.[119] LC-MS methods for quantitative analysis of hepcidin, a peptide hormone believed to be a master regulator of iron status, have been published.[120-122]

Genetic applications for clinical MS are beginning to emerge. For example, MALDI-TOF has been used for mutation detection in myeloproliferative disorders,[123] DNA methylation analysis,[124,125] and gene expression analysis.[126]

A promising genomic approach for pathogen identification uses polymerase chain reaction amplification of selected regions of a pathogen genome, followed by accurate mass measurement using ESI-TOF-MS. From the accurate mass information, a DNA base composition is computed, and the results are matched to pathogen DNA base compositions in a database.[127,128]

A burgeoning area in which MS plays a role is the emerging field of *metabolomics*. This scientific area involves the investigation and characterization of small molecules, including intermediates and products of metabolism, present in biological fluids under different conditions that include (1) normal homeostasis, (2) disease states, (3) stress, (4) dietary modification, (5) treatment protocols, and (6) aging. In a fashion similar to a mass spectrum providing a fingerprint signature for a specific molecule, it has been speculated that compounds identified and evaluated in metabolomic studies may provide a fingerprint signature for different physiologic states.

In practice, metabolites are identified through comparison with (1) known reference materials, (2) commercial or in-house developed mass spectral libraries or metabolite databases, (3) interpretation of mass spectra, or (4) ancillary techniques such as nuclear magnetic resonance. As with other applications of MS, both GC-MS and LC-MS have a place in such studies. GC-MS has some potential advantages that were described earlier. To use GC-MS, however, the metabolites in the sample must be volatile, or derivatization needs to be used to enhance detectability of a larger number of compounds.

LC-MS has its own usefulness in metabolomics because it has potentially wider applicability to polar and nonpolar compounds and allows the observation of the molecular or pseudomolecular ions. However, because reference materials or isotope-labeled internal reference materials do not exist for validating ionization efficiencies or recoveries for some of the biologically relevant compounds, the effects of ion suppression (discussed later) remain a potential confounding factor. Compared with proteomic research, metabolomics faces the added difficulties in that the MS/MS spectra are more difficult to interpret, scarce information is available in the MS/MS libraries, and DNA and protein sequence databases are of no use in interpreting the results.

Mass spectral imaging of tissue sections is another emerging technology that has potential for clinical applications and holds a great promise. The most common approach is to apply MALDI MS to image tissue sections.[129,130] A mass spectrum is acquired at each spot on a regularly spaced array across the sample. From these data, an image is constructed for each m/z. The images provide a spatial map of chemical composition (peptides or small molecules) from the sample. Another approach uses laser ablation, followed by ICP-MS, to provide a spatial map of the inorganic elemental composition of the sample.[131] This technique can be extended to immunohistochemical imaging by using metal-labeled antibodies.[132] With this scheme, it is possible to use different labels on different antibodies to do multiplexed imaging of several different targets on the same sample. However, it should be noted that at the present time mass spectral imaging is an extremely time-consuming process because the beam has to be rastered across the tissue by the laser many thousands of times and very large amounts of data need to be processed and analyzed. At this time this is mainly a research technique that is not used in routine diagnostic laboratories.

Practical Aspects of Mass Spectrometry: Logistics, Operations, and Quality

In many respects, the logistics, operations, quality control, and quality assurance processes for clinical MS laboratories follow the well-established clinical laboratory standards and guidelines. However, mass spectrometers are complex instruments and most manufacturers of instrumentation are still learning how to best support their clients in clinical laboratories. Consequently, the adoption of MS, and especially the more complex technologies such as LC-MS/MS, places added demands on training, competency, and manufacturers' support beyond those of more familiar and well-established technologies used in clinical laboratories.

In contrast to techniques such as optical spectrophotometry, mass spectrometers tend to require more frequent troubleshooting, tuning, calibration, and optimization, and the laboratory inspection checklist of the College of American Pathologists specifies that mass spectrometer performance should be verified daily.[133] In addition, the frequency of calibrations and optimizations needed to maintain instruments in fit-for-purpose condition will vary, depending on the requirements of the assays being performed, the instrumentation used, and other factors.

The term *calibration* in relation to MS is used in at least two distinct ways. The first is calibration of the *m/z* scale of the instrument, usually referred to as *mass calibration*. The other is calibration for quantitative analysis.

Schedules for mass calibration vary among laboratories, types of instruments used, and types of assays being run. For example, if accurate mass measurements are an important part of a method, as with many assays that employ TOF-MS, then very frequent mass calibration is typically required. In some cases, internal mass reference materials are included within each run, or even within each sample. For applications that are less dependent on mass accuracy, such as most quantitative methods performed on quadruple mass analyzers, mass calibration may be performed less frequently. For example, mass calibration may be performed every few weeks, with verification of mass calibration performed more frequently—as often as daily in some laboratories.

Similarly, based on validation results obtained in individual laboratories, schedules for calibration for quantitative analysis may vary. For example, some laboratories calibrate an assay daily, whereas others calibrate with every run.

One advantage of MS is that most methods in the MS laboratories avoid the use of highly specialized reagents such as commercial kits of reagents and antibodies. Consumables are mostly generic items, such as solvents, chromatographic columns, and sample vials or 96-well plates. This tends to buffer the laboratory from supply disruptions of specialized reagents, and in some cases can decrease consumable costs as well. However, consumables must be carefully selected and monitored for quality, because contaminated reagents and supplies may negatively affect performance of the methods; this problem is far too common. Solvent quality is of particular concern. For example, one study documented wide variations in methanol quality from different suppliers, which can lead to large differences in ionization efficiency and cause interferences in the analysis.[134]

Whenever possible, a quantitative method should use isotopically labeled internal standards, which typically differ from the analyte of interest by substitution of monoisotopic ions with isotope labels (typically deuterium, ^{13}C, or ^{15}N). However, it is not always possible to obtain isotopically labeled versions of each target analyte, in which case a closely related chemical analog should be selected as an internal standard.

MS provides several opportunities for enhancing analytical quality and therefore improving patient care. The high degree of selectivity of MS, particularly when included as a part of hyphenated techniques (GC-MS, LC-MS, LC-MS/MS, etc.) reduces the likelihood of interference compared with immunoassays or separation-based techniques (GC or LC) using nonspecific detection, particularly for small molecule analysis, where cross-reactivity is a concern.[135] Perhaps as important as a high degree of selectivity, MS provides a means to detect the presence of interferences when they occur. With methods that produce fragmentation, in the ion source (as in EI) or in a collision cell (as in MS/MS), fragment ions are produced with reproducible relative intensity (compared to the base peak). By monitoring one or more ratios of the ion fragments, and by comparing these ratios to ratios obtained from authentic reference materials measured in the same run, it is possible to detect the presence of interfering compounds on a sample-by-sample basis.[136]

In addition, accurate mass measurements are useful for detecting interferences if one is using an instrument capable of such measurements, such as a TOF mass spectrometer. To illustrate, the accurate mass of protonated cortisol ($C_{21}H_{31}O_5^+$) is 363.2166 Da, whereas the accurate mass of one of the isotope peaks of protonated molecule of the drug fenofibrate ($C_{20}H_{22}ClO_4^+$) is 363.1180 Da—a difference of 271 ppm. Therefore an interference of even a few percent by fenofibrate (drug used for treating patients with high cholesterol and high triglycerides) in a cortisol analysis would be detectable as a shift in mass of the observed peak on an instrument capable of low single-digit parts per million mass accuracy.[137]

Obviously, detection of interferences by accurate mass measurement alone becomes more difficult as the mass of the interfering compound approaches that of the target compound, but given the ability to detect interferences at a ≈ 20% level or better, and assuming a mass spectrometer of ≈ 3 ppm mass accuracy, a reasonable estimate is that interferences could be detected for all compounds with $|\Delta m/z|$ greater than 30 ppm relative to the target compound. Given the cortisol example discussed earlier, 22 chemical formulas for ions are within 30 ppm of the mass of protonated cortisol, provided we limit our list to the composition constraints listed earlier. Interferences from these would be difficult or impossible to detect by mass measurement alone. Thus accurate mass would likely detect the majority of possible interferences, but some of the potentially interfering compounds would be difficult or impossible to detect.

OPTIMIZATION OF INSTRUMENT CONDITIONS

When developing an MS-based method there are many parameters to optimize. This applies to both the mass spectrometer and the separation method.

Selection of Mass Transitions and Operating Conditions

MRM is the most commonly used type of acquisition in LC-MS/MS methods for targeted analysis. MRM-based methods allow sensitive and specific quantitation of analytes in samples with complex matrices. Typical MRM chromatograms contain one peak or a few peaks, which are easy to integrate, particularly if the sample preparation has been well designed.

When developing MRM methods, it is useful to start with identifying all analyte-specific mass transitions. The typical approach for selection and optimization of mass transitions is through infusion of solution of pure standard of the targeted compound using syringe pump. During the infusion, the signal can be optimized by adjusting the ion source conditions, declustering potential, the ion transmission conditions, and collisional fragmentation. When ions are transported from atmospheric pressure to the vacuum region, they typically exist in the form of clusters. Application of the declustering potential during the ion focusing causes low-energy collisions, which lead to declustering of the ions.

The MRM experiments are set by specifying m/z of the precursor ion (typically the molecular ion) of the targeted molecule and the m/z of the molecule specific fragments, produced by fragmentation in the collision cell. While developing a method, it is important to carefully assess which mass transitions to be used in the method. The best sensitivity and specificity are typically achieved using high-intensity unique fragment ions, which have minimal background noise and no interfering peaks coeluting with the analyte of interest. Use of fragment ions corresponding to the loss of water, ammonia, carbonyl (CO), and CO_2 groups generally should be avoided because they result in nonspecific mass transitions.

The optimal values of the voltages needed for declustering the molecular ion and the ion transmission are established by scanning the voltages and finding the apex values that correspond to the maximum signal intensity. Optimization of the collision energy (CE) is accomplished by scanning the CE used for fragmentation of the molecular ion, and plotting the abundances as a function of CE produces a profile, called a breakdown curve (Fig. 2.19).

In the majority of cases, the voltage corresponding to the apex is selected for use in the method. At this value of the CE there is maximum signal intensity. A second advantage of operating at the apex is that slight fluctuations in the instrument conditions do not result in large changes in the signal intensity, so the acquired signal is more stable. However, on some occasions, when unresolved chromatographic peaks could interfere with the analysis, it is beneficial to select CE on the leading slope or the trailing slope of the breakdown curve (more commonly on the leading slope) to improve specificity of analysis by avoiding fragmentation of the substance that potentially interferes with target analyte. This can be useful in cases in which the breakdown curve of an MS/MS transition of a potentially interfering compound partially overlaps that of the target compound. In such cases, by operating on the slope of the breakdown curve of the selected MS/MS transition of the target compound it may be possible to more strongly discriminate from the interfering substances.

In addition to evaluating breakdown curves, CE should be selected with the following principles in mind. If the CE is too low, ion fragmentation will be inefficient, and the signal abundance will be low. If the CE is too high, fragmentation can become too extensive; the number of peaks in the product ion spectrum may increase intensity of the peaks that are selective to the product ion of interest may have insufficient intensity to be useful, and the possibility of interfering peaks from coeluting isobars may increase.

Depending on the type of ionization source used, ionizability of molecules is influenced by their volatility, pKa, proton affinity, electronegativity, hydrophilicity/hydrophobicity, surfactant properties, solution pH, ionization energy, electron collision cross-section, etc. Of the ionization methods most commonly used in the MS applications in clinical laboratories, ESI is typically used for analysis of polar and ionizable molecules and APCI and APPI are used for nonpolar molecules. When a new method is being developed, available ionization techniques and polarity modes should be evaluated to assess effect of the conditions on the signal response and specificity of the detection.

The electrospray voltage and ion source conditions have a great effect on the ionization efficiency; the optimal voltage depends on the molecular structure, mobile phase composition, and flow rate. At higher electrospray voltages, greater fluctuation in the ionization efficiency can be observed and a larger number of impurities present in the sample may get ionized, potentially causing the loss of specificity and poor reproducibility. Therefore the lowest voltage resulting in an adequate sensitivity for the analyte is typically preferred. In general terms, there tends to be a threshold voltage, below which ionization is very inefficient; if the ESI voltage is too high, corona discharge or other undesirable effects may occur.

Online Two-Dimensional Separations

Two-dimensional chromatographic separation is a technique in which separation is performed using two HPLC columns with phases having different selectivity. The chromatographic columns are connected to each other through a switching valve, the sample is injected in the first column, and effluent of the column is directed to waste. At the time when the targeted peak is eluted from the the first column, the effluent is redirected into the second column, then during the column reconditioning and equilibration the switching valve is turned back to the original position. Using this approach, chromatographic columns with complementary (orthogonal

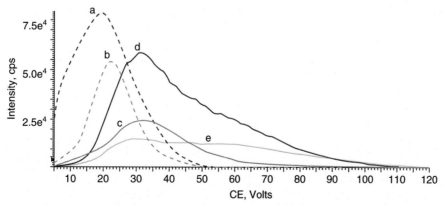

FIGURE 2.19 Breakdown curves for collision energy scans of cortisol. Curves correspond to mass transitions: a—m/z 363 → 345, b—m/z 363 → 327, c—m/z 363 → 171, d—m/z 363 → 121, e—m/z 363 → 97. *cps*, Counts per second; *CE*, collision energy.

or partially orthogonal) selectivity are typically used, so that peaks that are poorly resolved or unresolved by the first column would get separated on the second column.

In addition to the use of different stationary phases with complementary retention mechanisms, the selectivity may be modified through selection of the optimum for each separation mobile phase and temperature. Some advantages of well-designed 2D separations may include greater resolving power, faster analysis time (while the separation takes place on one column, the other column could be conditioned and reequilibrated), reduced contamination of the mass analyzer (major fraction of the effluent from the first column is directed to waste and not transferred into the second column), and ability to use for the second separation a mobile phase that is favorable for the optimal ionization efficiency. Various coupling strategies have been developed for the switching valve configuration and the peak transfer from the first to the second column; the choice of the specific strategy is method-dependent and would affect the robustness of the assay. Two-dimensional separations are more difficult to develop and troubleshoot, but in many cases the benefits in the methods' performance overweigh the drawbacks.

Conventional Versus Microflow Separations

As more LC-MS methods are developed, greater sensitivity and reduced sample volume are often required. This is especially true for analysis of novel biomarkers. Other trends in modern analytical laboratories are aimed at reducing the volume of solvents used, the cost of the used mobile phase disposal, and the costs of labor. The benefits of microflow separations in LC-MS analysis have been widely reported[138]; they include a higher sensitivity, greater efficiency of ion sampling, reduced solvent consumption and waste, and reduced contamination of the ion source. Despite the advantages, there are relatively few micro-flow based LC-MS methods currently used in routine laboratories. The main reasons are related to the fact that these separations historically were insufficiently rugged, required greater technical expertise of the staff, and caused frequent interruptions in the workflow of a laboratory. However, recent publications on comparison of the microflow and high-flow-rate traditional LC-MS/MS methods demonstrated the rugged method's performance with up to 10-fold gain in the signal-to-noise ratios and up to 20-fold reduction in the use of the solvents.[138]

Ion Suppression

Ion suppression is another quality issue that should be evaluated during method development and validation.[139] First described in 1993,[140,141] ion suppression is a matrix effect that results from the presence of coeluting nonvolatile or less volatile compounds (or compounds with greater proton affinity) that change the efficiency of spray droplet formation, ionic properties, and evaporation. These interfering substances, which include salts, ion-pairing agents, endogenous compounds, surfactants, drugs and/or metabolites, compete with analyte ions for access to the droplet surface or transfer to the gas phase, which, in turn, affects the number of charged ions in the gas phase that ultimately reach the detector.[142] Anions, such as phosphate or borate in buffers, also can neutralize the effective ionization of an analyte. Phospholipids present in biological samples and impurities introduced during the sample preparation and analysis have been demonstrated to be major contributors to ion suppression.[134,143]

Ion suppression refers to the effect of the constituents of the sample that suppresses ionization of the analyte of interest. Factors contributing to ion suppression include greater ionizability of the substances coeluting with the peak of interest, and concentration of the coeluting substance causing ion suppression. Ion suppression can have an adverse effect on the accuracy, precision, and sensitivity of the assay, particularly if the internal standard does not perfectly coelute with the targeted analyte, in which case the internal standard and the analyte may undergo different extents of ion suppression and thus compromise quantitative measurements. This is most likely to happen if the internal standard is highly deuterated. Ion suppression also may reduce the signal of the target analyte to nearly undetectable levels, in which case it is essentially impossible to obtain an accurate quantitative result. Of the different types of ionization techniques used in LC-MS methods, ESI tends to be most susceptible to ion suppression; methods using APCI are typically less prone to the effects of ion suppression, but the possibility of ion suppression in APCI should not be dismissed without validation.

Considering that biological samples contain a large number of endogenous molecules with concentrations ranging over a very wide dynamic range, ion suppression should be expected and the effects of ion suppression should be evaluated for all new or modified methods. The presence of ion suppression or other deleterious matrix effects can be evaluated via several experimental protocols.[1] One involves comparison of (1) the instrument response for reference materials (including any internal standards) injected directly in the mobile phase, and (2) the same amount of compound spiked into preextracted samples.[144] Data for the standard in the mobile phase provide a relative 100% response value. Data for the same amount of compound spiked into preextracted samples show the effects of sample matrix on MS response (ion suppression).

A second, more commonly used and preferred protocol involves postcolumn continuous infusion of compound into the MS detector, while analyzing samples (type intended for the evaluated method, eg, serum, plasma, urine) prepared according to the protocol of the evaluated method.[145-147] The instrumental setup includes a syringe pump connected via a tee to the column effluent (Fig. 2.20). Because the compound being tested is introduced into the ion source at a constant rate, a constant instrument response should be observed if no ionization suppression or enhancement occurs while analyzing biological specimens (Fig. 2.21, A). Typically there is suppression of the signal at the portion of the analysis that corresponds to the void volume of the HPLC column (see Fig. 2.21, B). The void volume is that portion of a chromatogram corresponding to no retention other than the time it takes for the mobile phase to flow through the column. Thus it represents a chromatographic time region, despite the word *volume* in the terminology.

The degree of ion suppression and the recovery time to full response can vary from assay to assay[145] and among the

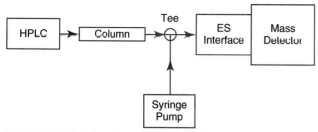

FIGURE 2.20 Postcolumn infusion system. Mobile phase or specimen extracts are injected into the high-performance liquid chromatography *(HPLC)* system. The analyte being evaluated is continuously infused, post column, and is mixed with the column effluent through a tee before entering the electrospray interface *(ES)*.

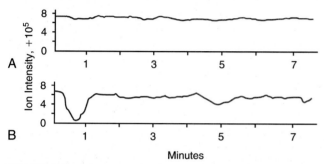

FIGURE 2.21 Infusion chromatograms for hypothetical analytes. A, Mobile-phase injection. B, Serum liquid-liquid extract injection. These profiles illustrate that ion suppression can be greater than 90%, that a recovery time may exist, and that suppression is not limited to the solvent front region. For a comprehensive presentation of these types of effects, the reader is referred to references 145 to 147.

samples and can be dependent on the sample preparation method, chromatographic column used, and LC separation conditions. Because endogenous compounds from the specimen matrix may elute at any time during the chromatographic run, ion suppression is not limited to the column void and not limited to the analysis time of the evaluated sample. In the case of strongly retained compounds, substances causing ion suppression may elute in subsequent injections. Considering this, the detector response should be monitored during analysis of multiple patient samples, to ensure that ion suppression will not affect subsequent injections. The observed degree of ion suppression also can be dependent on the sample volume aliquotted for the analysis, the injection volume, and the concentration of the analyte being monitored,[148] which is related to the matrix-to-analyte concentration ratio.[149] It should be noted that the degree of the matrix effect might differ among the samples of the same biological material, as has been ably shown by Matuszewski and colleagues.[150,151]

To control for ion suppression, it is highly desirable to use matrix-matched calibration standards and controls. It is important to evaluate ion suppression for all types of sample matrices intended for analysis by the method; in addition, considering the complexity of biological samples and the between-subject differences, there may be substantial fluctuations in the concentrations of the ion-suppressing species among samples. Because of this, a significant number of individual samples of all sample matrices intended for the method (serum, plasma, urine, etc.) should be used during the evaluation of the ion suppression.

In cases in which the isotope-labeled internal standards do not completely coelute with the analytes of interest, ion suppression cannot be completely compensated by the internal standard. This problem is particularly acute when using an internal standard that is highly deuterated, because these compounds are likely to not totally coelute with the native compound, and this can lead to significant quantitative errors in the analysis.[152] The ^{13}C- or ^{15}N-labeled compounds are chromatographically retained identically to the nonlabeled analogs and are not susceptible to the previously described problem.

Ion suppression is not limited to HPLC-MS or ESI/APCI ion sources. For MALDI analysis, arginine-containing peptides have been reported to dominate over the signal from other peptides in protein digests,[153] with the extent depending on the matrix used. The presence of ionic detergents, such as Triton X-100 and Tween 20, has also been shown to cause signal suppression in MALDI experiments, which can be countered by modifications to the matrix.[154]

Noise Reduction Techniques

In MS, background noise refers to the sum of electronic and chemical noise, which is independent of the data signal. Presence of the background noise interferes with the measurements and affects accuracy and specificity of analysis, especially at low concentrations. Reduction of chemical noise has been one of the aims for improvement since introduction of MS as an analytical technique. This is sometimes known as the "peak-at-every-mass" problem. The problem has long been known to mass spectrometrists, and it affects virtually all ionization methods to some degree.

Chemical noise is often dominant over electronic noise in MS. Background ions are inherent of atmospheric pressure ionization and related to the presence of impurities in the samples and in the mobile phases, residues accumulating on the surfaces of the ion source, and in part the high efficiency of atmospheric pressure ionization. Approaches used for noise reduction include optimizing the sample preparation, improving selectivity of ionization, optimization of the declustering conditions and ion transmission, maintaining cleanliness of the ion sources, and the flow path of the separation device.

One effective way for significant reduction of the effect of the background noise on the methods' performance is the use of MS/MS acquisition (MRM, neutral loss scan, product ion scan, and precursor ion scan), which allows substantial improvement of the detection specificity and reduction of the effects of chemical noise. Other approaches for reduction of the background noise include the use of mass analyzers with high resolving power (see Fig. 2.3); the use of multidimensional separations along with MS, such as ion mobility separations (IMS); high FAIMS; multidimensional chromatographic separations; and the incorporation of additional stages of fragmentation (MS/MS/MS). Software-based approaches (eg, dynamic background subtraction, active background noise reduction) also have been applied as noise reduction techniques. The previously mentioned techniques allow a reduction in the interference from the chemical background noise, but do not affect its cause. The best approach to the reduction of chemical background noise is the use of more extensive and efficient sample cleanup (as a way of minimizing introduction of contaminants into the ion source) and the use of high-purity solvents and additives for the mobile phases. This brings up the general issue of developing methods that are fit for purpose, but not so complex or expensive that their use will be impractical. Every laboratory needs to balance these factors in a way that is consistent with their goals, throughput, and constraints.

With regard to reducing background noise by using mass analyzers with high resolving power, it is important to understand the relationships among resolution, background noise (primarily chemical noise), electronic noise, and total signal level. A complete discussion is beyond the scope of the chapter, but a few general concepts can be useful to the clinical chemist without necessarily delving into all of the subtleties. For the sake of discussion, let us consider TOF detection. As mentioned earlier, this type of mass spectrometer acquires the full mass spectrum; that is, it is not possible to operate the instrument in SIM mode (or MRM mode in the case of a quadrupole TOF). However, it is possible to simulate SIM or MRM mode in postprocessing by integrating the mass spectrum over a limited m/z range and plotting the result as a function of spectrum number or chromatographic retention time.

If the m/z window is wide compared to the mass spectral peak width, the portion of the integrated signal arising from the targeted peak is independent of peak width. However, the signal arising from chemical noise generally increases with increasing width of the m/z window. This corresponds to the conditions in Figure 2.3, A—the FWHM of the TOF mass spectrometer was approximately 0.06 Da, whereas the window for generating the simulated SIM was 1 Da. Thus one would expect the chemical noise to decrease as the width of the integration window decreases.

The simulated SIM in Figure 2.3, B has the integration window set at 0.0072 Da, nearly an order of magnitude narrower than the peak width of the mass spectrometer. Under these conditions (ie, with the integration window much narrower than the peak width), narrowing the integration window still further affects chemical noise and target analyte signal nearly equally, so there is very little to be gained in terms of improving the signal-to-noise ratio by making the integration window narrower. Furthermore, in some cases there is a danger of increased statistical noise in the signal because ion numbers are quantized, which shows up as shot noise in the integrated signal. The signal-to-noise ratio in the shot noise-limit scales as 1 over the square root of the integration window width—that is, it gets worse as the window is made narrower. Discussion of the effect of the electronic noise on the noise budget is outside of the scope of this chapter, but it can be evaluated using methods analogous to those discussed previously.

Thus, in many cases, there is an optimum operating condition wherein the optimal integration window is often roughly equal to the peak width of the mass spectrometer. One additional comment is in order. Narrowing the integration window does nothing to reduce chemical noise if an interfering species is strictly isobaric with the targeted species. This implies that it is not at all useful if the goal is to reduce interferences from isomers. Furthermore, it is of very little usefulness if the mass offset of the interfering compound from the targeted compound is less than the peak width of the mass spectrometer. Nevertheless, as Figure 2.3 illustrates, a narrow window can be useful in improving the quality of chromatograms, and, as one can infer from the earlier discussion, this technique works best when using a high-resolution mass spectrometer.

POINTS TO REMEMBER

- MS is a highly sensitive and selective technique for analyzing a wide variety of clinically relevant analytes.
- MS relies on ionizing analytes in a sample, followed by separation of the ions according to their mass to charge ratios.
- When coupled with a separation technology such as gas chromatography or liquid chromatography, the hybrid technique is well suited for the quantitative analysis of clinically relevant molecules from bodily tissues and fluids.
- The use of MS for clinical applications has led to the development of highly specific assays that can overcome many of the issues faced when using immunoassays, such as cross-reactivity.
- Although highly selective, MS is not immune from interferences; molecules with same m/z (isomers and isobars) and similar fragmentation pattern may interfere with analysis

REFERENCES

1. Clinical Laboratory Standards Institute. *Mass spectrometry in the clinical laboratory. General principles and practice: approved guideline*, vol. C-50A. Wayne, Penn: Clinical Laboratory Standards Institute; 2007.
2. Clinical Laboratory Standards Institute. *Liquid chromatography-mass spectrometry methods*, vol. C-62A. Wayne, Penn: Clinical Laboratory Standards Institute; 2014.
3. Fenn JB. Electrospray wings for molecular elephants (Nobel lecture). *Angew Chem Int Ed Engl* 2003;**42**:3871—94.
4. Fenn JB, Mann M, Meng CK, et al. Electrospray ionization for mass spectrometry of large biomolecules. *Science* 1989;**246**:64—71.
5. Koy C, Mikkat S, Raptakis E, et al. Matrix-assisted laser desorption/ionization-quadrupole ion trap-time of flight mass spectrometry sequencing resolves structures of unidentified peptides obtained by in-gel tryptic digestion of haptoglobin derivatives from human plasma proteomes. *Proteomics* 2003;**3**:851—8.
6. Nakanishi T, Okamoto N, Tanaka K, et al. Laser desorption time-of-flight mass spectrometric analysis of transferrin precipitated with antiserum: a unique simple method to identify molecular weight variants. *Biol Mass Spectrom* 1994;**23**:230—3.
7. Tanaka K. The origin of macromolecule ionization by laser irradiation (Nobel lecture). *Angew Chem Int Ed Engl* 2003;**42**:3860—70.
8. Urry FM, Kushnir M, Nelson G, et al. Improving ion mass ratio performance at low concentrations in methamphetamine GC-MS assay through internal standard selection. *J Anal Toxicol* 1996;**20**:592—5.

9. Currie L. Nomenclature in evaluation of analytical methods including detection and quantification capabilities. *Pure Appl Chem* 1995;**67**:1699–723.
10. Calvert J. Glossary of atmospheric chemistry terms 1990. *Pure Appl Chem* 1990;**62**:2167–219.
11. Cech NB, Enke CG. Practical implications of some recent studies in electrospray ionization fundamentals. *Mass Spectrom Rev* 2001;**20**:362–87.
12. Ermer J, Vogel M. Applications of hyphenated LC-MS techniques in pharmaceutical analysis. *Biomed Chromatogr* 2000;**14**:373–83.
13. Glish GL, Vachet RW. The basics of mass spectrometry in the twenty-first century. *Nat Rev Drug Discov* 2003;**2**:140–50.
14. Huang C, Wachs T, Conboy J, et al. Atmospheric pressure ionization mass spectrometry. *Anal Chem* 1990;**62**:713A–25A.
15. Todd J. Recommendations for nomenclature and symbolism for mass spectrometry. IUPAC recommendations. *Pure Appl Chem* 1991;**63**:1541–66.
16. Whitehouse CM, Dreyer RN, Yamashita M, et al. Electrospray interface for liquid chromatographs and mass spectrometers. *Anal Chem* 1985;**57**:675–9.
17. Yamashita M, Fenn JB. Electrospray ion source: another variation on the free-jet theme. *J Phys Chem A* 1984;**88**:4451–9.
18. Chen J, Shiyanov P, Schlager JJ, et al. A pseudo MS3 approach for identification of disulfide-bonded proteins: uncommon product ions and database search. *J Am Soc Mass Spectrom* 2012;**23**:225–43.
19. Yamada N, Suzuki E, Hirayama K. Effective novel dissociation methods for intact protein: heat-assisted nozzle-skimmer collisionally induced dissociation and infrared multiphoton dissociation using a Fourier transform ion cyclotron resonance mass spectrometer equipped with a micromet electrospray ionization emitter. *Anal Biochem* 2006;**348**:139–47.
20. Rockwood AL, Busman M, Udseth HR, et al. Thermally induced dissociation of ions from electrospray mass spectrometry. *Rapid Commun Mass Spectrom* 1991;**5**:582–5.
21. Carroll D, Dizdic I, Stillwell R, et al. Atmospheric pressure ionization mass spectrometry: corona discharge ion source for use in liquid chromatograph-mass spectrometer-computer analytical system. *Anal Chem* 1975;**47**:2369–73.
22. Robb DB, Covey TR, Bruins AP. Atmospheric pressure photoionization: an ionization method for liquid chromatography-mass spectrometry. *Anal Chem* 2000;**72**:3653–9.
23. Syage J, Evans M, Hanold K. Photoionization mass spectrometry. *Am Lab* 2000;**12**:24–9.
24. Syage J, Hanning-Lee M, Hanold K. A man-portable, photoionization time-of-flight spectrometer. *Field Anal Chem Toxicol* 2000;**4**:204–15.
25. Elkadi M, Pillay A, Manuel J, et al. Sustainability study on heavy metal uptake in Neem Biodiesel using selective catalytic preparation and hyphenated mass spectrometry. *Sustainability* 2014;**6**:2413–23.
26. Becker JS. *Inorganic mass spectrometry: principles and applications*. Chichester, England: John Wiley & Sons; 2007. p. 181–7.
27. Takats Z, Wiseman JM, Gologan B, et al. Mass spectrometry sampling under ambient conditions with desorption electrospray ionization. *Science* 2004;**306**:471–3.
28. Cody RB, Laramee JA, Durst HD. Versatile new ion source for the analysis of materials in open air under ambient conditions. *Anal Chem* 2005;**77**:2297–302.
29. Wang H, Liu J, Cooks RG, et al. Paper spray for direct analysis of complex mixtures using mass spectrometry. *Angew Chem Int Ed Engl* 2010;**49**:877–80.
30. Manicke NE, Abu-Rabie P, Spooner N, et al. Quantitative analysis of therapeutic drugs in dried blood spot samples by paper spray mass spectrometry: an avenue to therapeutic drug monitoring. *J Am Soc Mass Spectrom* 2011;**22**:1501–7.
31. Price P. Standard definitions of terms relating to mass spectrometry: a report from the committee on measurements and standards of the American society for mass spectrometry. *J Am Soc Mass Spectrom* 1991;**2**:336–48.
32. Cooks RG, Rockwood AL. The "Thomson": a suggested unit for mass spectroscopists. *Rapid Commun Mass Spectrom* 1991;**5**:93.
33. de Hoffmann E, Stroobant V. *Mass spectrometry principles and applications*. 2nd ed. New York: John Wiley & Sons; 2001. p. 63–122. 132–155.
34. de Hoffmann E, Stroobant V. *Mass spectrometry principles and applications*. 2nd ed. New York: John Wiley & Sons; 2001. p. 361.
35. Siuzdak G. Mass analyzers and ion detectors. In: Siuzdak G, editor. *Mass analyzers and ion detectors*. San Diego: Academic Press; 1996. p. 32–55.
36. Dawson PH. *Quadrupole mass spectometry and its applications*. New York: Elsevier; 1976. p. 65–78.
37. Dahl D Simion 6.0 ion optics computer program.
38. Bucknall M, Fung KY, Duncan MW. Practical quantitative biomedical applications of MALDI-TOF mass spectrometry. *J Am Soc Mass Spectrom* 2002;**13**:1015–27.
39. Dawson J, Gilhaus M. Orthogonal-acceleration time-of-flight mass spectrometer. *Rapid Commun Mass Spectrom* 1989;**3**:155–9.
40. Lazar I, Lee E, Rockwood AL, et al. General considerations for optimizing a capillary electrophoresis-electrospray time-of-flight mass spectrometry system. *J Chromatogr A* 1998;**829**:279–88.
41. Sin C, Lee E, Lee M. Atmospheric pressure ionization time-of-flight mass spectrometry. *Anal Chem* 1991;**63**:2897–900.
42. Mamyrin B, Karataev V, Shmikk D, et al. Mass reflectron: new non-magnetic time-of-flight high-resolution mass spectrometer. *Zh Eksp Teor Fiz* 1973;**64**:82–9.
43. Rockwood AL. *An improved time of flight mass spectrometer*. Cincinnati, Ohio: 34th Annual American Society for Mass Spectrometry; 1986.
44. Bystrom CE, Sheng S, Clarke NJ. Narrow mass extraction of time-of-flight data for quantitative analysis of proteins: determination of insulin-like growth factor-1. *Anal Chem* 2011;**83**:9005–10.
45. Bystrom C, Sheng S, Zhang K, et al. Clinical utility of insulin-like growth factor 1 and 2: determination by high resolution mass spectrometry. *PLoS ONE* 2012;**7**:e43457.
46. Marin SJ, Hughes JM, Lawlor BG, et al. Rapid screening for 67 drugs and metabolites in serum or plasma by accurate-mass LC-TOF-MS. *J Anal Toxicol* 2012;**36**:477–86.
47. Makarov A, Scigelova M. Coupling liquid chromatography to orbitrap mass spectrometry. *J Chromatogr A* 2010;**1217**:3938–45.
48. Marshall AG, Hendrickson CL, Jackson GS. Fourier transform ion cyclotron resonance mass spectrometry: a primer. *Mass Spectrom Rev* 1998;**17**:1–35.
49. Kingdon K. A method for the neutralization of electron space charge by positive ionization at very low gas pressures. *Phys Rev* 1923;**21**:408–18.

50. Makarov A. Electrostatic axially harmonic orbital trapping: a high-performance technique of mass analysis. *Anal Chem* 2000;**72**:1156–62.
51. Makarov A, Denisov E, Kholomeev A, et al. Performance evaluation of a hybrid linear ion trap/orbitrap mass spectrometer. *Anal Chem* 2006;**78**:2113–20.
52. Makarov A, Denisov E, Lange O, et al. Dynamic range of mass accuracy in LTQ Orbitrap hybrid mass spectrometer. *J Am Soc Mass Spectrom* 2006;**17**:977–82.
53. Perry RH, Cooks RG, Noll RJ. Orbitrap mass spectrometry: instrumentation, ion motion and applications. *Mass Spectrom Rev* 2008;**27**:661–99.
54. Hu Q, Noll RJ, Li H, et al. The orbitrap: a new mass spectrometer. *J Mass Spectrom* 2005;**40**:430–43.
55. Erve JC, Gu M, Wang Y, et al. Spectral accuracy of molecular ions in an LTQ/Orbitrap mass spectrometer and implications for elemental composition determination. *J Am Soc Mass Spectrom* 2009;**20**:2058–69.
56. Rockwood AL, Erve JC. Mass spectral peak distortion due to Fourier transform signal processing. *J Am Soc Mass Spectrom* 2014;**25**:2163–76.
57. Chace DH. Mass spectrometry-based diagnostics: the upcoming revolution in disease detection has already arrived. *Clin Chem* 2003;**49**:1227–8. author reply 1228–1229.
58. Kanu AB, Dwivedi P, Tam M, et al. Ion mobility-mass spectrometry. *J Mass Spectrom* 2008;**43**:1–22.
59. Shvartsburg AA. *Differential ion mobility spectrometry: nonlinear ion transport and fundamentals of FAIMS*. Boca Raton, Fla: CRC Press; 2008.
60. Eiceman G, Karpas Z, Hill H. *Ion mobility spectrometry: ion mobility spectrometry*. 3rd ed. Boca Raton, Fla: CRC Press; 2013. p. 444.
61. Ray JA, Kushnir MM, Yost RA, et al. Performance enhancement in the measurement of 5 endogenous steroids by LC-MS/MS combined with differential ion mobility spectrometry. *Clin Chim Acta* 2015;**438**:330–6.
62. Caulfield MP, Li S, Lee G, et al. Direct determination of lipoprotein particle sizes and concentrations by ion mobility analysis. *Clin Chem* 2008;**54**:1307–16.
63. Krauss RM, Pinto CA, Liu Y, et al. Changes in LDL particle concentrations after treatment with the cholesteryl ester transfer protein inhibitor anacetrapib alone or in combination with atorvastatin. *J Clin Lipidol* 2015;**9**:93–102.
64. Rosenson RS, Davidson MH, Le NA, et al. Underappreciated opportunities for high-density lipoprotein particles in risk stratification and potential targets of therapy. *Cardiovasc Drugs Ther* 2015;**29**:41–50.
65. Mallard WG, Reed J. *Automated mass spectrometry deconvolution and identification system (AMDIS) user guide*. Gaithersburg, MD: US Department of Commerce-Technology Administration, National Institute of Standards and Technology (NIST), Standard Reference Data Program; 1997.
66. Rodriguez H, Tezak Z, Mesri M, et al. Analytical validation of protein-based multiplex assays: a workshop report by the NCI-FDA interagency oncology task force on molecular diagnostics. *Clin Chem* 2010;**56**:237–43.
67. Kelley RI. Diagnosis of Smith-Lemli-Opitz syndrome by gas chromatography/mass spectrometry of 7-dehydrocholesterol in plasma, amniotic fluid and cultured skin fibroblasts. *Clin Chim Acta* 1995;**236**:45–58.
68. Maurer HH. Role of gas chromatography-mass spectrometry with negative ion chemical ionization in clinical and forensic toxicology, doping control, and biomonitoring. *Ther Drug Monit* 2002;**24**:247–54.
69. Stellaard F, ten Brink HJ, Kok RM, et al. Stable isotope dilution analysis of very long chain fatty acids in plasma, urine and amniotic fluid by electron capture negative ion mass fragmentography. *Clin Chim Acta* 1990;**192**:133–44.
70. Knapp D. *Handbook of analytical derivatization reactions*. New York: John Wiley & Sons; 1979.
71. Drees JC, Stone JA, Olson KR, et al. Clinical utility of an LC-MS/MS seizure panel for common drugs involved in drug-induced seizures. *Clin Chem* 2009;**55**:126–33.
72. Maurer HH. Current role of liquid chromatography-mass spectrometry in clinical and forensic toxicology. *Anal Bioanal Chem* 2007;**388**:1315–25.
73. Maurer HH. Mass spectrometric approaches in impaired driving toxicology. *Anal Bioanal Chem* 2009;**393**:97–107.
74. Wu AH, Gerona R, Armenian P, et al. Role of liquid chromatography-high-resolution mass spectrometry (LC-HR/MS) in clinical toxicology. *Clin Toxicol (Phila)* 2012;**50**:733–42.
75. Annesley T, Majzoub J, Hsing A, et al. Mass spectrometry in the clinical laboratory: how have we done, and where do we need to be? *Clin Chem* 2009;**55**:1236–9.
76. Yang Z, Wang S. Recent development in application of high performance liquid chromatography-tandem mass spectrometry in therapeutic drug monitoring of immunosuppressants. *J Immunol Methods* 2008;**336**:98–103.
77. de Jong WH, Graham KS, van der Molen JC, et al. Plasma free metanephrine measurement using automated online solid-phase extraction HPLC tandem mass spectrometry. *Clin Chem* 2007;**53**:1684–93.
78. Knox S, Harris J, Calton L, et al. A simple automated solid-phase extraction procedure for measurement of 25-hydroxyvitamin D3 and D2 by liquid chromatography-tandem mass spectrometry. *Ann Clin Biochem* 2009;**46**:226–30.
79. Koal T, Burhenne H, Romling R, et al. Quantification of antiretroviral drugs in dried blood spot samples by means of liquid chromatography/tandem mass spectrometry. *Rapid Commun Mass Spectrom* 2005;**19**:2995–3001.
80. Remmel RP, Kawle SP, Weller D, et al. Simultaneous HPLC assay for quantification of indinavir, nelfinavir, ritonavir, and saquinavir in human plasma. *Clin Chem* 2000;**46**:73–81.
81. de Castro A, Concheiro M, Quintela O, et al. LC-MS/MS method for the determination of nine antidepressants and some of their main metabolites in oral fluid and plasma: study of correlation between venlafaxine concentrations in both matrices. *J Pharm Biomed Anal* 2008;**48**:183–93.
82. Subramanian M, Birnbaum AK, Remmel RP. High-speed simultaneous determination of nine antiepileptic drugs using liquid chromatography-mass spectrometry. *Ther Drug Monit* 2008;**30**:347–56.
83. Wohlfarth A, Weinmann W, Dresen S. LC-MS/MS screening method for designer amphetamines, tryptamines, and piperazines in serum. *Anal Bioanal Chem* 2010;**396**:2403–14.
84. Kushnir MM, Komaromy-Hiller G, Shushan B, et al. Analysis of dicarboxylic acids by tandem mass spectrometry: high-throughput quantitative measurement of methylmalonic acid in serum, plasma, and urine. *Clin Chem* 2001;**47**:1993–2002.

85. Lakso HA, Appelblad P, Schneede J. Quantification of methylmalonic acid in human plasma with hydrophilic interaction liquid chromatography separation and mass spectrometric detection. *Clin Chem* 2008;**54**:2028–35.
86. Yue B, Rockwood AL, Sandrock T, et al. Free thyroid hormones in serum by direct equilibrium dialysis and online solid-phase extraction-liquid chromatography/tandem mass spectrometry. *Clin Chem* 2008;**54**:642–51.
87. Kushnir MM, Blamires T, Rockwood AL, et al. Liquid chromatography-tandem mass spectrometry assay for androstenedione, dehydroepiandrosterone, and testosterone with pediatric and adult reference intervals. *Clin Chem* 2010;**56**:1138–47.
88. Soldin SJ, Soldin OP. Steroid hormone analysis by tandem mass spectrometry. *Clin Chem* 2009;**55**:1061–6.
89. Chace DH, Kalas TA. A biochemical perspective on the use of tandem mass spectrometry for newborn screening and clinical testing. *Clin Biochem* 2005;**38**:296–309.
90. Chace DH, Kalas TA, Naylor EW. Use of tandem mass spectrometry for multianalyte screening of dried blood specimens from newborns. *Clin Chem* 2003;**49**:1797–817.
91. Chace DH, DiPerna JC, Mitchell BL, et al. Electrospray tandem mass spectrometry for analysis of acylcarnitines in dried postmortem blood specimens collected at autopsy from infants with unexplained cause of death. *Clin Chem* 2001;**47**:1166–82.
92. Hardy DT, Preece MA, Green A. Determination of plasma free carnitine by electrospray tandem mass spectrometry. *Ann Clin Biochem* 2001;**38**:665–70.
93. Casetta B, Tagliacozzi D, Shushan B, et al. Development of a method for rapid quantitation of amino acids by liquid chromatography-tandem mass spectrometry (LC-MSMS) in plasma. *Clin Chem Lab Med* 2000;**38**:391–401.
94. Chalcraft KR, Britz-McKibbin P. Newborn screening of inborn errors of metabolism by capillary electrophoresis-electrospray ionization-mass spectrometry: a second-tier method with improved specificity and sensitivity. *Anal Chem* 2009;**81**:307–14.
95. Chace DH. Mass spectrometry in newborn and metabolic screening: historical perspective and future directions. *J Mass Spectrom* 2009;**44**:163–70.
96. Rinaldo P, Cowan TM, Matern D. Acylcarnitine profile analysis. *Genet Med* 2008;**10**:151–6.
97. Gao S, Zhang ZP, Karnes HT. Sensitivity enhancement in liquid chromatography/atmospheric pressure ionization mass spectrometry using derivatization and mobile phase additives. *J Chromatogr B Analyt Technol Biomed Life Sci* 2005;**825**:98–110.
98. Sherman NE, Stefansson B, Fox JW, et al. *Pseudomonas aeruginosa* and a proteomic approach to bacterial pathogenesis. *Dis Markers* 2001;**17**:285–93.
99. Wang Z, Dunlop K, Long SR, et al. Mass spectrometric methods for generation of protein mass database used for bacterial identification. *Anal Chem* 2002;**74**:3174–82.
100. Eigner U, Holfelder M, Oberdorfer K, et al. Performance of a matrix-assisted laser desorption ionization-time-of-flight mass spectrometry system for the identification of bacterial isolates in the clinical routine laboratory. *Clin Lab* 2009;**55**:289–96.
101. Szajli E, Feher T, Medzihradszky KF. Investigating the quantitative nature of MALDI-TOF MS. *Mol Cell Proteomics* 2008;**7**:2410–8.
102. Batista BL, Grotto D, Rodrigues JL, et al. Determination of trace elements in biological samples by inductively coupled plasma mass spectrometry with tetramethylammonium hydroxide solubilization at room temperature. *Anal Chim Acta* 2009;**646**:23–9.
103. Beauchemin D. Inductively coupled plasma mass spectrometry. *Anal Chem* 2002;**74**:2873–93.
104. Mandal BK, Ogra Y, Suzuki KT. Speciation of arsenic in human nail and hair from arsenic-affected area by HPLC-inductively coupled argon plasma mass spectrometry. *Toxicol Appl Pharmacol* 2003;**189**:73–83.
105. Anderson NL. Counting the proteins in plasma. *Clin Chem* 2010;**56**:1775–6.
106. Kenyon GL, DeMarini DM, Fuchs E, et al. Defining the mandate of proteomics in the post-genomics era: workshop report. *Mol Cell Proteomics* 2002;**1**:763–80.
107. Carr SA, Anderson L. Protein quantitation through targeted mass spectrometry: the way out of biomarker purgatory? *Clin Chem* 2008;**54**:1749–52.
108. Oberrauch W, Bergman AC, Helander A. HPLC and mass spectrometric characterization of a candidate reference material for the alcohol biomarker carbohydrate-deficient transferrin (CDT). *Clin Chim Acta* 2008;**395**:142–5.
109. Asif AR, Oellerich M, Amstrong VW, et al. Proteome of conidial surface associated proteins of *Aspergillus fumigatus* reflecting potential vaccine candidates and allergens. *J Proteome Res* 2006;**5**:954–62.
110. Clarke NJ, Zhang Y, Reitz RE. A novel mass spectrometry-based assay for the accurate measurement of thyroglobulin from patient samples containing antithyroglobulin autoantibodies. *J Investig Med* 2012;**60**:1157–63.
111. Hoofnagle AN, Becker JO, Wener MH, et al. Quantification of thyroglobulin, a low-abundance serum protein, by immunoaffinity peptide enrichment and tandem mass spectrometry. *Clin Chem* 2008;**54**:1796–804.
112. Kushnir MM, Rockwood AL, Roberts WL, et al. Measurement of thyroglobulin by liquid chromatography-tandem mass spectrometry in serum and plasma in the presence of antithyroglobulin autoantibodies. *Clin Chem* 2013;**59**:982–90.
113. Bateman RH, Green BN, Morris M. Electrospray ionization mass spectrometric analysis of the globin chains in hemoglobin heterozygotes can detect the variants HbC, D, and E. *Clin Chem* 2008;**54**:1256–7.
114. Wild BJ, Green BN, Stephens AD. The potential of electrospray ionization mass spectrometry for the diagnosis of hemoglobin variants found in newborn screening. *Blood Cells Mol Dis* 2004;**33**:308–17.
115. Rockwood AL, Busman M, Smith RD. Coulombic effects in the dissociation of large highly charged ions. *Int J Mass Spectrom Ion Processes* 1991;**111**:103–29.
116. Ganem B, Li Y, Henion J. Detection of noncovalent receptor-ligand complexes by mass spectrometry. *J Am Chem Soc* 1991;**113**:6294–6.
117. Ganem B, Li Y, Henion J. Observation of noncovalent enzyme substrate and enzyme product complexes by ion spray mass spectrometry. *J Am Chem Soc* 1991;**113**:7818–9.
118. Apostol I. Assessing the relative stabilities of engineered hemoglobins using electrospray mass spectrometry. *Anal Biochem* 1999;**272**:8–18.
119. Jeppsson JO, Kobold U, Barr J, et al. Approved IFCC reference method for the measurement of HbA1c in human blood. *Clin Chem Lab Med* 2002;**40**:78–89.

120. Wolff F, Deleers M, Melot C, et al. Hepcidin-25: measurement by LC-MS/MS in serum and urine, reference ranges and urinary fractional excretion. *Clin Chim Acta* 2013;**423**:99–104.
121. Kobold U, Dulffer T, Dangl M, et al. Quantification of hepcidin-25 in human serum by isotope dilution micro-HPLC-tandem mass spectrometry. *Clin Chem* 2008;**54**:1584–6.
122. Rochat B, Peduzzi D, McMullen J, et al. Validation of hepcidin quantification in plasma using LC-HRMS and discovery of a new hepcidin isoform. *Bioanalysis* 2013;**5**:2509–20.
123. Fu JF, Shi JY, Zhao WL, et al. MassARRAY assay: a more accurate method for *JAK2V617F* mutation detection in Chinese patients with myeloproliferative disorders. *Leukemia* 2008;**22**:660–3.
124. Igarashi J, Muroi S, Kawashima H, et al. Quantitative analysis of human tissue-specific differences in methylation. *Biochem Biophys Res Commun* 2008;**376**:658–64.
125. Wang SC, Oelze B, Schumacher A. Age-specific epigenetic drift in late-onset Alzheimer's disease. *PLoS ONE* 2008;**3**:e2698.
126. Au WY, Lam V, Pang A, et al. Glucose-6-phosphate dehydrogenase deficiency in female octogenarians, nanogenarians, and centenarians. *J Gerontol A Biol Sci Med Sci* 2006;**61**:1086–9.
127. Ecker DJ, Sampath R, Massire C, et al. Ibis T5000: a universal biosensor approach for microbiology. *Nat Rev Microbiol* 2008;**6**:553–8.
128. Hofstadler S, Sampatha R, Blyna L, et al. TIGER: the universal biosensor. *Int J Mass Spectrom* 2005;**242**:23–41.
129. Chaurand P, Sanders ME, Jensen RA, et al. Proteomics in diagnostic pathology: profiling and imaging proteins directly in tissue sections. *Am J Pathol* 2004;**165**:1057–68.
130. Cornett DS, Reyzer ML, Chaurand P, et al. MALDI imaging mass spectrometry: molecular snapshots of biochemical systems. *Nat Methods* 2007;**4**:828–33.
131. Becker J. *Inorganic mass spectrometry: principles and applications*. Chichester, England: John Wiley & Sons; 2007. p. 366–75.
132. Seuma J, Bunch J, Cox A, et al. Combination of immunohistochemistry and laser ablation ICP mass spectrometry for imaging of cancer biomarkers. *Proteomics* 2008;**8**:3775–84.
133. College of American Pathologists. Accreditation checklists. *CHM* 2009;**18600**.
134. Annesley TM. Methanol-associated matrix effects in electrospray ionization tandem mass spectrometry. *Clin Chem* 2007;**53**:1827–34.
135. Hoofnagle AN, Wener MH. The fundamental flaws of immunoassays and potential solutions using tandem mass spectrometry. *J Immunol Methods* 2009;**347**:3–11.
136. Kushnir MM, Rockwood AL, Nelson GJ, et al. Assessing analytical specificity in quantitative analysis using tandem mass spectrometry. *Clin Biochem* 2005;**38**:319–27.
137. Meikle AW, Findling J, Kushnir MM, et al. Pseudo-Cushing syndrome caused by fenofibrate interference with urinary cortisol assayed by high-performance liquid chromatography. *J Clin Endocrinol Metab* 2003;**88**:3521–4.
138. Christianson CC, Johnson CJ, Needham SR. The advantages of microflow LC-MS/MS compared with conventional HPLC-MS/MS for the analysis of methotrexate from human plasma. *Bioanalysis* 2013;**5**:1387–96.
139. Annesley TM. Ion suppression in mass spectrometry. *Clin Chem* 2003;**49**:1041–4.
140. Kebarle P, Tang L. From ions to solution in the gas phase: the mechanism of eletrospray mass spectrometry. *Anal Chem* 1993;**65**:972A–86A.
141. Tang L, Kebarle P. Dependence of ion intensity in electrospray mass spectrometry on the concentration of the analytes in the electrosprayed solution. *Anal Chem* 1993;**65**:3654–68.
142. King R, Bonfiglio R, Fernandez-Metzler C, et al. Mechanistic investigation of ionization suppression in electrospray ionization. *J Am Soc Mass Spectrom* 2000;**11**:942–50.
143. Xia YQ, Jemal M. Phospholipids in liquid chromatography/mass spectrometry bioanalysis: comparison of three tandem mass spectrometric techniques for monitoring plasma phospholipids, the effect of mobile phase composition on phospholipids elution and the association of phospholipids with matrix effects. *Rapid Commun Mass Spectrom* 2009;**23**:2125–38.
144. Matuszewski BK, Constanzer ML, Chavez-Eng CM. Matrix effect in quantitative LC/MS/MS analyses of biological fluids: a method for determination of finasteride in human plasma at picogram per milliliter concentrations. *Anal Chem* 1998;**70**:882–9.
145. Bonfiglio R, King RC, Olah TV, et al. The effects of sample preparation methods on the variability of the electrospray ionization response for model drug compounds. *Rapid Commun Mass Spectrom* 1999;**13**:1175–85.
146. Hsieh Y, Chintala M, Mei H, et al. Quantitative screening and matrix effect studies of drug discovery compounds in monkey plasma using fast-gradient liquid chromatography/tandem mass spectrometry. *Rapid Commun Mass Spectrom* 2001;**15**:2481–7.
147. Muller C, Schafer P, Stortzel M, et al. Ion suppression effects in liquid chromatography-electrospray-ionisation transport-region collision induced dissociation mass spectrometry with different serum extraction methods for systematic toxicological analysis with mass spectra libraries. *J Chromatogr B Analyt Technol Biomed Life Sci* 2002;**773**:47–52.
148. van Hout MW, Hofland CM, Niederlander HA, et al. On-line coupling of solid-phase extraction with mass spectrometry for the analysis of biological samples. II. Determination of clenbuterol in urine using multiple-stage mass spectrometry in an ion-trap mass spectrometer. *Rapid Commun Mass Spectrom* 2000;**14**:2103–11.
149. Heller DN. Ruggedness testing of quantitative atmospheric pressure ionization mass spectrometry methods: the effect of co-injected matrix on matrix effects. *Rapid Commun Mass Spectrom* 2007;**21**:644–52.
150. Matuszewski BK. Standard line slopes as a measure of a relative matrix effect in quantitative HPLC-MS bioanalysis. *J Chromatogr B Analyt Technol Biomed Life Sci* 2006;**830**:293–300.

151. Matuszewski BK, Constanzer ML, Chavez-Eng CM. Strategies for the assessment of matrix effect in quantitative bioanalytical methods based on HPLC-MS/MS. *Anal Chem* 2003;**75**: 3019−30.
152. Wang S, Cyronak M, Yang E. Does a stable isotopically labeled internal standard always correct analyte response? A matrix effect study on a LC/MS/MS method for the determination of carvedilol enantiomers in human plasma. *J Pharm Biomed Anal* 2007;**43**:701−7.
153. Krause E, Wenschuh H, Jungblut PR. The dominance of arginine-containing peptides in MALDI-derived tryptic mass fingerprints of proteins. *Anal Chem* 1999;**71**:4160−5.
154. Gharahdaghi F, Kirchner M, Fernandez J, et al. Peptide-mass profiles of polyvinylidene difluoride-bound proteins by matrix-assisted laser desorption/ionization time-of-flight mass spectrometry in the presence of nonionic detergents. *Anal Biochem* 1996;**233**:94−9.

Sample Preparation for Mass Spectrometry Applications

David A. Wells

ABSTRACT

Background

Biological samples require one or more pretreatment steps before analysis and detection by mass spectrometry. This chapter discusses the different sample preparation steps performed in laboratories analyzing drugs and proteins in the clinical and research setting and introduces high-throughput applications.

Content

The general techniques discussed are dilution, centrifugation, sonication, and homogenization. Separation techniques are filtration and ultrafiltration, dialysis and microdialysis, desalting, buffer exchange, enzymatic hydrolysis, and acid-base digestion. The precipitation technique discussed is protein precipitation. Enrichment techniques are evaporation, solvent exchange, and derivatization. Extraction techniques reviewed are liquid-liquid extraction, solid-supported liquid-liquid extraction, and solid-phase extraction (off-line and online sample processing). The chromatographic techniques discussed are column-switching (single and dual column modes) for turbulent flow chromatography, restricted access media, monolithic columns, and immunoaffinity extraction. The evolving techniques described are dried blood spots, capillary microsampling, and tissue imaging.

The clinical chemistry laboratory encounters a wide variety of samples in its day-to-day operations of chemical analysis. Although an automated clinical chemistry analyzer can detect analytes directly from only a small volume of serum, keep in mind that the serum came from a biological fluid (whole blood) that required a sample preparation step before it was added to the sample queue. Therefore the act of centrifuging a whole blood sample in the collection tube to prepare either serum or plasma constitutes a sample preparation step. When red blood cells break within a whole blood sample and hemolysis occurs, that sample will require further sample cleanup because the cell contents are no longer separated out with the cells during centrifugation, and the resulting red coloration from hemoglobin and cell contents may interfere with detection. Urine may appear to the novice as a clean sample matrix because it is often translucent, but this matrix contains proteins and salts that must be removed or diluted before analysis. The objective of this chapter is to discuss the goals for sample preparation and describe the many different methodologies that are encountered in the clinical chemistry laboratory, with a focus on those procedures performed before quantitation by mass spectrometry (MS) instrumentation.

BACKGROUND

A mass spectrometer cannot accept repeated injections of a raw sample matrix of biological origin because endogenous components from the matrix are often present in high concentrations and are detrimental for many reasons. They may cause a rapid deterioration in the separation performance of the chromatographic column, suppress the ionization process, mask the analyte of interest chromatographically, accumulate in the mass spectrometer ion source with repeated injections, or clog frits or the liquid chromatography lines to cause an increased system back pressure and reduce overall system performance and/or result in mass spectrometer system downtime for cleaning the ionization source. As a result, some type of sample preparation step is always required before analysis by MS, whether it be a simple dilution or a multistep automated extraction scheme. Although MS analysis methods demonstrate speed, sensitivity, and specificity, a sample cleanup by sample preparation is not the only focus for the analyst—the success of an assay depends also on a well-chosen column and mobile phase to provide the desired chromatographic separation.[1] The importance of chromatography in the entire process cannot be overlooked; the column separates analyte(s) of interest while further removing unwanted matrix interferences present in samples that can potentially mask an analyte or introduce ion suppression. Sample preparation is an integral component of the assay, along with the chromatographic separation and the chosen detection mode of the mass spectrometer.

The primary goal of sample preparation is to isolate the analyte of interest from the sample matrix so that unwanted interferences (proteins, salts, metabolites, endogenous

> **BOX 3.1 Goals for Sample Preparation**
>
> - Remove unwanted interferences (proteins, salts, metabolites, endogenous substances)
> - Solubilize analytes
> - Remove particulates that may block chromatographic tubing
> - Concentrate analytes to achieve sensitivity gains
> - Dilute analyte concentrations or solvent strength when outside the range of detection
> - Modify pH to promote or counteract ionization
> - Exchange the solvent in which the analyte resides (eg, aqueous to a nonpolar solvent)
> - Derivatize or complex with a chemical species to improve detection sensitivity
> - Remove an unwanted metabolic functional group (eg, glucuronide conjugate)
> - Hydrolyze large proteins for peptide analysis
> - Disrupt analyte binding to protein or another component within sample matrix
> - Remove proteins and lipids that can cause ion suppression

substances) are eliminated, thus improving method specificity. Additional goals (Box 3.1) may include one or more of the following steps:

- solubilization of analyte,
- removing particulates that could block the tubing of the chromatographic system,
- concentrating the analyte to achieve a gain in sensitivity,
- dilution when the analyte concentration or solvent strength is too high,
- altering the pH to promote or counteract ionization,
- exchanging the solvent in which the analyte resides from an aqueous to a nonpolar solvent more compatible for injection into a chromatographic system,
- derivatization or complexation with a chemical species to improve detection sensitivity,
- enzymatic cleavage of a functional group,
- acid-base hydrolysis or sonication to release analyte bound to sample matrix,
- disruption of protein binding with the analyte,
- removal of proteins and lipids that can cause ion suppression,
- filtration to eliminate fine particulates.

The need for one or more of these additional steps is determined by the chemistry of the analyte, the characteristics of the sample matrix, and the choice of instrumentation; the end result is a purified analyte, ready for injection into the mass spectrometer.

Whole blood, serum, plasma, and urine are by far the most common sample matrices the clinical laboratory is asked to analyze, but many more sample types may potentially be encountered, such as dried blood spots (DBS), cerebrospinal fluid (CSF), saliva, tissue, bile, seminal fluid, vitreous humor, sputum, hair, nails, and meconium. Each sample matrix has an associated sample preparation complexity factor; that is, whole blood is very complex considering it contains cells, cell membranes, protein, and hemoglobin that must be removed during sample preparation, whereas CSF and urine are less complex because cells and the majority of proteins are removed on isolation (the presence of proteins in high concentrations indicates an abnormality). Plasma contains an anticoagulant that prevents the proteins in solution from clotting; serum is the solution isolated after blood is clotted. Tissue, hair, and nails are solid samples and commonly require a preliminary digestion or homogenization step to remove the analyte from the sample matrix. DBS can require a solubilization step in an organic solvent. Feces is a very complex sample matrix that may require lyophilization to remove water and convert the matrix to a solid for subsequent sample preparation steps; meconium, the first fecal excretion of a newborn child, is more often encountered in the toxicology laboratory. Saliva, sputum, and vitreous humor are of aqueous origin but are more viscous than water and present their own challenges to analysis as a result of their composition. The exact procedure chosen by the analyst will take into account the most desirable chemical and/or physical means to isolate the particular analyte from the sample matrix and enrich and purify it so that sensitivity needs of the assay can be met.

Many different choices of sample preparation techniques are available to an analyst. The different methods vary in several regards, such as simplicity (number of steps); the time required for the overall procedure; cost per sample considering all materials involved; cost of analysis, which includes material and labor expenses; expertise or proficiency of the analyst; ease of method development; ability of the assay to be automated; the level of concentration factor attained; and cleanliness of the final extract. A sample preparation procedure is selected when considering such factors with regard to the particular needs of the assay. A rate-limiting step in method development is time. The amount of time required to develop and validate a method that results in a very clean extract is often much greater than the time required to select a simple off-the-shelf sample preparation method. Therefore a trade-off always exists between the choice of sample preparation method in regard to cleanliness, cost, and time.

Automation is a productivity and performance solution that is beneficial when the number of samples assayed exceeds the number that can be processed manually within a given time period or working shift. Some sample preparation methods comprise only liquid handling steps and can be fully automated in a very straightforward manner, whereas others are more difficult to automate because they include one or more manual or labor-intensive steps. Laboratories are motivated to adopt an automation solution for different reasons. Generally, automation is desired to achieve higher throughput. Although throughput considerations can meet established goals and allow for unattended overnight operation, they do not always meet every expectation. Automation is not always faster than manual sample processing, but it can reduce hands-on time significantly and free an analyst to perform other tasks in the laboratory. Freeing an individual from a hazardous (eg, radioactivity or toxic chemical usage) environment is another reason for implementing automation. The automation of a number of individual processes has been shown to add reproducibility and quality to the results and makes it extremely likely that an assay can be transferred or duplicated. Sometimes the specific sample preparation task is miniaturized or complex, and allowing the software on an instrument to calculate and perform repetitive functions allows a lower skilled operator to monitor the work. Automation also can be viewed as a strategic investment for the future, allowing the rapid implementation of new technology. A very important goal for implementing automation into a laboratory workflow is greater employee job satisfaction. Specific automation considerations for sample preparation are discussed within this text and are reviewed within book chapters.[2-6]

> **POINTS TO REMEMBER**
>
> **Factors in Choosing a Sample Preparation Method**
> - Number of individual steps
> - Time required for overall procedure
> - Cost per sample including materials and labor
> - Expertise of the analyst
> - Ease of method development
> - Automation considerations
> - Concentration factor
> - Cleanliness of final extract

GENERAL TECHNIQUES

Dilution

Sample dilution is an important pretreatment step in separation methods for the following reasons: (1) reduces the concentration of salts and endogenous materials in a sample matrix, (2) reduces the viscosity or ionic strength of the sample, (3) enhances the compatibility of the sample with the mobile phase, (4) reduces the concentration of the sample to fall within the range of the standard curve, and (5) protects the analytical column from overloading.[7] Sample dilution is most commonly performed with urine because drug concentrations are fairly high and dilution can be accomplished without a detrimental effect on sensitivity. A simple dilution approach is preferred when the primary goal is minimal time spent preparing samples for analysis. However, dilution also can be one component within a multistep sample preparation protocol that provides for additional sample cleanup. Although the amount of protein in a sample is usually a concern, under normal physiologic conditions the protein concentrations in urine are almost negligible; however, repeated injections over time can result in protein buildup within the chromatographic separation column as well as the ion source. When diluted urine is injected repeatedly, system downtime must be anticipated to clean the ion source.

A very simple protocol involves the dilution of urine with an aqueous or organic solvent and subsequent injection of an aliquot onto a chromatographic column with MS detection; this technique is sometimes referred to as "dilute and shoot." However, this approach is not without challenges because large amounts of endogenous components within urine may potentially coelute with a target analyte. Although they are not seen in the selected ion monitoring mode, the presence of interferences may significantly affect the efficiency and reproducibility of the ionization process within the mass spectrometer interface. In a report by Fu and associates,[8] the ionization behavior of indinavir in human urine is presented as an example in which endogenous matrix components were found to interfere with the ionization of the target analyte, leading to increased variation in tandem MS (MS/MS) responses. The sample preparation for this method used a volume of 1 mL urine diluted with 650 μL acetonitrile so that the resulting concentration of organic in the sample was equal to or less than that of the mobile phase. An aliquot of 6 μL was injected into a liquid chromatography (LC)-MS/MS system. The characterization of this effect and methods to overcome these interferences are described in the report. Deventer and associates[9] discuss urine analysis in doping control, in which 24-hour turnaround times are mandatory in major sports events and minimal time for sample preparation is a major objective. Whether dilution is acceptable within the analytical method depends on two factors: (1) ionization behavior of the target analyte(s) and (2) detection level requirement. Analytes that are efficiently ionized are more likely to be diluted before injection and examples include opiates and stimulants (the basic nitrogen atom is easily ionized) and diuretics (sulfonamide groups, carboxylic acids, or basic amines can be protonated or deprotonated). The minimum limits of detection (cutoffs) depend on the drug class, and this factor is a major concern before no preconcentration step is employed. When the cutoff cannot be met using diluted urine, an extraction step must be included. Dilution is not the only sample preparation step used in doping control because crystallization of salts (calcium phosphate, calcium oxalate) and proteinuria (which can be 10 times higher in performance athletes) can clog the chromatographic system if not removed; therefore centrifugation is first employed to sediment these materials to the bottom of the sample tube and yield a supernatant free of particulates and sediment. Filtration is an alternative to centrifugation but involves additional cost for the disposable filter unit. The analytical considerations of ionization behavior and minimum detectable limits are also important in analytical toxicology including workplace drug testing.

Advances in LC column technology using monolithic silica rods (single piece of porous silica) have recently made dilute and shoot sample preparation more attractive. Columns made with this monolithic stationary phase having a defined pore structure (macropores and mesopores) can accommodate the injection of diluted urine samples because they have high permeability, porosity, and capacity, as well as faster equilibration times. The total porosity of monoliths is approximately 15% higher than that of conventional particulate high-performance liquid chromatography (HPLC) columns, and so the resulting column pressure drop is therefore much lower, allowing operation at higher flow rates, including flow gradients.[10] The chemistry of organic, inorganic, and hybrid (organic-inorganic) monoliths and the application of these columns for sample preparation are reviewed by Nema and associates.[11] In addition to outlining the chemical synthesis and variety of monolithic LC columns, Nema and associates describe the fabrication of silica monoliths into polypropylene syringe barrels, spin columns, and pipette tips to replace traditional solid-phase extraction (SPE) media. Koyuturk and colleagues[12] illustrate the use of carbon-18 (^{18}C)-bonded monolithic columns used in a method for the quantification of irbesartan and hydrochlorothiazide in urine, in which diluted urine samples were injected directly into the chromatographic system. The sample preparation method involves centrifugation of 10 mL of urine at 4000 rpm for 10 minutes to pellet particulates and removal of 2 mL of supernatant that is then filtered through a 0.22-μm polyvinylidene difluoride (PVDF) membrane. A portion (1 mL) of the filtrate is diluted to 10 mL with water, and an aliquot (5 μL) is injected into the chromatographic system. The ease of sample preparation in this approach provided an advantage, and the column was reported to have a long lifetime. After hundreds of dilute and shoot injections, the authors reported there was nearly no change in the retention, resolution, and peak morphology of the analytes.

Dilution followed by injection, in an off-line sample preparation approach, is rarely used for plasma because analyte concentrations are not as high as in urine, so any dilution negatively affects sensitivity. Also, this sample matrix contains high amounts of protein that must be greatly reduced. However, knowing that dilution can be very attractive for the minimal effort and total time involved, it is possible to dilute and inject an aliquot from plasma when the sensitivity needs are met, small injection volumes are used, and proper choice of LC column is made. A report by McCauley-Myers describes a method in which plasma samples are first centrifuged, pipetted into wells of a microplate, and then placed on a liquid-handling workstation where a volume of 15 μL plasma supernatant is removed and diluted with 485 μL of a solution of water/methanol/formic acid (70:30:0.1, v/v/v) containing internal standard.[13] The samples were sealed and mixed; the dilution resulted in a slightly viscous solution with no observed precipitation. An aliquot of 5 μL was injected into an LC-MS/MS system. The lower limit of quantitation for the dilution assay (2 ng/mL) was 400 times higher than that of a more selective procedure that also concentrated the analyte (liquid-liquid extraction [LLE]; 5 pg/mL lower limit of quantitation). However, the advantage offered by the dilution procedure was 50 times greater throughput. In this case, throughput was a more important consideration than analyte sensitivity. The preferred method for direct injection of neat plasma samples is by online procedures for LC analysis that employ specialty columns constructed from restricted-access materials or system approaches such as turbulent flow chromatography (TFC) or online SPE; these procedures are discussed in the "Chromatographic Techniques" section.

In the analytical laboratory, the process of preparing samples in one or more 96-well plates using manual pipetting with a dilution protocol can be labor-intensive and time-consuming. Liquid-handling workstations that automate this entire process are ideally suited for these tasks. Jiang and co-workers[14] describe a fully automated and validated sample dilution and preparation process using a Tecan Freedom EVO150 liquid-handling workstation (Tecan, Research Triangle Park, North Carolina) equipped with an eight-channel liquid-handler arm and a 96-multichannel arm. A robotic sample preparation program was validated that contains a dilution calculation spreadsheet and a Visual Basic macro to automatically transform sample information from a laboratory information management system work list into executable comma-separated values (CSV) work lists that contain each sample's dilution scheme, source well positions, destination well positions, and liquid classes. A preprogrammed robotic script within the Freedom EVO software having several executable commands transforms CSV files to executable work lists and then executes the specified sample pipetting and dilution scheme. The dynamic dilution range is reported to be 1- to 1000-fold and is divided into three dilution steps: 1- to 10-, 11- to 100-, and 101- to 1000-fold. The entire process is accomplished within 1 hour for two racks of samples (96 samples/rack) and includes pipetting samples, diluting samples, and adding internal standard(s).

Centrifugation

The process of centrifugation allows for the sedimentation and separation of particulates in solution or as part of heterogeneous mixtures when centrifugal force is applied in the radial direction for a given time. Particles in solution, when subjected to a given centrifugal force, will migrate away from the axis of rotation at a rate that depends on size and density. Factors such as the rotational speed and the distance from the axis of rotation influence this process. The centrifugal force applied to the sample can be much greater than the force of gravity, which allows for even very fine particulates to settle out of the solution. Centrifugation can separate or fractionate fragile components such as platelets and does not damage their function or induce their activation.[15] In its simplest application, centrifugation will sediment large particles or molecules to the bottom of a tube as a visible pellet. Centrifugation at a combination of different forces and times will allow the differential fractionation of materials, such as separation of nucleus cytoskeleton from mitochondria, plasma membranes, and ribosomes because endogenous materials have varying intrinsic densities.[16] Adding trichloroacetic acid or acetonitrile to a sample of plasma precipitates proteins out of solution and a subsequent centrifugation procedure pellets the proteins to the bottom while an aliquot of the solution above the pellet is nearly (>95%) protein-free and is analyzed.[17] Sample preparation methods for urine that are useful to produce the metabolic profiles required for metabolomic research usually involve only centrifugation and dilution. Typical dilution factors with water range between 1:1 and 1:10 before LC-MS analysis. Mild centrifugation (1000 to 3000 relative centrifugal force × 5 minutes) immediately after sample collection is recommended to remove cellular components. LC-MS—based methods for blood, plasma, serum, CSF, other biofluids, and tissues require more complex procedures than those for urine.[18,19]

Density gradient separations involve a gradient of high concentrations of a small molecule, such as sucrose, which are distributed along the axis of the centrifugally generated force. Low concentrations are at the top of the tube, and high concentrations are at the bottom. When concentrations of the molecule are high enough, the density and viscosity of the solution vary along with the gradient. A dense molecule such as cesium chloride (CsCl) actually forms a density gradient during the process of centrifugation. When a sample undergoes centrifugation in a density gradient, the different components within the sample will sediment toward the bottom if their buoyant density is greater than that of the solution or they will float to the top if their density is lower than that of the solution. In another approach, centrifugation in the gradient is performed to the point at which the solution density of the molecule equals its buoyant density; then the molecule does not sediment or float, and separation occurs by differences in their buoyant densities. Sucrose density gradient ultracentrifugation is a valuable technique for fractionating DNA, RNA, and proteins.[20,21] Jasinski and colleagues[20] describe methodology for the large-scale purification of RNA prepared by in vitro transcription using T7 RNA polymerase by CsCl equilibrium density gradient ultracentrifugation and large-scale purification of RNA nanoparticles by sucrose gradient ultracentrifugation.[20]

Sonication

The action of directing sound waves (mechanical vibrations) toward a liquid or solid causes molecular disruption and is useful as a sample preparation step to remove an analyte from or disrupt its binding to a sample matrix.[22] Sonication is also referred to as ultrasound-assisted extraction. Typically, sample preparation methods use sonication more for solids than liquids, although sonication applied to a liquid sample does aid in analyte dissolution. Sound waves are delivered by an ultrasonic probe placed beside or in proximity with the sample, within an ultrasonic bath, or using a cup horn, and the overall effect achieved is removal of analyte from its association with the sample matrix. A frequent application of sonication is leaching minerals, chemicals, and natural compounds from plant tissues, soils, fruits, and vegetables. Applications involving biological matrices are also encountered, such as the extraction of mercury from human urine[23] and cobalamins from animal tissues and biological fluids.[24] Overviews on ultrasound-assisted extraction for the pretreatment of solid samples are summarized by Bendicho and associates.[25,26] Applications are presented, and the variables that influence ultrasound-assisted extraction are discussed, such as sonication time, ultrasound amplitude, extraction solvent, particle size, and solid concentration in liquid. Ultrasound is also useful as an adjunct to assist with tissue homogenization[27] and digestion procedures[28] in the clinical or bioanalytical setting.

Homogenization

The process of homogenization involves aggressive mixing and blending of two immiscible materials (eg, liquid and tissues) to create a homogenous composition. Note that homogenization is a disruptive procedure for the sample matrix and may involve crushing, grinding, pulverizing, and/or cutting processes to breakdown the solid matrix. The goals of the homogenization process are to provide uniform and repeatable results, use instrumentation that is easy to clean, be able to accommodate a range of sample sizes, and be automatable if at all possible.[29] Solid and semisoft sample matrices such as tissues and fecal samples are commonly homogenized with a buffer solution of known pH to disrupt or release analyte bound to the solid matrix material and create a suspension of extremely small particles distributed uniformly throughout a liquid. After a filtration or centrifugation step to remove particulates, an aliquot of the liquid is analyzed or undergoes additional sample preparation, depending on the cleanliness and sensitivity needs of the assay.

The classic method of homogenization is the use of a mortar and pestle, which can homogenize small amounts of sample; however, the procedure is manual and limited in scale. When a larger quantity of tissue requires homogenization, an instrument such as a blender is appropriate, but this approach is also a time-consuming and manual procedure. Instrumentation has been introduced to provide for homogenization of multiple samples to remove the manual component, whether adapted within a laboratory or a commercial product such as the Autogizer (Tomtec, Hamden, Connecticut).[30,31] An application using a liquid-handling workstation is described that transfers homogenized brain tissue samples from individual test tubes into a 96-well microplate for further sample purification and analysis.[32]

Typical applications requiring homogenization techniques are the isolation of RNA, DNA, and proteins from tissues for proteomic and genomic analyses,[33] oligopeptide analysis from tissues,[34] metabolomics (fecal samples provide insight to gut microbiota)[35] and screening of drugs from various tissues.[32,36] A sample preparation method for analysis of cell morphology in sputum samples, which are viscous and not miscible in buffer solution, employs homogenization using mucolytic agents such as N-acetylcysteine or dithiothreitol.[37] Semisoft matrices such as feces can sometimes use only water for the homogenization procedure, as in the isolation and purification of metabolites of the cytotoxic drug paclitaxel.[38] Solid biological tissues are commonly homogenized in saline or various buffers at a temperature of 4°C to retain structural integrity of proteins and/or analyte stability. Liquid organic acids such as trifluoroacetic acid or trichloroacetic acid are sometimes added in a 10% concentration as a protein precipitating agent. The end result of the homogenization process is a uniform mixture of tiny fragments of tissues, ready for subsequent sample preparation before analysis.[34]

SEPARATION TECHNIQUES

Filtration

Filtration as a sample preparation method is a pressure-driven process that effectively removes particulates from biological fluids that can potentially foul LC lines, column frits, and/or MS interface. Liquid-handling workstations and pipetting systems also benefit from filtration so that all liquids are free of materials that may introduce pipetting challenges, especially plasma clots. Filtration for bioanalysis applications is more appropriately named microfiltration and employs membrane pore sizes that are typically in the range of 0.1 to 10 μm.[39] Within this text and the published bioanalytical literature, *filtration* is the common terminology, however. Filtration is attractive as a first or second step in sample preparation because the sample concentration is not diluted and no organic solvents are used; filtration is often followed by additional techniques that concentrate the sample. Applications that commonly benefit from a filtration step include the separation of tissues from liquid after homogenization, removal of a mass of precipitated proteins or cellular debris from neat plasma before use with any of the traditional sample preparation techniques, and direct injection techniques (TFC, restricted access media [RAM], and online SPE). Filtration at the end of a sample preparation method is also effective, such as to clarify eluates and/or reconstituted extracts before chromatographic analysis. Methods for precipitating proteins from plasma use

TABLE 3.1 Comparison of Pressure-Driven Membrane Processes

	Microfiltration	Ultrafiltration	Nanofiltration
Microns	0.1–10 micron	0.01–0.1 micron	0.001–0.01 micron
MW (approximate)	>500,000	1000–500,000	<1000
Components retained by membrane	Intact cells Cell debris Bacteria **Membrane**	Proteins	Nucleic acids Antibodies
Components passed through membrane	Colloids Viruses Proteins Nucleic acids Sugars Salts	Nucleic acids Surfactants Sugars Salts	Salts Water

Reprinted by permission from Zydney AL. Membrane. In: Wilson ID, editor. *Encyclopedia of separation science.* Oxford: Academic Press; 2000: 1748–1755.

microfiltration, and these procedures are discussed in the "Precipitation Techniques" section.

Membranes used for filtration are generally composed of a porous polymer such as polyethersulfone, PVDF, polypropylene, or a cellulose derivative (eg, nitrocellulose) and are available in a range of porosities for exclusion by pore size. Filtration membranes are configured as single-use disposable units that accommodate from one sample at a time (syringe filters or spin filters) to 96 and 384 samples in the microplate format. The term *ultrafiltration* typically refers to a pore size less than 0.1 μm, whereas *nanofiltration* specifies a pore size less than 0.01 μm and a molecular weight cutoff of approximately 1000 Da[40] (Table 3.1).

Passive filtration through a membrane using gravity can be performed, but vacuum is a more practical approach, except for protein-containing and viscous samples. When using vacuum to pull proteinaceous samples through the membrane, the force achieved is not always sufficient to completely pull liquids through a fine-porosity membrane and residual liquid can be left above the membrane or adsorbed onto the plastic inside wall, below the membrane but above the collected liquid. Centrifugation is superior to vacuum because it attains higher forces and effectively passes the full volume of liquid through a fine-porosity exclusion filter. Passive ultrafiltration was compared with centrifugal ultrafiltration in a study by Blanco and coworkers.[41] Detection of *Legionella pneumophila* antigen in urine by enzyme immunoassay or immunochromatographic testing is specific for diagnosing legionellosis, and the use of concentrated urine obtained by selective ultrafiltration has been shown to significantly improve the sensitivity of antigen detection. When antigen from 4 mL urine is concentrated by centrifugal ultrafiltration (Amicon Ultra-4, Millipore, Bedford, Massachusetts) using a force of 3000 × g for 15 minutes, the sensitivity of antigen detection was shown to be equal to that obtained using traditional passive ultrafiltration (Urifil-10, Millipore), which requires 1 to 3 hours. Therefore great time savings results from the use of centrifugation as part of the sample preparation method.

Ultrafiltration uses a semipermeable membrane to separate molecules in liquids having a molecular weight range of 1000 to 500,000 Daltons.[42] The membrane is made of either polymeric or inorganic materials and contains pores of a defined size distribution. Note that the molecular weight cutoff (MWCO) rating for a membrane is a nominal number that is not absolute in terms of capability to exclude proteins by size. Ultrafiltration is commonly used to clarify protein solutions, and in proteomics the technique filter-aided sample preparation (FASP) describes the on-filter cleanup and digestion of protein samples for MS analyses. This method allows for gelfree processing of biological samples solubilized with detergents for proteomic analysis by MS. Using FASP, detergents are removed by ultrafiltration, and after protein digestion, peptides are separated from undigested material and recovered in the eluate. Manza[43] described the use of commercially available microcentrifugation devices (cellulose spin filters with a 5000 MWCO) for this purpose.[43] The protein sample is added to the upper chamber of a spin filter with a MWCO membrane, and contaminating species are washed away. Proteins are resuspended in a buffer compatible with digestion. Proteins are then reduced, alkylated, and digested on the filter, and the resulting peptides are isolated in the eluate by centrifugation. The method significantly reduces the complexity and the time required for sample preparation and minimizes the loss of sample. The results obtained are reported as equivalent to those for in-solution digestions.

In clinical proteomics, the urine supernatant obtained after centrifugation is devoid of particulates and endogenous materials, including the pellet fraction. However, the pellet fraction also can be important to analyze when looking for whole human cells that may have been shed into the urine from proximal tissues and organs or viruses and any microbes that may have infected the urogenital tract. The detection of microbes in the isolated pellet after centrifugation provides diagnostic information that is complementary to traditional cell culture–based laboratory tests. Yu and Pieper[44] describe a FASP method used in shotgun proteomics that effectively lyses cells present in urinary pellets isolated after centrifugation, solubilizes the majority of proteins resulting from microbial and human cells, and yields protein mixtures that are compatible with enzymatic digestion. The subsequent use of desalting procedures yields a peptide fraction suitable for analysis by LC-MS/MS. The methodology is scaled to higher throughput using parallel sample preparation in 96-well plates, peptide separation using nano-LC in one dimension, and analysis via a Q-

Exactive benchtop quadrupole Orbitrap MS (Thermo Scientific, Waltham, Massachusetts). Using this technique, it was shown that more than 1000 distinct microbial proteins and 1000 distinct human proteins can be identified from a single experiment. Michalski and associates[45] discuss the technical and performance advantages of coupling of a quadrupole mass filter to an orbitrap analyzer and its application to MS-based proteomics.[45]

A porous hollow fiber membrane is an attractive alternative to conventional flat membranes because it increases the membrane area for diffusion (ie, high surface-to-volume ratio) and is easy to incorporate into flow streams.[40] An application of centrifugal ultrafiltration based on molecular weight separation uses a hollow fiber membrane for the measurement of the drug amoxicillin in human plasma; in this case the membrane acts as a dialysis membrane to isolate free (non–protein bound) drug for analysis.[46] The separation device is constructed using a slim glass tube (6 cm × 5 mm) and a U-shaped hollow fiber membrane. After addition of a plasma sample followed by centrifugation (eg, 1.25×10^3 g for 10 min), small molecules pass through the ultrafiltration membrane without the use of additional buffer or high centrifugal force. The filtrate is withdrawn from the hollow fiber, and an aliquot is free of proteins and suitable for direct injection into a mass spectrometer.

POINTS TO REMEMBER

Filtration
- Does not dilute or enrich the sample concentration
- No organic solvents used
- Removes cellular debris and particulates
- May be followed by subsequent steps that concentrate the sample
- Clarifies eluates before injection but does not remove proteins unless they have been precipitated from solution in a previous step
- Performed in single disposable units and 96-well microtiter plate formats

Dialysis and Microdialysis

Dialysis, also known as equilibrium dialysis, is a concentration-driven separation technique that allows the selective diffusion of low molecular weight solutes from larger macromolecules through a semipermeable membrane made from a synthetic polymer or cellulose. The MWCO of the membrane is determined by the size of its pores, and the analyst chooses the appropriate specification for the separation. Dialysis is commonly used to separate and measure the concentrations of protein-bound drug and free drug within a liquid sample. There is no enrichment or concentration factor involved in the dialysis process. The analytes become diluted because the driving force of the mass transfer process is simply the difference in concentration across a semipermeable membrane. The sample is placed on one side of the membrane, and a buffer solution or dialysate is placed on the opposite side. Molecules such as protein-bound drug that are larger than the pores of the membrane are retained, and smaller molecules pass freely through the membrane.

Over time, small molecules approach equilibrium with the entire dialysate volume. The classic dialysis technique is clearly a useful separation technology to capture non–protein-bound (free) drug from protein-bound drug, but it is not commonly used as the only sample preparation step before MS analysis because the analyte concentration is greatly diluted. An aliquot of the liquid volume representing free drug concentration is then subjected to an enrichment technique such as SPE and then is amenable to chromatography and MS detection. Dialysis is also useful for desalting and exchanging buffer, techniques that are useful with gel filtration chromatography because they also use MWCO limits for separation. Gel filtration is much faster than dialysis (a few minutes compared with longer than 1 hour) and is able to remove contaminants from small sample volumes, as long as the column size and format are matched appropriately to the sample. Dialysis techniques are generally focused on determining plasma protein binding for drug analytes under physiologic conditions.[47,48] Note that the choice of buffer used in the equilibrium dialysis method plays an important role in the quality of the data obtained using this technique. A major disadvantage of dialysis is the dilution of the sample, thus reducing analyte concentration; also, in some cases, analyte binding to the chosen membrane may be observed.

Microdialysis is a particular application of dialysis that measures free (unbound) analyte and is used for the real-time monitoring of physiologic events occurring within the interstitial fluid in living tissues. A great advantage of microdialysis is that the technique is able to be connected directly to online MS analysis. The dialysis membrane is contained within a probe that is placed in proximity to tissues in vitro or in vivo. Microdialysis does not add or remove any fluids, because only molecules equilibrate across the semipermeable membrane in response to a concentration gradient, enabling the diffusion of substances from the interstitial space into the dialysis probe. This technique is valuable for serial sampling or continuous monitoring in animals in which the probe is mechanically positioned so that they can move freely without sedation or impairment in any way. The inlet of the probe is connected to a micro syringe pump that can provide ultralow flow rates; the other end of the probe is connected to either a collection vial or an LC micro valve that enables online analysis. There are different shapes, sizes, and varieties of the probes depending on the tissue and the region being investigated. The use of hollow fiber membranes increases the surface area available for diffusion.[49] Microdialysis methodology has been shown useful for serial sampling of the extracellular fluid for drug analysis applications in pharmacokinetics, as well as for the continuous monitoring of brain chemicals such as neurotransmitters and peptides and other small molecule substances such glucose, lactate, pyruvate, and glycerol.[50] The dialysate requires no further sample cleanup and is analyzed directly by LC-MS/MS.

Tang and colleagues[51] describe the dynamic, continuous, and simultaneous analysis of multiple neurotransmitters to research the complex interactions between neuronal and intercellular communications. Online microdialysis is coupled with hydrophilic interaction chromatography–MS/MS for the simultaneous measurement of the transmitters acetylcholine, serotonin, dopamine, norepinephrine, glutamate, γ-aminobutyric acid (GABA), and glycine, toward the goal of understanding transmitter release from embryonal carcinoma stem cells in vitro. The limit of detection was determined as the concentration of analytes with a signal-to-noise ratio of at least 3 and are described in pg of microdialysates; limits are reported as 2 pg for acetylcholine, serotonin, and glutamate and 10 pg for dopamine, norepinephrine, GABA, and glycine.

Microdialysis sampling in the brain is summarized by Ducey and colleagues,[52] who discuss approaches to sampling and calibration, as well as the methods employed for the analysis of the microdialysis samples. Many applications of off-line and online microdialysis systems, and some microfluidic applications, are published for the monitoring of pharmacokinetics or biological events using LC separation (many with MS detection).[53-61] Korf and associates[62] review the current status of microdialysis and microfiltration, which can be combined easily with other analytical techniques, and focus on the use of small volumes with ultraslow sampling to provide advantages. Applications discussed are quantitative pharmacokinetics, glucose metabolism in the brain, cytokines, and proteomics (tumor secretomes), both in vivo and in vitro. Although the sampling of interstitial fluid using microdialysis has become common, the small sample volumes, low concentration of analytes, and many different low molecular weight molecules assayed all present great analytical challenges. A review by Guihen and O'Connor[63] highlights these challenges and discusses the importance of proper evaluation of conditions and needs to select the most suitable analytical methodology for chromatography and detection. Microdialysis tools have been adapted for protein, oligopeptide, and peptide analyses in proteomics sample preparation as well, using sample volumes from 10 to 100 µL.[34,64]

Desalting and Buffer Exchange

The process of desalting separates salts and small molecules from larger proteins in a sample so they are removed from subsequent analysis steps. Buffer exchange replaces the original equilibration buffer pair of the sample with a different buffer pair, one more suitable for subsequent processing or analysis by ion exchange or affinity chromatography. Desalting and buffer exchange represent a particular type of size exclusion chromatography using porous particles that are characterized by a maximum effective pore size. Size exclusion chromatography is also referred to as gel permeation chromatography, gel filtration, or molecular sieve chromatography.[65] The larger proteins in a sample are excluded by pore size of the resin beads and quickly flow around the beads and exit the column first, along with the matrix components. Salts and small molecules enter the resin beads and flow through, but their migration is slowed and they elute from the column at a later time with the column equilibration liquid or buffer. The choice of effective pore size in resin beads can be varied to target certain analytes; for example, peptides can be separated from proteins using resin beads having larger pore sizes.[66] The technique can be useful for sequential fractionation based on molecular size, in addition to simply removing larger molecules from smaller molecules. A wide selection of resins or gels is available to accomplish a range of sample preparation needs.

The common formats for performing gel filtration on the laboratory bench are gravity-flow columns and centrifuge columns, also known as spin columns. Sample is loaded onto the top of a gravity-flow column, and it slowly passes through the resin bed; the column is positioned upright. The sequential addition of one or more buffer solutions (ionic strength and pH are varied) to the column, during wash and elution steps, applies gentle pressure to move existing liquids slowly through the gel matrix and out. As fractions elute from the bottom of the column, they can be collected to isolate proteins or macromolecules of interest. Gravity-flow columns can be prepared individually with chromatographic resin or they can be purchased already configured. A variation of the gravity-flow column is one that is a closed system and the sample and buffer are sequentially forced through the cartridge using pressure from a syringe. Centrifuge or spin columns use a much greater centrifugal force to push a sample through the gel filtration matrix, and this procedure takes far less time than using the gravity-flow format. In the case of centrifugation, the sample is not diluted because no additional liquids are needed to force it through the gel matrix. A 96-well plate configuration of gel filtration columns is commercially available for higher throughput applications.

Chromatographic techniques are also used in an online mode with LC and MS for desalting and buffer exchange of proteins using size exclusion chromatography. Four different primary approaches are used: (1) TFC, (2) RAM, (3) monolithic columns, and (4) immunoaffinity extraction (IAE). These techniques are discussed further in the "Chromatographic Techniques" section.

Enzymatic Hydrolysis
Phase II Metabolic Conjugation

The body metabolizes drugs and xenobiotics by pathways called phase I and phase II. Phase I reactions alter the structure of the molecule and may involve oxidation, reduction, hydrolysis, or dealkylation. During a phase II reaction, enzymes catalyze the addition of an endogenous conjugate to a molecule to make it more water soluble and promote its excretion into urine. Common conjugates include glucuronic acid, a sugar molecule, as well as glutathione and sulfate. For example, the drug morphine is metabolized by a phase II pathway that creates morphine-3-glucuronide and morphine-6-glucuronide.[67] Sample preparation for a conjugated analyte may involve the removal of a glucuronic acid or sulfate group using the enzymes β-glucuronidase or sulfatase, respectively. A solution of the enzyme in buffer is incubated with a solution containing the metabolite at 37°C for 18 to 24 hours to cleave the glucuronic acid or sulfate conjugate from the analyte. Free (unconjugated) concentrations of analyte can then be determined. An analytical method for the determination of free and glucuronide-conjugated female steroid hormones (progesterone, pregnenolone, estradiol, estriol, and estrone) in urine at the pg mL−1 level is published by Alvarez Sanchez and colleagues.[68] Following incubation of urine with enzyme, centrifugation is performed before a subsequent sample preparation step (SPE) for additional cleanup. Separation and detection were carried out by LC electrospray ionization (ESI) and tandem MS (LC-ESI-MS/MS) with a triple quadrupole mass detector.

Tissue Digestion

Enzymatic tissue digestion provides an alternative to the manually intensive tissue homogenization in some cases. The enzymes collagenase and proteinase K are known to degrade connective tissue and promote its dissolution. A typical procedure involves adding a known mass of tissue with an aqueous suspension containing the enzyme collagenase. Buffer is added to the homogenate and the mixture is incubated at 37°C for 24 hours. The digestion reaction is quenched by addition of a buffer, and the suspension is then centrifuged at room temperature to pellet the particulates. The analytes desipramine and fluoxetine were analyzed in dog and rat brain tissue using this enzymatic approach, and detection was made by LC-MS/MS.[69] Enzymatic digestion also can be performed before homogenization. It has been reported that mouse lung and heart tissues were best treated with collagenase first and then homogenized by a bead beater; however, mouse brain, bone marrow, kidney, spleen, and liver tissues are effectively homogenized using only a mechanical bead beater.[36]

Proteolysis

Enzymatic digestion of proteins is referred to as proteolysis. Intact proteins larger than 10 kDa are not compatible with quantification by LC-MS/MS; therefore sample preparation commonly involves proteolytic digestion using proteases, most commonly trypsin, to cleave the protein into smaller peptides. One peptide, the signature peptide, is commonly used for quantification as a surrogate for the larger protein.[70] MS-based protein quantification studies rely on the use of peptide-centered analytical methods; therefore efficient protein digestion protocols for sample preparation are very important.[71] An unbiased digestion is particularly important for the "shotgun approach," in which the proteome is digested, followed by peptide sequencing using MS/MS. Shotgun proteomics has become a routine method for identifying, characterizing, and quantifying proteins on a large scale. Fonslow and Yates[72] provide a summary of the proteases used for digestion of proteins for shotgun proteomics and the types of experiments for which they are best employed.[72] Complete digestion of proteins is also essential for the field of quantitative clinical chemistry proteomics. Beyond complete protein digestion, adequate calibration using commutable calibrators that undergo all sample preparation steps is relevant for absolute and accurate quantitation of serum proteins.[73]

Many protocols for enzyme digestion of protein mixtures are published in the literature. Traditional methods require 6 to 48 hours to complete and often are not effective for all types of proteins. Therefore the rate-limiting step in protein analysis is the digestion method. Microwave energy is one approach that has been used to accelerate protein sample preparation for proteome analysis. Enzyme- or chemical-induced proteolysis can be accomplished in minutes with the assistance of microwave irradiation. A chapter by Wang and Li[74] reviews the applications of microwave technologies to facilitate both enzymatic digestion and chemical digestion. Approaches of microwave-assisted enzymatic digestion, particularly trypsin digestion, adapted to both in-solution and in-gel based protein analysis are discussed. Histology-directed microwave-assisted enzymatic protein digestion for matrix-assisted laser desorption ionization (MALDI) MS analysis of mammalian tissue is discussed by Taverna and coworkers.[75] The method uses hydrogel disks (1 mm diameter) embedded with trypsin solution applied to a tissue section, directing enzymatic digestion to a spatially confined area.

The application of microwave radiation allows protein digestion to occur in 2 minutes on tissue, and the extracted peptides are analyzed by MALDI MS and LC-MS/MS.

Acid-Base Digestion

Trace elements in blood are effectively analyzed using inductively coupled plasma mass spectrometry (ICP-MS). The sample preparation step is critical to remove proteins, salts, and other interferences so that matrix and spectral interferences are reduced. Blood samples can be digested in acid or base solutions to perform this cleanup. A comparison of acid versus alkali sample preparation methods for multiple elements in human blood and serum was performed by Lu and coworkers.[76] Aliquots (0.2 or 0.5 mL) of human whole blood and serum samples, including reference samples, were subjected to alkali dilution (ammonia solution) or acid digestion (nitric acid), and then the samples were analyzed for their concentrations of 25 metals, using a quadrupole ICP-MS instrument equipped with a collision/reaction cell. The results suggest that the alkali dilution method is suitable for the determination of lithium (Li), boron (B), manganese (Mn), cobalt (Co), copper (Cu), zinc (Zn), arsenic (As), selenium (Se), rubidium (Rb), and strontium (Sr) in whole blood and serum; molybdenum (Mo), cadmium (Cd), and lead (Pb) in whole blood; and antimony (Sb) in serum by ICP-MS. Acid digestion is preferred for iron (Fe) and for low concentrations of cesium (Cs).

PRECIPITATION TECHNIQUES

Protein Precipitation

Most biological matrices contain protein to varying extents; that is, plasma contains much more protein than CSF. The characteristics of protein binding do influence drug-drug interactions in patients. Among all plasma proteins, serum albumin is the most widely studied and is known to be the most important carrier for drugs. The presence of proteins in the biological samples assayed creates problems because they must be removed before the process of chromatographic separation and detection begins. When proteins are not removed and are injected into a chromatographic system, precipitation occurs as soon as the protein interacts with organic solvents used in LC mobile phases. Over time, these precipitated masses build up inside the column inlet, thereby reducing LC column lifetime and increasing the backpressure of the total system. The protein that is not removed and that is carried through the analytical system will foul the mass spectrometer interface; the end results include downtime for the system and cleaning.

A common approach to remove proteins from a biological sample is the addition of an organic solvent (methanol, ethanol, or acetonitrile). Typically, a volume of sample matrix (1 part) is diluted with a volume of precipitating agent (2 to 4 parts), followed by vortex mixing and then centrifugation or filtration to isolate or remove the mass of precipitated protein. An aliquot of the supernatant or filtrate is injected for analysis (Fig. 3.1). Note that acids, salts, and metal ions also can be used to precipitate proteins, such as saturated aqueous ammonium sulfate or zinc sulfate heptahydrate (10%, w/v), trichloroacetic acid (10%, w/v), perchloric acid (6%, w/v), and metaphosphoric acid (5% w/v). Among these choices, acetonitrile, trichloroacetic

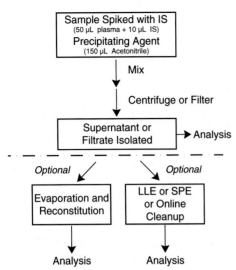

FIGURE 3.1 Schematic diagram of the protein precipitation technique for bioanalysis. Optional steps are shown. *IS*, Internal Standard; *LLE*, liquid-liquid extraction; *SPE*, solid-phase extraction. (Reprinted by permission from Wells DA. *High throughput bioanalytical sample preparation: methods and automation strategies.* Amsterdam: Elsevier; 2003.)

acid, and zinc sulfate are reported to be optimal at removing protein (>96, 92, and 91% protein precipitation efficiency at a 2:1 ratio of precipitant to plasma, respectively).[77]

Protein precipitation is a popular choice for sample preparation because it involves simply a one-step addition of solvent to sample and is considered a universal procedure (requiring no method development). The speed of this technique is very attractive to analysts, as well as its low cost. The high resolving power of LC-MS/MS analytical methods generally overcomes the nonselective cleanup procedure. Typical sample matrices that are used with protein precipitation techniques are plasma, serum, whole blood, tissue homogenates, and in vitro incubation mixtures.

There are several disadvantages of the protein precipitation technique to consider. Samples are diluted by a factor of 3 or more from the addition of organic solvent; therefore precipitation is a useful technique only when analyte concentrations are relatively high and the detection limits allow adequate quantitation. To allow for concentration of analyte, the supernatant may be evaporated with nitrogen and heat; however, an evaporation step requires an additional transfer step (with possible transfer loss of analyte) and added time for the dry-down and reconstitution procedures. If an analyte is volatile and labile to heat, an evaporation step has the potential to reduce recovery of that analyte.

Matrix components are not completely removed during a protein precipitation procedure, and small amounts will be contained in the isolated supernatant or filtrate. In MS/MS detection systems, matrix contaminants have been shown to reduce the efficiency of the ionization process and their effect has been studied and methods developed to reduce this problem.[8,9,78-84] The end result of not sufficiently removing matrix components is a loss in response that is referred to as ionization suppression.[85] This effect leads to decreased reproducibility and accuracy for an assay and failure to reach the desired limit of quantitation. Additionally, the efficiency of protein removal using organic solvents is not complete, ranging from 98.7 to 99.8%.[17] Repeated injections over time will accumulate tiny amounts of protein and foul the ionization source of a mass spectrometer, requiring system downtime for cleaning. However, as a means to overcome this problem, the use of rapid gradient LC-MS/MS for analysis coupled with flow diversion of the solvent front has been shown to permit the introduction of protein-precipitated samples into the mass spectrometer without the necessity for source cleaning.[86] In addition to the presence of proteins, phospholipids in blood also can cause matrix effects and ion suppression, but the precipitation technique described here does not affect their removal before analysis.[87] The evaluation and minimization of matrix effects caused by phospholipids in LC-MS analysis of biological samples are described.[88,89]

When the number of samples is reasonable and can be accommodated within a large sample capacity centrifuge to pellet precipitated proteins, precipitation in this case can be performed using individual tubes, but the technique is labor-intensive and does not satisfy the needs for high throughput. The microplate format is the preferred approach for faster sample processing because this format is used in autosamplers for injection into the chromatographic system. The following two general approaches are common for performing protein precipitation in the high-throughput microplate format:

1. Use a collection plate or microtube rack as the source container, pellet the precipitated protein at the bottom of the wells by centrifugation, and aspirate the supernatant for analysis. Automation in the form of a multiple-tip liquid-handling workstation can be used to transfer aliquots of supernatant to a clean microplate for placement into the autosampler.
2. Use a filtration microplate to trap the precipitated protein on top of a filter and collect and analyze an aliquot of the filtrate contained within a mated collection plate. Samples can be processed through the filters either by vacuum or positive pressure (centrifugation).

Each of these approaches is practical for performing protein precipitation methods in a high-throughput format. Considerations when an analyst would select the collection plate format over the filtration plate would include the extent of available hardware and automation accommodating the microplate format, total cost of materials and thus the cost per sample, number of physical manipulations, and the degree of transfer loss determined to be acceptable. To gain sensitivity, it is possible to dry-down the supernatant or filtrate and reconstitute before injection. Protein precipitation in the 96-well plate format is a preferred sample preparation procedure before LC-MS/MS analysis. Many published reports of protein precipitation applications are available using either the collection plate[90-95] or the filtration plate,[96-99] and several of these papers discuss the automation of these formats to achieve high throughput.

ENRICHMENT TECHNIQUES

Evaporation and Solvent Exchange

An enrichment process is required when the analyte to be concentrated is present in amounts or concentrations that are too low to be measured directly. Typically a dilute solution of analyte(s) is encountered in an isolated filtrate or supernatant and the injection of an aliquot will not present enough mass of the analyte for detection. Analyte is also diluted during an LLE procedure. Evaporation of the isolated solution under nitrogen and/or heat is performed to remove volatile solvent having a low vapor pressure. Evaporation times are longer for aqueous-containing solutions because their vapor pressure is higher than that of a straight organic solvent. An important consideration for performing this process is the thermal stability of the analyte; if the analyte is labile under the evaporation conditions chosen, loss will result. A procedure to determine analyte recovery using spiked solutions can determine whether evaporative loss is occurring. Once the solution is evaporated, an appropriate reconstitution liquid is added and the sample is vortex mixed before analysis. A solvent exchange occurs using this approach when the sample before evaporation and reconstitution is primarily in an aqueous solution and after evaporation is solubilized in an organic solvent or a solvent of high organic composition.

Derivatization

The process of derivatizing a drug, that is, introducing a particular functional group onto an analyte by chemical reaction to change its chemical and physical properties, is used to improve the sensitivity of detection. Historically, analytical methods using detection by gas chromatography (GC)-MS benefited from derivatization, but recent applications demonstrate an improvement in the detection of analytes poorly ionized by atmospheric pressure ionization techniques, such as ESI used in LC-MS/MS. For example, derivatization can be used to introduce charged functional groups to analytes poorly ionized in positive ESI-MS by adding quaternary ammonium, pyridinium, or phosphonium moieties.[100] The most often targeted functional groups of analytes that are modified are carbonyl, hydroxyl, carboxyl, amine, and thiol groups. The key advantages of integrating derivatization with LC-MS analysis include[101]:

1. improvement of selectivity and separation
2. enhancement of ionization efficiency
3. improvement of structural elucidation
4. removal of endogenous interference
5. facilitation of isomer separation

A comprehensive review of derivatization-based LC-MS analysis is provided by Qi and colleagues,[101] which summarizes the reaction mechanisms of representative derivatization reagents and the strategy used for their selection for established reaction methodologies. Applications are provided in peptide and protein analysis, metabolite analysis, pharmaceutical analysis, and MS imaging studies. In peptide and protein analysis, derivatization methods include[101]:

1. Isotope-coded affinity tag (iCAT) method, which derivatizes cysteine residues by iCAT reagent;
2. Isobaric tags for relative and absolute quantification (iTRAQ) method, which derivatizes the N-terminus of peptides by iTRAQ reagent;
3. Stable-isotope dimethyl labeling method, which derivatizes the N-terminal and ε-amino group of lysine through reductive amination by formaldehyde.

Turfus and associates[100] report an application in metabolite analysis, specifically the enhanced detection of glucuronide conjugates of benzodiazepines by derivatization with tris(trimethoxyphenyl) phosphonium propylamine, which introduces a quaternary cation functionality. This technique was shown beneficial also for conjugates of testosterone, epitestosterone, 5α-dihydrotestosterone, androsterone, morphine, and paracetamol. Benzodiazepine glucuronides were studied in greatest detail; after positive mode ESI-MS, average improvements to peak areas after derivatization were 67-, 6-, and 7-fold for temazepam, oxazepam, and lorazepam glucuronides, respectively. Average improvements to the signal-to-noise ratios for temazepam, oxazepam, and lorazepam glucuronides were 1336-, 371-, and 217-fold, respectively. Although this work was performed with neat standards, the presence of endogenous compounds competing for the derivatization agent in biological sample matrices can be a concern. However, this technique was shown to be applicable to human urine containing benzodiazepine glucuronides.

In the clinical analysis of drugs, the goal of derivatization is to improve chromatographic retention on reversed-phase columns and obtain higher ionization efficiency in MS. Derivatization also can provide satisfactory chromatographic separation for some chiral drugs.[101] Anabolic-androgen steroids are often abused in professional sports to increase performance, and derivatization of these compounds is a useful approach to enhance their detection in urine.[102] MS imaging studies, reviewed by Chughtai and Heeren,[103] are challenged by the need to detect low-abundant endogenous compounds in biological tissues, especially those with poor ionization yields during MS analysis. In this respect, the recent innovation of chemical derivatization–coupled MS imaging has proved to be a powerful tool to improve detection.[101]

EXTRACTION TECHNIQUES

Liquid-Liquid Extraction

LLE is a technique used to separate analytes from interferences in a sample matrix by the act of partitioning between two immiscible liquids; under the optimal conditions, the analyte in solution will preferentially migrate into the organic solvent. The fundamental procedure (Fig. 3.2) is as follows: an aliquot of aqueous sample matrix (eg, plasma) containing analyte is mixed with a volume of buffer at a known pH (or a strongly acidic or basic solution that modifies pH) which maintains the analyte in its un-ionized (uncharged) state. The resulting solution is then vigorously mixed (~10 minutes) with several ratio volumes of a water-immiscible organic solvent or a mixture of two or more solvents from choices such as hexane, diethyl ether, methyl tert-butyl ether (MTBE) or ethyl acetate. Constituents in the sample matrix distribute between the aqueous and organic liquid phases and partition preferentially into the organic phase when analytes are un-ionized (uncharged) and demonstrate solubility in that organic solvent. The organic phase is transferred from the sample well and placed into a clean receiving well, by a technique such as aspiration at room temperature, aspiration after freezing of the aqueous component using dry ice/acetone, or use of a molded gasket to transfer contents from one microplate to another.[104] Once isolated, the organic solution is evaporated to dryness and then reconstituted in a mobile phase–compatible solvent. The method provides efficient sample cleanup as well as sample enrichment.

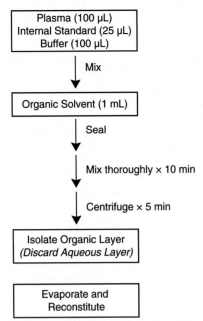

FIGURE 3.2 Schematic diagram of the liquid-liquid extraction technique for bioanalysis. Optional steps are shown. (Reprinted by permission from Wells DA. *High throughput bioanalytical sample preparation: methods and automation strategies.* Amsterdam: Elsevier; 2003.)

LLE is a popular technique for sample preparation; it is widely applicable for many drug compounds because of hydrophobic functional groups and is a relatively inexpensive procedure, requiring only the cost of the solvents and sample tubes. When the optimal organic solvent is chosen and the sample pH is adjusted, very clean extracts can be obtained with good selectivity for the target analytes. Inorganic salts are insoluble in the solvents commonly used for LLE and remain behind in the aqueous phase along with proteins and water-soluble endogenous components. Some disadvantages of LLE include its labor-intensive nature with several disjointed steps such as vortex, mix, and centrifugation. The organic solvents used are volatile and present hazards to worker safety. Also, emulsions can be formed without warning and can result in loss of sample. Regardless of these constraints, LLE is often one of the top three sample preparation methods in use and various ways to semi-automate the process have been developed.

Instead of using individual sample tubes, LLE can be performed in collection microplates to provide for high-throughput sample preparation.[105-109] The use of liquid-handling workstations reduces the hands-on analyst time required for this technique and offers semi-automation to an otherwise labor-intensive task. The plate sealing and vortex mixing steps still require manual intervention, but the pipetting steps are all performed by the workstation. An application for the LLE of cyclosporine in human blood is described by Brignol and associates.[109] After addition of internal standard and ammonium hydroxide to the blood sample (0.3 mL), extraction was performed twice with MTBE. The organic extract was evaporated to dryness and reconstituted in mobile phase. Pipetting automation used a 96-tip liquid-handling workstation. Samples were analyzed by ESI-LC-MS/MS. A lower limit of quantitation of 5.23 ng/mL was achieved, and the calibration curves covered the range 5.24 to 1748 ng/mL. Throughput achieved was reported as four 96-well plates (384 samples) extracted and analyzed within 28 hours.

> **POINTS TO REMEMBER**
>
> **Liquid-Liquid Extraction**
> - Analyte(s) in aqueous phase must be in the un-ionized form by adjusting pH.
> - Several ratio volumes of a water-immiscible organic solvent are added.
> - Analyte(s) must demonstrate solubility in the organic solvent used.
> - Common organic solvents are MTBE, ethyl acetate, hexane, and diethyl ether.
> - Procedure can be performed in (1) individual tubes or microplates by vigorous shaking or (2) using a solid particle support (diatomaceous earth) in which analytes first adsorb and then desorb when organic solvent is passed through the particle bed
> - On isolation of organic solvent, evaporation followed by reconstitution yields a concentrated extract for analysis.

Solid-Supported Liquid-Liquid Extraction

The many disjointed mixing and centrifugation steps of traditional LLE can be eliminated by performing LLE in a flow-through column filled with inert diatomaceous earth particles. This technique is referred to as solid-supported LLE (SS-LLE). The high surface area provided by the diatomaceous earth facilitates efficient, emulsion-free interactions between the aqueous sample and the organic solvent. Essentially, the diatomaceous earth with its treated aqueous phase behaves as the aqueous phase of a traditional LLE, yet it has the characteristics of a solid support.

Using the SS-LLE method, a mixture of sample (eg, plasma), internal standard, and buffer solution is prepared. This aqueous mixture is then added directly to the dry particle bed without a conditioning or pretreatment step. The mixture is allowed to partition for about 3 to 5 min on the particle surface via gravity flow. The analyte in aqueous solution is now spread among the particles in a high surface area. A hydrophobic filter on the bottom of each column or well prevents the aqueous phase from breaking through into the collection vessel placed underneath. Organic solvent is then added to the column, and as it slowly flows through the particle bed via gravity the analyte partitions from the adsorbed aqueous phase into the organic solvent the eluate is collected into an appropriately sized receiving column, tube, or microplate. A second addition of organic solvent is performed and the combined eluates are evaporated to dryness. The residue is reconstituted in a mobile phase—compatible solution, and an aliquot is injected into the chromatographic system for analysis.

Solid-supported LLE can be performed in microplates rather than individual columns for high throughput and is fully automatable with liquid-handling workstations because each step is simply a liquid transfer or addition. The many capping and mixing steps as used for traditional LLE are unnecessary, so less hands-on analyst time is required. However, there are a few precautions in using SS-LLE that deserve mention. The diatomaceous earth particles used in these procedures demonstrate a capacity limit for adsorption of aqueous sample matrix. It is possible to overload the particle with too much sample volume; the end result of overloading the column is incomplete recovery. Also, method transfer from an established LLE procedure to one that uses SS-LLE is not always straightforward. An investment in time may be required to optimize the method in terms of sample volume loaded onto the particle bed, the mass of particle in each well or column, the volume of organic solvent required, and/or the mechanics of collecting and working with the isolated organic solvent. LC-MS applications using SS-LLE as a sample preparation method before MS detection include the analyses in human plasma of the following analytes: aldosterone[110]; a novel topoisomerase I inhibitor[111]; simvastatin[112]; and a β3-adrenergic receptor agonist.[113] Additional applications include the analysis of benzodiazepines in antemortem and postmortem whole blood[114] and a urine screening assay of 189 doping agents in sports drug testing.[115]

Solid-Phase Extraction
Methodology
Solid-phase extraction (SPE) is a procedure in which an analyte, contained in a liquid phase, comes in contact with a solid phase (sorbent particles in a column or disk) and is selectively attracted to the surface of that solid phase. All other materials not adsorbed by chemical attraction or affinity remain in the liquid phase and go to waste. A wash solution is then passed through the sorbent bed to remove any loosely adsorbed contaminants from the sample matrix, yet retain the analyte of interest on the solid phase. Finally, an eluting solvent (usually an organic solvent such as methanol or acetonitrile that may be modified with acid or base) is added to the sorbent bed. This solvent disrupts the attraction between the analyte and solid phase, causing desorption or elution from the sorbent. Liquid processing through the sorbent bed can be accomplished by vacuum or positive displacement. Solvent exchange (evaporation followed by reconstitution) or dilution is followed by analysis on a chromatographic system. The general scheme for performing SPE is shown in Fig. 3.3.

SPE is often performed independently of the chosen analytical detection system; online SPE refers to coupling the extraction with the chosen detection system. An advantage to performing sample preparation in an online manner is higher throughput. Although the sample preparation, chromatographic separation, and MS detection steps are fundamentally sequential, significant time savings can be realized by positioning the extraction and separation steps in parallel; therefore, while the LC-MS/MS system is analyzing one sample, the online sample preparation component is programmed to extract the next sample and prepare it for injection as soon as the system is ready to accept it.

SPE is an attractive sample preparation technique for the following reasons:
- Very selective extracts can be obtained (reducing the potential for ionization suppression from matrix materials).
- A wide variety of sample matrices are accepted.
- Analytes are able to be concentrated at the end of the procedure.
- High recoveries are often achieved with good reproducibility.

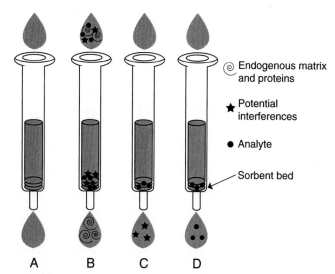

FIGURE 3.3 The basic steps for solid-phase extraction. *A*, Conditioning the sorbent bed. *B*, Loading analytes. *C*, Washing away interferences. *D*, Selective elution for further workup and analysis. The procedure is similar regardless of solid-phase extraction format. (Reprinted by permission from Wells DA. *High throughput bioanalytical sample preparation: methods and automation strategies.* Amsterdam: Elsevier; 2003.)

- Improved throughput is demonstrated by parallel processing.
- Low solvent volumes are accommodated.
- It is able to be fully automated.
- Emulsions are not formed as with LLE.
- Elution solvent when compatible with mobile phase can be injected directly, thus eliminating a dry-down step.
- A large number of product formats and chemistry choices are available to meet nearly all extraction requirements.

In spite of its many advantages, SPE may not be the preferred sample preparation technique primarily for its higher cost per extraction compared with other methods such as protein precipitation and LLE. Additional considerations against the use of SPE may be that it consists of several sequential steps, its perceived difficulty to master use of the many sorbent chemistries, and a historical perception that method development takes too much time.

There have been many advances in SPE over the years that have made it a more attractive choice for sample preparation. The technology has been improved with the introduction of additional solid sorbent chemistries that allow for generic methods applicable for many different classes of analytes, multiple types of disk-based SPE devices, smaller bed mass sorbent loading, and a multitude of 96-well plate formats to accommodate a range of sample volumes from microliters to milliliters. Liquid-handling workstations (from 4- to 96-tip capacity) also have been improved to offer completely unattended automation for performing the SPE procedure in microplates.

Sorbents and Sorbent Chemistries
Many choices are available for SPE sorbent chemistries and they can be categorized as reversed phase (in varying relative degrees from strongly to slightly hydrophobic), ionic (anionic or cationic), mixed phase (reversed phase and ionic), and hydrophilic. The two major types of base sorbent particles used

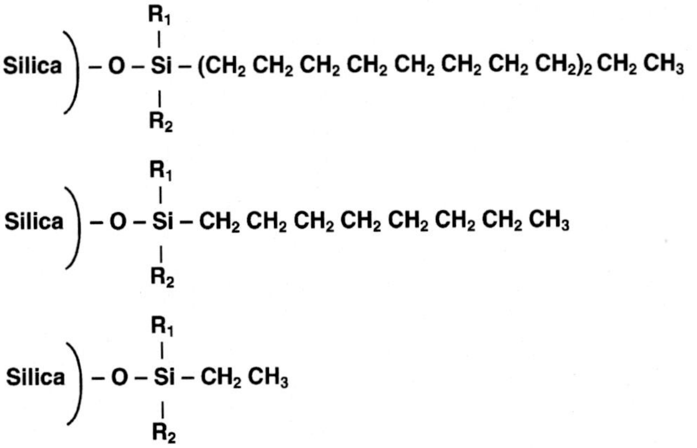

Silica⟩–O–Si(R₁)(R₂)–(CH₂CH₂CH₂CH₂CH₂CH₂CH₂CH₂)₂CH₂CH₃

Silica⟩–O–Si(R₁)(R₂)–CH₂CH₂CH₂CH₂CH₂CH₂CH₂CH₃

Silica⟩–O–Si(R₁)(R₂)–CH₂CH₃

FIGURE 3.4 Illustration of silica bonded with alkyl chains of varying lengths: C18 (octadecyl), C8 (octyl), and C2 (ethyl) bonded silica, shown *top* to *bottom*, respectively. (Reprinted by permission from Wells DA. *High throughput bioanalytical sample preparation: methods and automation strategies.* Amsterdam: Elsevier; 2003.)

in SPE are silica and polymer. Raw, underivatized silica is an amorphous, porous solid that contains polysiloxane (Si-O-Si) and silanol groups (Si-OH). The presence of these silanol groups allows polar adsorption sites and makes the surface weakly acidic. Silica is covalently bonded to a functional group through a chemical process. These groups are commonly alkyl chains of varying length, such as octadecyl (C18), octyl (C8), butyl (C4), ethyl (C2), and methyl (C1), as shown in Fig. 3.4. The following characteristics are known to vary among silica sources from different manufacturers: particle size distribution; pore size distribution; surface area; mono-, di-, or trifunctional bonding process; percent carbon loading; and the type and degree of end capping (which in turn influences the amount of residual silanols).

A typical polymer sorbent for SPE is a cross-linked poly(-styrene divinylbenzene [STB]). The advantages of a polymer sorbent include:
- 100% organic composition (chemically synthesized)
- stable across the entire pH range from 0 to 14
- predictable attraction (no silanols)
- more consistent manufacturing process batch to batch
- greater capacity per gram than bonded silica

The sorbent poly(styrene divinylbenzene) imparts a different selectivity than C18-bonded silica by nature of the aromatic rings and vinyl groups in its chemical structure (Fig. 3.5). Generally, it is able to adsorb a wider range of analytes; some more polar analytes are able to be captured

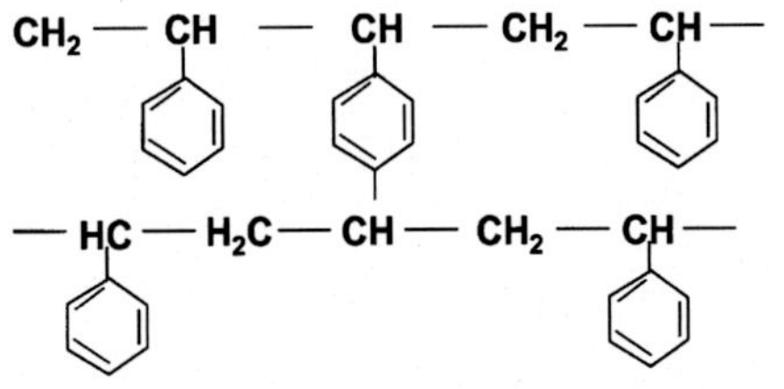

FIGURE 3.5 An original polymer sorbent useful for solid-phase extraction is poly(styrene divinylbenzene). (Reprinted by permission from Wells DA. *High throughput bioanalytical sample preparation: methods and automation strategies.* Amsterdam: Elsevier; 2003.)

compared with traditional reversed phase—bonded silica sorbents. A modified and improved version of the SDB polymer sorbent was introduced in 1996 as Oasis HLB (Waters Corporation, Milford, Massachusetts). This sorbent is a synthetic polymer of divinylbenzene and N-vinylpyrrolidone. The letters *HLB* refer to hydrophilic lipophilic balance, which describes hydrophilicity in terms of its wetting properties and lipophilicity for analyte retention. The Oasis sorbent has shown great utility for the SPE of a wide variety of analytes having diverse chemical structures. Although this sorbent can perform selective extractions when a method has been optimized for particular analytes, its great appeal is its ability to extract analytes with success using a generic method (condition with methanol then water, load sample, wash using 5% methanol in water, and elute with methanol).[116]

Specialty sorbent chemistries that are bonded to silica sorbents and provide unique selectivities have been synthesized. An example of a specialty bonded phase is phenylboronic acid, which is very specific for isolating coplanar vicinal hydroxyl molecules such as are found in catecholamines (eg, epinephrine, norepinephrine, and dopamine). Under slightly alkaline load conditions (pH 8.7), these catecholamines are extracted onto the phenylboronic acid; a selective retention occurs secondary to the formation of a cyclic ester between the boronic acid bound to the stationary phase and the 1,2-diol functional group on the catecholamines. After a wash step to remove residual components, the pH is made acidic, causing the cyclic boronic acid ester to hydrolyze and the catecholamines to be eluted.[117,118] In addition to phenylboronic acid, alumina as a solid-phase sorbent has been shown useful for catecholamine extraction.[119,120]

The amount of published information available on SPE is considerable, and further coverage of its theory and chemistry is outside the scope of this text. The reader is referred to book chapters on the subject by this author, which review the theory of SPE, sorbent chemistries, product formats, high-throughput applications, method development and optimization procedures, and automation methodologies.[5,121-122] In addition, there are several informative reviews of the established SPE sorbents, their modes of interaction with analytes, the method development and optimization process, and mathematical predictive models.[123-126]

> **POINTS TO REMEMBER**
>
> **Solid-Phase Extraction**
> - Performed off-line in a batch mode
> - Fully automatable off-line to yield an extract that (1) is ready for the evaporation step or (2) can be injected directly if mobile phase—compatible solution was used for elution
> - Can obtain very clean extracts
> - Accepts a wide variety of sample matrices
> - Products available in a wide range of formats, sorbent chemistries, and capacities to meet nearly all extraction requirements
> - Multiple steps required for the overall procedure
> - Can be performed in high-throughput mode using 96-well microplates
> - Higher cost per sample than filtration and LLE

High-Throughput Methods

High-throughput SPE is performed in the 96-well microplate format, and many sorbent chemistries and formats (particle bed mass, well volume, disks) are available to meet most high-throughput sample preparation needs. The great number of published applications for quantitative analysis using 96-well plates (or 96-tips) confirms the wide acceptance of this technique.[116,127-135] Some reports have used a higher density 384-well format; however, use of this configuration is not in common practice.[136,137] Although SPE microplates can be processed manually using hand-held pipettors and a vacuum workstation, liquid-handling workstations are preferred. The use of 96-tip semi-automated workstations are popular for high throughput, while 8- and 12-tip workstations are able to perform additional functions such as reformat samples from tubes to plates and offer full automation. The automation of SPE in microplates is covered in detail in literature reviews and book chapters[6,138,139]

Processing Approaches: Off-Line Versus Online

SPE is commonly performed in the off-line or batch mode, in which a batch of samples is processed independently from the analytical detection system; once aliquots are ready for analysis, they are placed into the autosampler for MS analysis. SPE sample processing in this manner uses sorbent particles packed in columns, cartridges, pipette tips, or microplates, and one analyst can typically perform up to 100 samples per day. With the addition of a liquid-handling workstation to introduce semi-automation to the process, several hundred samples per day can be prepared and typically the samples are analyzed overnight.

Online SPE is performed in a totally automated manner using disposable extraction cartridges that are processed on-line with the chromatographic system. An example of a commercial system for online SPE using disposable extraction cartridges is named Symbiosis (Spark Holland, Emmen, the Netherlands). The SPE cartridge used for online applications has a standard dimension of 10 × 2 mm and can withstand LC system pressures to 300 bar; a complete set of sorbent chemistries is available in this format. Typical particle sizes used are 40 µm, although the HySphere cartridges (Spark Holland) specifically use a smaller particle size, less than 10 µm. Sorbent mass per cartridge is typically between 20 and 45 mg, depending on particle chemistry. The standard capacity of the system is 192 cartridges (two trays containing 96 cartridges each); an optional feeder mechanism allows access up to 960 cartridges (10 trays). The total system is integrated, complete with autosampler. A more introductory system is also offered, the Symbiosis Fixed SPE, which can be connected to any LC, ultra-high-performance LC (UHPLC), and LC-MS configuration. The system works with the Spark 10 mm SPE cartridges or with a third-party SPE column.

The use of online SPE with LC analysis in a serial mode is straightforward; one sample is processed after the previous sample has finished. With the advent of parallel and staggered parallel LC systems, throughput of online SPE is similarly increased. The use of two SPE cartridges in staggered parallel fashion, one being eluted while the other is being extracted, becomes an important feature in helping to meet throughput needs in clinical laboratories. Note that to achieve maximum throughput, the cycle time for online SPE should be faster than the LC run time.

An important feature and advantage of online SPE, compared with off-line SPE, is direct elution of the analyte from the SPE cartridge into the mobile phase of the LC system. The time-consuming off-line steps of evaporation, reconstitution, and preparation for injection are eliminated, making the online SPE more efficient and fully automated. The entire volume of eluate is analyzed, and thus maximum sensitivity for detection is obtained. Some other advantages of this online approach include:

1. Samples and SPE cartridges are processed in a completely enclosed system protected against light and air.
2. The operator is protected from working with hazardous and/or volatile organic solvents.
3. Less handling and manipulation are involved, with no transfer loss of analyte.

Online SPE presents the advantage of automated method development. This process involves the examination of several extraction variables, such as the sorbent chemistry; the composition of the load, wash, and elution solvents; sorbent particle size; solvent volumes; and flow rates, with subsequent optimization. Typically, raw sample matrix (combined with internal standard) is used for injection onto the SPE cartridge, although sometimes a pretreatment sample preparation step such as filtration may precede this online analysis.

Many applications in clinical chemistry and drug analysis have been published using online SPE. In proteomics, high-throughput SPE analysis for determination of peptide and protein profiles was demonstrated by processing over 1000 samples in 24 hours using the Symbiosis cartridge-based method.[140] Clinical applications include the extraction and analysis of cortisol and cortisone in saliva,[141] antiretroviral drugs in human plasma,[142] the opioid tilidine and two metabolites in human urine,[143] 6-acetylmorphine in human urine,[144] and fluconazole in human plasma.[145]

POINTS TO REMEMBER

Online Solid-Phase Extraction
- Performed online in synchronization with chromatographic system
- Fully automated
- Analyte eluted from the cartridge directly into the LC mobile phase
- Maximum sensitivity for detection obtained by analysis of entire eluate volume
- No off-line steps required for evaporation, reconstitution, or preparation for injection
- Samples and SPE cartridges protected from light and ambient air
- No operator exposure to hazardous and/or volatile organic solvents
- Method development can be fully automated
- Wide range of sorbent chemistries available

CHROMATOGRAPHIC TECHNIQUES

Column-Switching: Single Column Mode

Column-switching refers to the use of a switching valve (6-, 10- or 12-port) placed in series with the LC system so that the liquid flowing through the extraction column can be directed away from the system to waste or flow through and reach the detector. This valve, also called a flow divert valve, is commonly placed after the column in the simplest case when only one column is used; in dual-column systems, it is often placed between the extraction column and the analytical column. In single column mode, this column is both the extraction column and the analytical column (Fig. 3.6). The biological sample matrix is injected directly into the LC system, and the analyte(s) are adsorbed onto the extraction column, but unwanted matrix components are not retained or are poorly retained. The valve placed after the column allows the mobile phase to flow to waste and not foul the detector. After the extraction procedure, the valve is switched using software and the analytes on the extraction column are eluted using a different mobile phase composition and they reach the detector. The extraction column is then reequilibrated so it can accept the next sample. The particle composition within the extraction column is the important specification that allows this method to be successful. Four primary approaches for direct injection in single column mode are (1) TFC, (2) monolithic columns, (3) IAE, and (4) RAM.

Turbulent Flow Chromatography

Turbulent flow liquid chromatography (TFC) is a fast technique that allows for the direct injection of plasma or serum samples after only a filtration pretreatment step, at most. TFC is considered fast because it is performed at higher flow rates (~5 mL/min) than laminar flow chromatography and uses a single short column (1 × 50 mm) containing particles of a larger size (30 to 50 μm). When a biological sample is injected into the LC system, it passes through the column at a high flow rate and encounters a highly aqueous solvent (typically 100% water with pH or salt additives as needed) that is diverted to waste for the load and wash period, lasting less than a minute. With the use of a reversed-phase sorbent in the extraction column, the analyte is retained by partitioning while hydrophilic components in the sample matrix, including proteins, are washed off the column and go to waste. Following the load and wash steps, the flow is changed to a high-percentage organic solvent and the divert valve is switched to direct the flow into a splitter; a portion of the split flow enters the mass spectrometer. Depending on the type of ion source and the ionization mode used, the flow is usually split from approximately 5 mL/min down to 0.3 to 1.0 mL/min. The retained analyte is eluted from the column with some limited separation and is detected by the mass spectrometer. After elution, the flow composition is changed back to high aqueous to equilibrate the extraction column in preparation for the next sample. The entire online extraction and analysis procedure can be accomplished in approximately 3 minutes per sample. This single-column mode provides the distinct advantage of high throughput and simplicity. However, carryover between samples may be a potential concern if the equilibration procedure is not adequate.

The single-column mode of TFC yields the shortest injection cycle, but it may not provide sufficient cleanup; when this is the case, an analytical column is placed in series downstream from the extraction column. This dual-column mode provides the advantage of improved separation performance, as well as improved sensitivity that results from chromatographic focusing. Another advantage of the dual-column configuration mode is that the extraction and separation processes can be performed simultaneously because they are driven by two separate LC pumps. While the sample is running on the analytical separation column, the extraction column is in use with the next sample.

Many applications of clinical interest have been published using TFC for the analysis of 25-hydroxyvitamin D and 3-EPI-25-hydroxyvitamin D_3,[146] tricyclic antidepressants,[147] hydroxychloroquine,[148] immunosuppressants,[149] risperidone,[150] and catechins.[151] The original theory of TFC was presented by Pretorius and Smuts[152]; more recent reviews on the theory and history of TFC have been published by Edge (2003)[153] and Couchman (2012).[154]

Monolithic Columns

The use of monolith sorbent packed into LC columns has been demonstrated to be a viable alternative to porous particle packed columns.[10,155] Their unique biporous structure and smaller pores (mesopores, diameter ~12 or 13 nm) provide an enhanced surface area and sufficient capacity; larger pores (macropores, diameter ~2 μm) on the silica skeleton

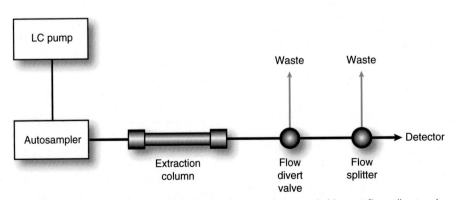

FIGURE 3.6 A single extraction column configuration is shown using a switching or flow divert valve between the extraction column and the detector; a flow splitter is optional. *LC,* Liquid chromatography. (Reprinted by permission from Wells DA. LC automation. In: Pawliszyn J, editor. *Comprehensive sampling and sample preparation: analytical techniques for scientists,* vol 2. Amsterdam: Academic Press; 2012:613–648.)

provide channels of high permeability that permit higher flow rates (up to 10 mL/min) without generating high backpressure as is typical in conventional columns. Their biporous structure offers the unique advantage of direct injection of the sample matrix; small- to medium-sized analytes achieve good partitioning and separation through mesopores, whereas large molecules and particulates in the sample matrix pass directly through macropores.

The advantages and disadvantages of monolith columns, as summarized by Cabrera and colleagues,[10] are as follows. The advantages are low back pressure, fast analyses, high efficiency, possibility to couple many columns, and compatibility with conventional instruments. The disadvantages are high solvent consumption, low pH resistance of silica monoliths, low temperature stability, lack of narrow-bore columns, and method transfer is not straightforward. Over the years since their introduction, monoliths have been demonstrated to be convenient for routine use and applications in sample preparation are reviewed by Nema and associates[156] and Namera and colleagues.[157] Monolithic materials are also packaged in the 96-tip format and Skoglund and coworkers[132] report the extraction of busulfan and cyclophosphamide from whole blood samples and their determination by LC-MS.

Immunoaffinity Extraction

Immunoaffinity extraction (IAE) uses antibody-antigen interactions to provide a very high specificity to attract molecules of interest. Antibodies (polyclonal or monoclonal) are immobilized onto a pressure-resistant solid support and then packed into a precolumn or column for use in LC applications. These antibodies can remove a specific analyte or class of analytes from among all other materials in a sample, and their recognition is based on a particular chemical structure, rather than on a general attraction such as occurs with SPE. IAE can be performed off-line but online coupling to LC is preferred for higher throughput. The detailed fundamental theory of immunosorbent-based sample preparation is provided by Hennion and Pichon.[158] This technique is also referred to as immunoaffinity SPE and the particles as immunosorbents.

The usage of immunosorbents is similar to that of a typical SPE sequence except that the immunosorbent is stored in wet media, such as phosphate-buffered saline solution. The binding of antibody to antigen is a reversible process, and its equilibrium can be influenced by manipulating the solution pH and aqueous/organic composition. The affinity between antibodies and analytes is strongly dependent on the nature of the antibodies as well as the amounts of antibodies.[159] There exists a cross-reactivity of certain antibodies that can be exploited to yield immunosorbents that isolate mixtures of structurally related analytes rather than a single specific analyte.

The coupling of IAE with LC allows a high degree of selectivity because extraction, concentration, and isolation can be achieved simultaneously. The single-column mode can be used with immunoaffinity techniques, but it is often not used because large volumes of aqueous buffer at low pH are commonly required for the elution step; the sample is in a dilute solution and a chromatographic focusing, or concentration step at the head of an analytical column, is required to provide more sensitivity for detection. Also, passing totally aqueous solutions through the IAE column is desirable in prolonging its lifetime; in a dual-column mode this aqueous solution can be easily diverted to waste without adversely affecting the performance of the analytical column, which prefers some organic concentration in the mobile phase. Flow diversion also permits a larger IAE column to be used as the larger volume passed through it also can be directed to waste.

An overview of chromatographic immunoassays, antibodies, and various binding immunoassays is published by Hage,[160] and another paper introduces the concept of immunosorbents as applied to both off-line and online techniques.[161] A discussion of immunoaffinity chromatography in clinical analysis is published by Hage and Clarke.[162] Several papers discuss advances in analytical applications of immunoaffinity chromatography.[163-165]

Restricted Access Media

Another technique that allows for the direct injection of plasma or serum online with the chromatographic system uses an analytical column containing particles referred to as restricted access media (RAM). The RAM particles are designed to prevent or restrict large macromolecules from accessing the inner adsorption sites of the bonded phase. Commercially available RAM columns, all silica based, include internal surface reversed phase, semipermeable surface, and a hydrophobic shielded phase named Hisep (Supelco, Bellefonte, Pennsylvania). The most popular column variety used in bioanalysis is the internal surface reversed-phase column. The internal surface is covered with a bonded reversed-phase material, and the external surface is covered with a nonadsorptive but hydrophilic material. This dual-phase column permits effective separation of the analyte of interest from macromolecules in the sample matrix; drugs and other small molecules enter the pores of the hydrophobic reversed phase to partition and be retained, whereas proteins and larger matrix components are excluded. When serum or plasma is injected onto a RAM column, the proteins are excluded by the outer, hydrophilic layer and pass through to waste.

The dual-phase nature of these RAM materials allows the direct injection of the biological sample matrix onto the column without pretreatment. Some disadvantages with the use of RAM columns are that retention times can be long (>10 minutes), the column must be washed between injections, and the mobile phases are not always compatible with some ionization techniques used in LC-MS/MS. Dual-column RAM techniques are also used. These methods use an analytical separation column placed in series downstream from the RAM column. A general overview of the use of restricted access materials in LC has been published in two parts.[166,167] A full and detailed compilation of RAM applications, and history of RAM column development, is summarized in the papers by Souverain and associates,[168] Mullett,[169] and Cassiano and colleagues.[170-172]

Column-Switching: Dual Column Mode

Although the single-column mode of column switching has been shown to be useful in bioanalysis, the assay is often prone to matrix effects and interferences because of the limited chromatographic resolving power offered by only one extraction column.[173] To improve the separation capability, a second column is introduced into the system, which is a typical analytical column and the primary technique for connecting the two columns is column switching.

In the dual-column configuration (Fig. 3.7), an electrically actuated six-port valve is used to connect the two columns and

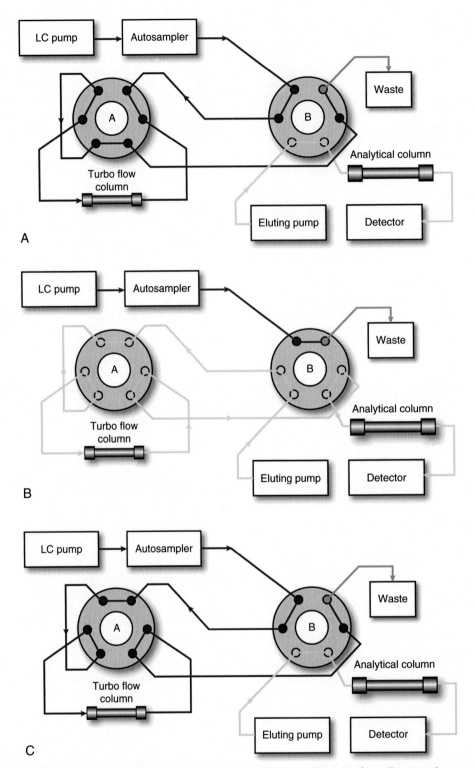

FIGURE 3.7 Schematic diagrams illustrating a column switching valve configuration for online sample preparation using dual columns, consisting of a turbulent flow column for extraction and an analytical column for separation. **A,** Loading step: Turbulent flow sweeps debris from sample matrix through the TurboFlow column (Thermo Fisher Scientific, Waltham, Massachusetts) while the analyte is retained. **B,** Transfer step: Gradient mobile phase elutes analytes back out of TurboFlow column to analytical column. **C,** Eluting step: Analytes are separated through eluting from analytical column to detector (see text and reference 184). (Reprinted by permission from Xu Y, Willson KJ, Musson DG. Strategies on efficient method development of on-line extraction assays for determination of MK-0974 in human plasma and urine using turbulent-flow chromatography and tandem mass spectrometry. *J Chromatogr B* 2008;863:64–73.)

perform the column switching function; two LC pumps are required. LC pump 1 is used for most of the sample extraction functions (load, wash, and equilibration steps) except for elution. During these extraction steps, the switching valve is set to position 1, which directs the effluent from the extraction column to waste and allows LC pump 2 to perform separation on the analytical column. The valve is switched to position 2 for the elution step. LC pump 2 now delivers flow through the extraction column, eluting the analyte, which passes on to the analytical column for further separation.

The dual-column mode provides the advantage of improved separation as well as detection sensitivity resulting from chromatographic focusing.[174] Another advantage of dual-column configuration is that the extraction and separation processes can be performed simultaneously because they are now driven by two independent LC pumps. While the sample is running on the separation column, the extraction column can be used to extract the next sample. Therefore all or part of the time that the system is used for online extraction can be buried in the run time of the separation column.

Most commonly used online sample preparation methodologies are based on the principle of a dual-column configuration. These approaches include TFC, RAM, SPE using a disposable cartridge, and IAE. The major differences among these approaches are the type of extraction column or cartridge, the solvent composition, and the flow rate of the chromatographic system.

EVOLVING TECHNIQUES

Dried Blood Spots

In the 1960s, phenylketonuria was identified from dried blood spots (DBS) that were placed on paper cards and mailed to the laboratory. An extracting solution dissolved the amines from the paper and allowed separation and detection by GC. Since the advent of more sensitive LC-MS/MS techniques, the analysis of drugs, amino acids, and other analytes from DBS (obtained from a finger or heel prick) has been revisited and become an active area of investigation. This sampling method is preferable because it is far less invasive than collecting venous blood into a tube, and from a logistical perspective a paper card is easier to ship than volumes of biological fluid in tubes under refrigeration. However, blood as a sample matrix contains more potential interferences than serum, for example, so the sample preparation and detection steps must be selective enough to meet the requirements of the assay. Over time, the investigation of this technique has resulted in many more questions raised beyond that of matrix effects, such as hematocrit concentration of blood, which affects size of the blood spot, spot volume, point of addition of the internal standard, homogeneity of spots, carryover between spots, and analyte stability under conditions of storage and transport.[175] Recent technologic developments in DBS include automation, online extraction, and direct analysis by MS.[176] Many representative DBS-LC-MS/MS applications are summarized and the important aspects of developing and validating a rugged DBS-LC-MS/MS method are presented by Li and Tse[177] and Jager associates.[175] A review by Taneja and colleagues discusses the applicability and relevance of DBS in the quantitative assessment of antimalarial drugs, the advantages and drawbacks of DBS, and the difficulties encountered during the implementation of this methodology.[178] High-throughput DBS analysis (sample-to-sample run times of 75 seconds) has been demonstrated in a screening method of the five most common lysosomal storage disorders in newborn DBS using TFC and UHPLC.[179]

Capillary Microsampling

An alternative to DBS is the technique of capillary microsampling to yield dried plasma spots (DPS) which overcomes the area bias and homogeneity issues associated with conventional DBS samples when a sub-punch is taken.[180] Bowen and associates describe a novel approach to capillary plasma microsampling for quantitative bioanalysis.[181] This published method involves collection of blood into a plastic-wrapped, ethylenediaminetetraacetic acid-coated capillary tube that contains a small amount of a thixotropic gel and a porous plug. The blood-filled capillary is placed into a secondary labeled container suitable for centrifugation and plasma is generated; during this process, the thixotropic gel isolates the plasma from the red blood cells and creates a physical barrier between the two matrices. The plasma is then dispensed from the capillary tube into a separate container for storage or processing. Applications are appearing in the literature, such as the evaluation of plasma microsampling for DPS in quantitative LC-MS/MS bioanalysis using ritonavir[182] and guanfacine[183] as model compounds.

REFERENCES

1. Jemal M, Xia Y-Q. The need for adequate chromatographic separation in the quantitative determination of drugs in biological samples by high performance liquid chromatography with tandem mass spectrometry. *Rapid Commun Mass Spectrom* 1999;**13**:97–106.
2. Wells DA. Automation tools and strategies for bioanalysis. In: *High throughput bioanalytical sample preparation: methods and automation strategies*. Amsterdam: Elsevier; 2003. p. 135–97.
3. Wells DA. Protein precipitation: automation strategies. In: *High throughput bioanalytical sample preparation: methods and automation strategies*. Amsterdam: Elsevier; 2003. p. 255–76.
4. Wells DA. Liquid-liquid extraction: automation strategies. In: *High throughput bioanalytical sample preparation: methods and automation strategies*. Amsterdam: Elsevier; 2003. p. 327–60.
5. Wells DA. Solid-phase extraction: automation strategies. In: *High throughput bioanalytical sample preparation: methods and automation strategies*. Amsterdam: Elsevier; 2003. p. 485–504.
6. Wells DA, Lloyd TL. Automation of sample preparation for pharmaceutical and clinical analysis. In: Pawliszyn J, editor. *Sampling and sample preparation for field and laboratory*, vol. 37. Amsterdam: Elsevier; 2002. p. 837–68.
7. Zacharis CK, Verdoukas A, Tzanavaras PD, et al. Automated sample preparation coupled to sequential injection chromatography: on-line filtration and dilution protocols prior to separation. *J Pharm Biomed Anal* 2009;**49**:726–32.
8. Fu I, Woolf EJ, Matuszewski BK. Effect of the sample matrix on the determination of indinavir in human urine by HPLC with turbo ion spray tandem mass spectrometric detection. *J Pharm Biomed Anal* 1998;**18**:347–57.

9. Deventer K, Pozo OJ, Verstraete AG, et al. Dilute-and-shoot-liquid chromatography-mass spectrometry for urine analysis in doping control and analytical toxicology. *TrAC Trends Anal Chem* 2014;**55**:1–13.
10. Cabrera K, Lubda D, Eggenweiler HM, et al. A new monolithic-type HPLC column for fast separations. *J High Res Chromatogr* 2000;**23**:93–9.
11. Nema T, Chan ECY, Ho PC. Applications of monolithic materials for sample preparation. *J Pharm Biomed Anal* 2014;**87**:130–41.
12. Koyuturk S, Can NO, Atkosar Z, et al. A novel dilute and shoot HPLC assay method for quantification of irbesartan and hydrochlorothiazide in combination tablets and urine using second generation C18-bonded monolithic silica column with double gradient elution. *J Pharm Biomed Anal* 2014;**97**:103–10.
13. McCauley-Myers DL, Eichhold TH, Bailey RE, et al. Rapid bioanalytical determination of dextromethorphan in canine plasma by dilute-and-shoot preparation combined with one minute per sample LC-MS/MS analysis to optimize formulations for drug delivery. *J Pharm Biomed Anal* 2000;**23**:825–35.
14. Jiang H, Ouyang Z, Zeng J, et al. A user-friendly robotic sample preparation program for fully automated biological sample pipetting and dilution to benefit the regulated bioanalysis. *J Lab Autom* 2012;**17**:211–21.
15. Bagamery K, Kvell K, Barnet M, et al. Are platelets activated after a rapid, one-step density gradient centrifugation? Evidence from flow cytometric analysis. *Clin Lab Haematol* 2005;**27**:75–7.
16. Harford JB, Bonifacino JS. Subcellular fractionation and isolation of organelles. In: *Current protocols in cell biology*. John Wiley & Sons; 2009.
17. Blanchard J. Evaluation of the relative efficacy of various techniques for deproteinizing plasma samples prior to high-performance liquid chromatographic analysis. *J Chromatogr* 1981;**226**:455–60.
18. Wilson ID, Michopoulos F, Theodoridis G. Sampling and sample preparation for LC-MS-based metabonomics/metabolomics of samples of mammalian origin. In: Pawliszyn J, editor. *Comprehensive sampling and sample preparation*. Oxford: Academic Press; 2012. p. 339–57.
19. Vuckovic D. Sample preparation in global metabolomics of biological fluids and tissues. In: Issaq HJ, Veenstra TD, editors. *Proteomic and metabolomic approaches to biomarker discovery*. Waltham, MA: Academic Press; 2013. p. 52–77.
20. Jasinski DL, Schwartz CT, Haque F, et al. Large scale purification of RNA nanoparticles by preparative ultracentrifugation. *Methods Mol Biol* 2015;**1297**:67–82.
21. Masek T, Valasek L, Pospisek M. Polysome analysis and RNA purification from sucrose gradients. *Methods Mol Biol* 2011;**703**:293–309.
22. Priego-Capote F, Luque De Castro MD. Analytical uses of ultrasound. I. Sample preparation. *TrAC Trends Anal Chem* 2004;**23**:644–53.
23. Capelo JL, dos Reis CD, Maduro C, et al. Tandem focused ultrasound (TFU) combined with fast furnace analysis as an improved methodology for total mercury determination in human urine by electrothermal-atomic absorption spectrometry. *Talanta* 2004;**64**:217–23.
24. Viñas P, López-Erroz C, Balsalobre N, et al. Speciation of cobalamins in biological samples using liquid chromatography with diode-array detection. *Chromatographia* 2003;**58**:5–10.
25. Bendicho C, De La Calle I, Pena F, et al. Ultrasound-assisted pretreatment of solid samples in the context of green analytical chemistry. *TrAC Trends Anal Chem* 2012;**31**:50–60.
26. Bendicho C, Lavilla I. EXTRACTION | *Ultrasound extractions. Reference module in chemistry, molecular sciences and chemical engineering*. Elsevier; 2013.
27. Smith KM, Xu Y. Tissue sample preparation in bioanalytical assays. *Bioanalysis* 2012;**4**:741–9.
28. Priego-Capote F, de Castro L. Ultrasound-assisted digestion: a useful alternative in sample preparation. *J Biochem Biophys Methods* 2007;**70**:299–310.
29. Lacorte S, Bono-Blay F, Cortina-Puig M. Sample homogenization. In: Pawliszyn J, editor. *Comprehensive sampling and sample preparation*. Oxford: Academic Press; 2012. p. 65–84.
30. Ilyin SE, Plata-Salamán CR. An efficient, reliable and inexpensive device for the rapid homogenization of multiple tissue samples by centrifugation. *J Neurosci Methods* 2000;**95**:123–5.
31. Size A, Sharon A, Sauer-Budge A. An automated low cost instrument for simultaneous multi-sample tissue homogenization. *Robotics and Computer-Integrated Manufacturing* 2011;**27**:276–81.
32. Xu S, Zheng S, Shen X, et al. Automated sample preparation and purification of homogenized brain tissues. *J Pharm Biomed Anal* 2007;**44**:581–5.
33. Pena-Llopis S, Brugarolas J. Simultaneous isolation of high-quality DNA, RNA, miRNA and proteins from tissues for genomic applications. *Nat Protoc* 2013;**8**:2240–55.
34. Poliwoda A, Wieczorek PP. Sample pretreatment techniques for oligopeptide analysis from natural sources. *Anal Bioanal Chem* 2009;**393**:885–97.
35. Deda O, Gika HG, Wilson ID, et al. An overview of fecal sample preparation for global metabolic profiling. *J Pharm Biomed Anal* 2015;**113**:137–50.
36. Liang X, Ubhayakar S, Liederer BM, et al. Evaluation of homogenization techniques for the preparation of mouse tissue samples to support drug discovery. *Bioanalysis* 2011;**3**:1923–33.
37. Saraswathy Veena V, Sara George P, Jayasree K, et al. Comparative analysis of cell morphology in sputum samples homogenized with dithiothreitol, N-acetyl-l cysteine, Cytorich® red preservative and in cellblock preparations to enhance the sensitivity of sputum cytology for the diagnosis of lung cancer. *Diagn Cytopathol* 2015;**43**:551–8.
38. Sparreboom A, Huizing MT, Boesen JJB, et al. Isolation, purification, and biological activity of mono- and dihydroxylated paclitaxel metabolites from human feces. *Cancer Chemother Pharmacol* 1995;**36**:299–304.
39. Huisman IH. Membrane separations: microfiltration. In: Wilson ID, editor. *Encyclopedia of separation science*. Oxford: Academic Press; 2000. p. 1764–77.
40. van Bruijnsvoort M, Schoenmakers PJ. Membrane preparation: hollow-fibre membranes. In: Wilson ID, editor. *Encyclopedia of separation science*. Oxford: Academic Press; 2000. p. 3312–9.
41. Blanco S, Prat C, Pallares MA, et al. Centrifugal ultrafiltration method for rapid concentration of *Legionella pneumophila* urinary antigen. *J Clin Microbiol* 2004;**42**:4410.
42. Cheryan M. Membrane separations: ultrafiltration. In: *Reference module in chemistry: molecular sciences and chemical engineering*. Elsevier; 2013.
43. Manza LL, Stamer SL, Ham AJ, et al. Sample preparation and digestion for proteomic analyses using spin filters. *Proteomics* 2005;**5**:1742–5.

44. Yu Y, Pieper R. Urinary pellet sample preparation for shotgun proteomic analysis of microbial infection and host-pathogen interactions. *Methods Mol Biol* 2015;**1295**:65–74.
45. Michalski A, Damoc E, Hauschild JP, et al. Mass spectrometry-based proteomics using Q Exactive, a high-performance benchtop quadrupole Orbitrap mass spectrometer. *Mol Cell Proteomics* 2011;**10**. M111.011015.
46. Dong WC, Hou ZL, Jiang XH, et al. A simple sample preparation method for measuring amoxicillin in human plasma by hollow fiber centrifugal ultrafiltration. *J Chromatogr Sci* 2013; **51**:181–6.
47. Huang Y, Chen H, He F, et al. Simultaneous determination of human plasma protein binding of bioactive flavonoids in *Polygonum orientale* by equilibrium dialysis combined with UPLC-MS/MS. *J Pharm Anal* 2013;**3**:376–81.
48. Wang C, Wang Q, Yuan Z, et al. Drug-protein-binding determination of stilbene glucoside using cloud-point extraction and comparison with ultrafiltration and equilibrium dialysis. *Drug Dev Ind Pharm* 2010;**36**:307–14.
49. Papadoyannis IN, Samanidou VF. Sample pretreatment in clinical chemistry. In: Aboul-Enein HY, editor. *Separation techniques in clinical chemistry*. New York, NY: Marcel Dekker; 2003.
50. Lanckmans K, Sarre S, Smolders I, et al. Quantitative liquid chromatography/mass spectrometry for the analysis of microdialysates. *Talanta* 2008;**74**:458–69.
51. Tang YB, Sun F, Teng L, et al. Simultaneous determination of the repertoire of classical neurotransmitters released from embryonal carcinoma stem cells using online microdialysis coupled with hydrophilic interaction chromatography-tandem mass spectrometry. *Anal Chim Acta* 2014:70–9.
52. Ducey MW, Regel AR, Nandi P, et al. Microdialysis sampling in the brain: analytical approaches and challenges. In: Janusz P, editor. *Comprehensive sampling and sample preparation*. Oxford: Academic Press; 2012. p. 535–57.
53. Behrens HL, Li L. Monitoring neuropeptides in vivo via microdialysis and mass spectrometry. *Methods Mol Biol* 2010; **615**:57–73.
54. Nandi P, Lunte SM. Recent trends in microdialysis sampling integrated with conventional and microanalytical systems for monitoring biological events: a review. *Anal Chim Acta* 2009; **651**:1–14.
55. Jin G, Cheng Q, Feng J, et al. On-line microdialysis coupled to analytical systems. *J Chromatogr Sci* 2008;**46**:276–87.
56. Shackman HM, Shou M, Cellar NA, et al. Microdialysis coupled on-line to capillary liquid chromatography with tandem mass spectrometry for monitoring acetylcholine in vivo. *J Neurosci Methods* 2007;**159**:86–92.
57. Musteata FM. Recent progress in in-vivo sampling and analysis. *TrAC Trends Anal Chem* 2013;**45**:154–68.
58. Musteata FM. Pharmacokinetic applications of microdevices and microsampling techniques. *Bioanalysis* 2009;**1**:171–85.
59. Hsieh YC, Zahn JD. On-chip microdialysis system with flow-through glucose sensing capabilities. *J Diabetes Sci Technol* 2007;**1**:375–83.
60. Moon BU, de Vries MG, Cordeiro CA, et al. Microdialysis-coupled enzymatic microreactor for in vivo glucose monitoring in rats. *Anal Chem* 2013;**85**:10949–55.
61. Li Q, Zubieta JK, Kennedy RT. Practical aspects of in vivo detection of neuropeptides by microdialysis coupled off-line to capillary LC with multistage MS. *Anal Chem* 2009;**81**: 2242–50.
62. Korf J, Huinink KD, Posthuma-Trumpie GA. Ultraslow microdialysis and microfiltration for in-line, on-line and off-line monitoring. *Trends Biotechnol* 2010;**28**:150–8.
63. Guihen E, O'Connor WT. Current separation and detection methods in microdialysis the drive towards sensitivity and speed. *Electrophoresis* 2009;**30**:2062–75.
64. Maischak H, Tautkus B, Kreusch S, et al. Proteomic sample preparation by microdialysis: easy, speedy, and nonselective. *Anal Biochem* 2012;**424**:184–6.
65. Hagel L. Gel filtration: size exclusion chromatography. In: Janson J-C, editor. *Protein purification: principles, high resolution methods, and applications*vol. 149. Hoboken, NJ: Wiley; 2011. p. 51.
66. Hedlund H. Desalting and buffer exchange of proteins using size-exclusion chromatography. *CSH Protoc* 2006;**2006**(1). https://doi.org/10.1101/pdb.prot4199.
67. Slawson MH, Crouch DJ, Andrenyak DM, et al. Determination of morphine, morphine-3-glucuronide, and morphine-6-glucuronide in plasma after intravenous and intrathecal morphine administration using HPLC with electrospray ionization and tandem mass spectrometry. *J Anal Toxicol* 1999;**23**:468–73.
68. Álvarez Sánchez B, Capote FP, Jiménez JR, et al. Automated solid-phase extraction for concentration and clean-up of female steroid hormones prior to liquid chromatography–electrospray ionization–tandem mass spectrometry: an approach to lipidomics. *J Chromatogr A* 2008;**1207**:46–54.
69. Yu C, Penn LD, Hollembaek J, et al. Enzymatic tissue digestion as an alternative sample preparation approach for quantitative analysis using liquid chromatography-tandem mass spectrometry. *Anal Chem* 2004;**76**:1761–7.
70. Bronsema KJ, Bischoff R, van de Merbel NC. High-sensitivity LC-MS/MS quantification of peptides and proteins in complex biological samples: the impact of enzymatic digestion and internal standard selection on method performance. *Anal Chem* 2013;**85**:9528–35.
71. Leon IR, Schwammle V, Jensen ON, et al. Quantitative assessment of in-solution digestion efficiency identifies optimal protocols for unbiased protein analysis. *Mol Cell Proteom* 2013; **12**:2992–3005.
72. Fonslow BR, Yates 3rd JR. Proteolytic digestion methods for shotgun proteomics. In: Pawliszyn J, editor. *Comprehensive sampling and sample preparation*. Oxford: Academic Press; 2012. p. 261–76.
73. Lehmann S, Hoofnagle A, Hochstrasser D, et al. Quantitative clinical chemistry proteomics (qCCP) using mass spectrometry: general characteristics and application. *Clin Chem Lab Med* 2013;**51**:919–35.
74. Wang N, Li L. Microwave digestion of protein samples for proteomics applications. In: Pawliszyn J, editor. *Comprehensive sampling and sample preparation*. Oxford: Academic Press; 2012. p. 277–90.
75. Taverna D, Norris JL, Caprioli RM. Histology-directed microwave assisted enzymatic protein digestion for MALDI MS analysis of mammalian tissue. *Anal Chem* 2015; **87**:670–6.
76. Lu Y, Kippler M, Harari F, et al. Alkali dilution of blood samples for high throughput ICP-MS analysis-comparison with acid digestion. *Clin Biochem* 2015;**48**:140–7.
77. Polson C, Sarkar P, Incledon B, et al. Optimization of protein precipitation based upon effectiveness of protein removal and ionization effect in liquid chromatography-tandem mass

spectrometry. *J Chromatogr B Analyt Technol Biomed Life Sci* 2003;**785**:263−75.
78. Matuszewski BK, Constanzer ML, Chavez-Eng CM. Matrix effect in quantitative LC/MS/MS analyses of biological fluids: a method for determination of finasteride in human plasma at picogram per milliliter concentrations. *Anal Chem* 1998;**70**:882−9.
79. Li F, Ewles M, Pelzer M, et al. Case studies. The impact of nonanalyte components on LC-MS/MS-based bioanalysis: strategies for identifying and overcoming matrix effects. *Bioanalysis* 2013;**5**:2409−41.
80. Singleton C. Recent advances in bioanalytical sample preparation for LC-MS analysis. *Bioanalysis* 2012;**4**:1123−40.
81. Marchi I, Viette V, Badoud F, et al. Characterization and classification of matrix effects in biological samples analyses. *J Chromatogr A* 2010;**1217**:4071−8.
82. King R, Bonfiglio R, Fernandez-Metzler C, et al. Mechanistic investigation of ionization suppression in electrospray ionization. *J Am Soc Mass Spectrom* 2000;**11**:942−50.
83. Bergeron A, Garofolo F. Importance of matrix effects in LC−MS/MS bioanalysis. *Bioanalysis* 2013;**5**:2331−2.
84. Cote C, Bergeron A, Mess JN, et al. Matrix effect elimination during LC-MS/MS bioanalytical method development. *Bioanalysis* 2009;**1**:1243−57.
85. Furey A, Moriarty M, Bane V, et al. Ion suppression; a critical review on causes, evaluation, prevention and applications. *Talanta* 2013;**115**:104−22.
86. Watt AP, Morrison D, Locker KL, et al. Higher throughput bioanalysis by automation of a protein precipitation assay using a 96-well format with detection by LC-MS/MS. *Anal Chem* 2000;**72**:979−84.
87. Tulipani S, Llorach R, Urpi-Sarda M, et al. Comparative analysis of sample preparation methods to handle the complexity of the blood fluid metabolome: when less is more. *Anal Chem* 2013;**85**:341−8.
88. Janusch F, Kalthoff L, Hamscher G, et al. Evaluation and subsequent minimization of matrix effects caused by phospholipids in LC−MS analysis of biological samples. *Bioanalysis* 2013;**5**:2101−14.
89. Guo X, Lankmayr E, Guzzetta AA, et al. Phospholipid-based matrix effects in LC-MS bioanalysis. *Bioanalysis* 2011;**3**:349−52.
90. Ma J, Shi J, Le H, et al. A fully automated plasma protein precipitation sample preparation method for LC-MS/MS bioanalysis. *J Chromatogr B* 2008;**862**:219−26.
91. Murphy AT, Berna MJ, Holsapple JL, et al. Effects of flow rate on high-throughput quantitative analysis of protein-precipitated plasma using liquid chromatography/tandem mass spectrometry. *Rapid Commun Mass Spectrom* 2002;**16**:537−43.
92. Bakhtiar R, Lohne J, Ramos L, et al. High-throughput quantification of the anti-leukemia drug STI571 (Gleevec) and its main metabolite (CGP 74588) in human plasma using liquid chromatography-tandem mass spectrometry. *J Chromatogr B Analyt Technol Biomed Life Sci* 2002;**768**:325−40.
93. Sadagopan N, Cohen L, Roberts B, et al. Liquid chromatography-tandem mass spectrometric quantitation of cyclophosphamide and its hydroxy metabolite in plasma and tissue for determination of tissue distribution. *J Chromatogr B Biomed Sci Appl* 2001;**759**:277−84.
94. Shou WZ, Bu HZ, Addison T, et al. Development and validation of a liquid chromatography/tandem mass spectrometry (LC/MS/MS) method for the determination of ribavirin in human plasma and serum. *J Pharm Biomed Anal* 2002;**29**:83−94.
95. Ramos L, Brignol N, Bakhtiar R, et al. High-throughput approaches to the quantitative analysis of ketoconazole, a potent inhibitor of cytochrome P450 3A4, in human plasma. *Rapid Commun Mass Spectrom* 2000;**14**:2282−93.
96. Walter RE, Cramer JA, Tse FL. Comparison of manual protein precipitation (PPT) versus a new small volume PPT 96-well filter plate to decrease sample preparation time. *J Pharm Biomed Anal* 2001;**25**:331−7.
97. Biddlecombe RA, Pleasance S. Automated protein precipitation by filtration in the 96-well format. *J Chromatogr B Biomed Sci Appl* 1999;**734**:257−65.
98. Berna M, Murphy AT, Wilken B, et al. Collection, storage, and filtration of in vivo study samples using 96-well filter plates to facilitate automated sample preparation and LC/MS/MS analysis. *Anal Chem* 2002;**74**:1197−201.
99. Rouan MC, Buffet C, Marfil F, et al. Plasma deproteinization by precipitation and filtration in the 96-well format. *J Pharm Biomed Anal* 2001;**25**:995−1000.
100. Turfus SC, Halket JM, Parkin MC, et al. Signal enhancement of glucuronide conjugates in LC-MS/MS by derivatization with the phosphonium propylamine cation tris(trimethoxyphenyl) phosphonium propylamine, for forensic purposes. *Drug Test Anal* 2014;**6**:500−5.
101. Qi B-L, Liu P, Wang Q-Y, et al. Derivatization for liquid chromatography-mass spectrometry. *TrAC Trends Anal Chem* 2014;**59**:121−32.
102. Persichilli S, Gervasoni J, Cocci A, et al. Anabolic steroids by LC-MSMS. *Clin Chem Lab Med* 2012;**50**. eA68.
103. Chughtai K, Heeren RMA. Mass spectrometric imaging for biomedical tissue analysis. *Chem Rev* 2010;**110**:3237−77.
104. Hoofnagle AN, Laha TJ, Donaldson TF. A rubber transfer gasket to improve the throughput of liquid-liquid extraction in 96-well plates: application to vitamin D testing. *J Chromatogr B Analyt Technol Biomed Life Sci* 2010;**878**: 1639−42.
105. Zhang N, Hoffman KL, Li W, et al. Semi-automated 96-well liquid-liquid extraction for quantitation of drugs in biological fluids. *J Pharm Biomed Anal* 2000;**22**:131−8.
106. Jemal M, Teitz D, Ouyang Z, et al. Comparison of plasma sample purification by manual liquid-liquid extraction, automated 96-well liquid-liquid extraction and automated 96-well solid-phase extraction for analysis by high-performance liquid chromatography with tandem mass spectrometry. *J Chromatogr B Biomed Sci Appl* 1999;**732**:501−8.
107. Ramos L, Bakhtiar R, Tse FL. Liquid-liquid extraction using 96-well plate format in conjunction with liquid chromatography/tandem mass spectrometry for quantitative determination of methylphenidate (Ritalin) in human plasma. *Rapid Commun Mass Spectrom* 2000;**14**:740−5.
108. Shen Z, Wang S, Bakhtiar R. Enantiomeric separation and quantification of fluoxetine (Prozac) in human plasma by liquid chromatography/tandem mass spectrometry using liquid-liquid extraction in 96-well plate format. *Rapid Commun Mass Spectrom* 2002;**16**:332−8.
109. Brignol N, McMahon LM, Luo S, et al. High-throughput semi-automated 96-well liquid/liquid extraction and liquid chromatography/mass spectrometric analysis of everolimus (RAD 001) and cyclosporin a (CsA) in whole blood. *Rapid Commun Mass Spectrom* 2001;**15**:898−907.
110. Owen LJ, Keevil BG. Supported liquid extraction as an alternative to solid phase extraction for LC-MS/MS aldosterone analysis? *Ann Clin Biochem* 2013;**50**(Pt 5):489−91.

111. Wang AQ, Zeng W, Musson DG, et al. A rapid and sensitive liquid chromatography/negative ion tandem mass spectrometry method for the determination of an indolocarbazole in human plasma using 96-well diatomaceous earth plates for solid-liquid extraction [correction of using internal standard (IS) 96-well diatomaceous earth plates for solid-liquid extraction]. *Rapid Commun Mass Spectrom* 2002;**16**: 975–81.
112. Zhao JJ, Xie IH, Yang AY, et al. Quantitation of simvastatin and its beta-hydroxy acid in human plasma by liquid-liquid cartridge extraction and liquid chromatography/tandem mass spectrometry. *J Mass Spectrom* 2000;**35**:1133–43.
113. Wang AQ, Fisher AL, Hsieh J, et al. Determination of a beta(3)-agonist in human plasma by LC/MS/MS with semi-automated 48-well diatomaceous earth plate. *J Pharm Biomed Anal* 2001; **26**:357–65.
114. Sauve EN, Langodegard M, Ekeberg D, et al. Determination of benzodiazepines in ante-mortem and post-mortem whole blood by solid-supported liquid-liquid extraction and UPLC-MS/MS. *J Chromatogr B Analyt Technol Biomed Life Sci* 2012; **883-884**:177–88.
115. Dominguez-Romero JC, Garcia-Reyes JF, Molina-Diaz A. Comparative evaluation of seven different sample treatment approaches for large-scale multiclass sport drug testing in urine by liquid chromatography-mass spectrometry. *J Chromatogr A* 2014;**1361**:34–42.
116. Cheng Y-F, Neue UD, Bean L. Straightforward solid-phase extraction method for the determination of verapamil and its metabolite in plasma in a 96-well extraction plate. *J Chromatogr A* 1998;**828**:273–81.
117. Mazzeo J, Krull I. Immobilized boronates for the isolation and separation of bioanalytes. *Biochromatography* 1989;**4**: 124–30.
118. Martin P, Leadbetter B, Wilson I. Immobilized phenylboronic acids for the selective extraction of β-blocking drugs from aqueous solution and plasma. *J Pharm Biomed Anal* 1993;**11**:307–12.
119. Maycock P, Frayn K. Use of alumina columns to prepare plasma samples for liquid-chromatographic determination of catecholamines. *Clin Chem* 1987;**33**:286–7.
120. Wu A, Gornet TG. Preparation of urine samples for liquid-chromatographic determination of catecholamines: bonded-phase phenylboronic acid, cation-exchange resin, and alumina adsorbents compared. *Clin Chem* 1985;**31**:298–302.
121. Wells DA. Solid-phase extraction: high throughput techniques. In: *High throughput bioanalytical sample preparation: methods and automation strategies*. Amsterdam: Elsevier; 2003. p. 361–432.
122. Wells DA. Solid-phase extraction: strategies for method development and optimization. In: *High throughput bioanalytical sample preparation: methods and automation strategies*. Amsterdam: Elsevier; 2003. p. 433–84.
123. Gilar M, Bouvier ES, Compton BJ. Advances in sample preparation in electromigration, chromatographic and mass spectrometric separation methods. *J Chromatogr A* 2001;**909**: 111–35.
124. Hennion MC, Cau-Dit-Coumes C, Pichon V. Trace analysis of polar organic pollutants in aqueous samples: tools for the rapid prediction and optimisation of the solid-phase extraction parameters. *J Chromatogr A* 1998;**823**:147–61.
125. Hennion MC. Solid-phase extraction: method development, sorbents, and coupling with liquid chromatography. *J Chromatogr A* 1999;**856**:3–54.
126. Pawliszyn J. Theory of extraction. In: Janusz P, editor. *Comprehensive sampling and sample preparation*. Oxford: Academic Press; 2012. p. 1–25.
127. Janiszewski J, Schneider RP, Hoffmaster K, et al. Automated sample preparation using membrane microtiter extraction for bioanalytical mass spectrometry. *Rapid Commun Mass Spectrom* 1997;**11**:1033–7.
128. Grouzmann E, Dunand M, Gubian D, et al. High throughput and sensitive quantitation of plasma catecholamines by UPLC-tandem mass spectrometry using a solid phase microwell extraction plate. *Clin Chem Lab Med* 2013;**51**:eA54–A55.
129. Hempenius J, Wieling J, Brakenhoff JP, et al. High-throughput solid-phase extraction for the determination of cimetidine in human plasma. *J Chromatogr B Biomed Sci Appl* 1998;**714**: 361–8.
130. Wang W, Qin S, Li L, et al. An optimized high throughput clean-up method using mixed-mode SPE plate for the analysis of free arachidonic acid in plasma by LC-MS/MS. *Int J Anal Chem* 2015;**2015**:374819.
131. Teo CC, Chong WPK, Tan E, et al. Advances in sample preparation and analytical techniques for lipidomics study of clinical samples. *TrAC Trends Anal Chem* 2015;**66**:1–18.
132. Skoglund C, Bassyouni F, Abdel-Rehim M. Monolithic packed 96-tips set for high-throughput sample preparation: determination of cyclophosphamide and busulfan in whole blood samples by monolithic packed 96-tips and LC-MS. *Biomed Chromatogr* 2013;**27**:714–9.
133. Yu Y, Suh M-J, Sikorski P, et al. Urine sample preparation in 96-well filter plates for quantitative clinical proteomics. *Anal Chem* 2014;**86**:5470–7.
134. Li KM, Rivory LP, Clarke SJ. Solid-phase extraction (SPE) techniques for sample preparation in clinical and pharmaceutical analysis: a brief overview. *Curr Pharm Anal* 2006;**2**:95–102.
135. Song Q, Junga H, Tang Y, et al. Automated 96-well solid phase extraction and hydrophilic interaction liquid chromatography-tandem mass spectrometric method for the analysis of cetirizine (ZYRTEC®) in human plasma: with emphasis on method ruggedness. *J Chromatogr B Analyt Technol Biomed Life Sci* 2005;**814**:105–14.
136. Rule G, Chapple M, Henion JA. 384-well solid-phase extraction for LC/MS/MS determination of methotrexate and its 7-hydroxy metabolite in human urine and plasma. *Anal Chem* 2001;**73**:439–43.
137. Biddlecombe RA, Benevides C, Pleasance S. A clinical trial on a plate? The potential of 384-well format solid phase extraction for high-throughput bioanalysis using liquid chromatography/tandem mass spectrometry. *Rapid Commun Mass Spectrom* 2001;**15**:33–40.
138. Li M, Chou J, Jing J, et al. MARS: bringing the automation of small-molecule bioanalytical sample preparations to a new frontier. *Bioanalysis* 2012;**4**:1311–26.
139. Rossi DT, Zhang N. Automating solid-phase extraction: current aspecs and future prospects. *J Chromatogr A* 2000;**885**:97–113.
140. Bladergroen MR, Derks RJ, Nicolardi S, et al. Standardized and automated solid-phase extraction procedures for high-throughput proteomics of body fluids. *J Proteomics* 2012;**77**:144–53.
141. Jones RL, Owen LJ, Adaway JE, et al. Simultaneous analysis of cortisol and cortisone in saliva using XLC-MS/MS for fully automated online solid phase extraction. *J Chromatogr B Analyt Technol Biomed Life Sci* 2012;**881-882**:42–8.
142. Koal T, Sibum M, Koster E, et al. Direct and fast determination of antiretroviral drugs by automated online solid-phase

142. extraction-liquid chromatography-tandem mass spectrometry in human plasma. *Clin Chem Lab Med* 2006;**44**:299−305.
143. Kohler C, Grobosch T, Binscheck T. Rapid quantification of tilidine, nortilidine, and bisnortilidine in urine by automated online SPE-LC-MS/MS. *Anal Bioanal Chem* 2011;**400**:17−23.
144. Robandt PV, Bui HM, Scancella JM, et al. Automated solid-phase extraction-liquid chromatography-tandem mass spectrometry analysis of 6-acetylmorphine in human urine specimens: application for a high-throughput urine analysis laboratory. *J Anal Toxicol* 2010;**34**:470−5.
145. Mitchell RJ, Christian R, Hughes H, et al. The application of fully automated on-line solid phase extraction in bioanalysis. *J Pharm Biomed Anal* 2010;**52**:86−92.
146. Couchman L. Automated, high-throughput workflow for the analysis of 25-hydroxyvitamin D and 3-EPI-25-hydroxyvitamin D3 by multiplexed turboflow LC-tandem MS. *Biochim Clin* 2013;**37**:S81.
147. Breaud AR, Harlan R, Di Bussolo JM, et al. A rapid and fully-automated method for the quantitation of tricyclic antidepressants in serum using turbulent-flow liquid chromatography-tandem mass spectrometry. *Clin Chim Acta* 2010;**411**:825−32.
148. Füzéry AK, Breaud AR, Emezienna N, et al. A rapid and reliable method for the quantitation of hydroxychloroquine in serum using turbulent flow liquid chromatography-tandem mass spectrometry. *Clin Chim Acta* 2013;**421**:79−84.
149. Ceglarek U, Lembcke J, Fiedler GM, et al. Rapid simultaneous quantification of immunosuppressants in transplant patients by turbulent flow chromatography combined with tandem mass spectrometry. *Clin Chim Acta* 2004;**346**:181−90.
150. Kousoulos C, Dotsikas Y, Loukas YL. Turbulent flow and ternary column-switching on-line clean-up system for high-throughput quantification of risperidone and its main metabolite in plasma by LC-MS/MS. Application to a bioequivalence study. *Talanta* 2007;**72**:360−7.
151. Takino M, Daishima S, Yamaguchi K, et al. Quantitative liquid chromatography-mass spectrometry determination of catechins in human plasma by automated on-line extraction using turbulent flow chromatography. *Analyst* 2003;**128**:46−50.
152. Pretorius V, Smuts TW. Turbulent flow chromatography: a new approach to faster analysis. *Anal Chem* 1966;**38**:274−81.
153. Edge T. Turbulent flow chromatography in bioanalysis. In: Ian DW, editor. *Handbook of analytical separations bioanalytical separations*, vol. 4. Amsterdam: Elsevier Science; 2003. p. 91−128.
154. Couchman L. Turbulent flow chromatography in bioanalysis: a review. *Biomed Chromatogr* 2012;**26**:892−905.
155. Iberer G, Hahn R, Jungbauer A. Monoliths as stationary phases for separating biopolymers: fourth-generation chromatography sorbents. *LC GC* 1999;**17**:998−1005.
156. Nema T, Chan EC, Ho PC. Applications of monolithic materials for sample preparation. *J Pharm Biomed Anal* 2014;**87**:130−41.
157. Namera A, Saito T. Advances in monolithic materials for sample preparation in drug and pharmaceutical analysis. *TrAC Trends Anal Chem* 2013;**45**:182−96.
158. Hennion MC, Pichon V. Immuno-based sample preparation for trace analysis. *J Chromatogr A* 2003;**1000**:29−52.
159. Hage DS, Nelson MA. Chromatographic immunoassays. *Anal Chem* 2001;**73**:199A−205A.
160. Hage DS. Survey of recent advances in analytical applications of immunoaffinity chromatography. *J Chromatogr B Biomed Sci Appl* 1998;**715**:3−28.
161. De Frutos M, Regnier FE. Tandem chromatographic-immunological analyses. *Anal Chem* 1993;**65**:17A−25A.
162. Hage DS, Clarke W. Immunoaffinity chromatography in clinical analysis. In: Hage DS, editor. *Handbook of affinity chromatography*. New York, NY: Taylor & Francis; 2006. p. 361.
163. Tsikas D. Quantitative analysis of biomarkers, drugs and toxins in biological samples by immunoaffinity chromatography coupled to mass spectrometry or tandem mass spectrometry: a focused review of recent applications. *J Chromatogr B Analyt Technol Biomed Life Sci* 2010;**878**:133−48.
164. Stevenson D. Immuno-affinity solid-phase extraction. *J Chromatogr B Biomed Sci Appl* 2000;**745**:39−48.
165. Lord HL, Rajabi M, Safari S, et al. A study of the performance characteristics of immunoaffinity solid phase microextraction probes for extraction of a range of benzodiazepines. *J Pharm Biomed Anal* 2007;**44**:506−19.
166. Boos KS, Rudolphi A. The use of restricted-access media in HPLC. I. Classification and review. *LC GC* 1997;**15**:602−11.
167. Rudolphi A, Boos KS. The use of restricted-access media in HPLC. II. Applications. *LC GC* 1997;**15**:814−22.
168. Souverain S, Rudaz S, Veuthey JL. Restricted access materials and large particle supports for on-line sample preparation: an attractive approach for biological fluids analysis. *J Chromatogr B Analyt Technol Biomed Life Sci* 2004;**801**:141−56.
169. Mullett WM. Determination of drugs in biological fluids by direct injection of samples for liquid-chromatographic analysis. *J Biochem Biophys Methods* 2007;**70**:263−73.
170. Cassiano NM, Barreiro JC, Moraes MC, et al. Restricted access media supports for direct high-throughput analysis of biological fluid samples: review of recent applications. *Bioanalysis* 2009;**1**:577−94.
171. Cassiano NM, Lima VV, Oliveira RV, et al. Development of restricted-access media supports and their application to the direct analysis of biological fluid samples via high-performance liquid chromatography. *Anal Bioanal Chem* 2006;**384**:1462−9.
172. Cassiano NM, Lima VV, Oliveira RV, et al. Development of restricted-access media supports and their application to the direct analysis of biological fluid samples via high-performance liquid chromatography. *Anal Bioanal Chem* 2006;**385**:1580.
173. Jemal M, Yuan Q, Higan DB. The use of high-flow high performance liquid chromatography coupled with positive and negative ion electrospray tandem mass spectrometry for quantitative bioanalysis via direct injection of the plasma/serum samples. *Rapid Commun Mass Spectrom* 1998;**12**:1389−99.
174. Zeng H, Wu JT, Unger SE. The investigation and the use of high flow column-switching LC/MS/MS as a high-throughput approach for direct plasma sample analysis of single and multiple components in pharmacokinetic studies. *J Pharm Biomed Anal* 2002;**27**:967−82.
175. Jager NGL, Rosing H, Schellens JHM, et al. Procedures and practices for the validation of bioanalytical methods using dried blood spots: a review. *Bioanalysis* 2014;**6**:2481−514.
176. Meesters RJ, Hooff GP. State-of-the-art dried blood spot analysis: an overview of recent advances and future trends. *Bioanalysis* 2013;**5**:2187−208.
177. Li W, Tse FL. Dried blood spot sampling in combination with LC-MS/MS for quantitative analysis of small molecules. *Biomed Chromatogr* 2010;**24**:49−65.

178. Taneja I, Erukala M, Raju KS, et al. Dried blood spots in bioanalysis of antimalarials: relevance and challenges in quantitative assessment of antimalarial drugs. *Bioanalysis* 2013;5:2171–86.
179. Herman JL, Shushan B, De Jesus VR, et al. The application of multiplexed, multi-dimensional ultra high pressure liquid chromatography/tandem mass spectrometry to the high throughput screening of lysosomal storage disorders in newborn dried bloodspots. *J Inherit Metab Dis* 2010:S126.
180. Denniff P, Spooner N. Volumetric absorptive microsampling: a dried sample collection technique for quantitative bioanalysis. *Anal Chem* 2014;**86**:8489–95.
181. Bowen CL, Licea-Perez H, Karlinsey MZ, et al. A novel approach to capillary plasma microsampling for quantitative bioanalysis. *Bioanalysis* 2013;**5**:1131–5.
182. Li W, Doherty J, Favara S, et al. Evaluation of plasma microsampling for dried plasma spots (DPS) in quantitative LC-MS/MS bioanalysis using ritonavir as a model compound. *J Chromatogr B* 2015;**991**:46–52.
183. Li Y, Henion J, Abbott R, et al. The use of a membrane filtration device to form dried plasma spots for the quantitative determination of guanfacine in whole blood. *Rapid Commun Mass Spectrom* 2012;**26**:1208–12.
184. Xu Y, Willson KJ, Musson DG. Strategies on efficient method development of on-line extraction assays for determination of MK-0974 in human plasma and urine using turbulent-flow chromatography and tandem mass spectrometry. *J Chromatogr B Analyt Technol Biomed Life Sci* 2008;**863**:64–73.

Mass Spectrometry Applications in Infectious Disease and Pathogens Identification

*Phillip Heaton and Robin Patel**

ABSTRACT

Background
Matrix-assisted laser desorption ionization time-of-flight mass spectrometry (MALDI-TOF-MS) is a powerful tool in the clinical microbiology laboratory enabling accurate identification of bacteria, fungi, and mycobacteria. First adopted in European microbiology laboratories, its ease of use, accuracy, rapid turnaround times, and low cost have led to its widespread adoption in microbiology laboratories worldwide. In contrast, polymerase chain reaction electrospray ionization—mass spectrometry (PCR-ESI-MS) is an emerging technology for clinical microbiology with the potential for direct-from-sample testing and actionable results in a few hours.

Content
This first half of this chapter briefly discusses the history of MALDI-TOF-MS leading to its commercialization and adoption in clinical microbiology laboratories. Identification of aerobic and anaerobic organisms as well as mycobacteria and fungi is discussed. Additional applications, such as direct identification from blood and urine cultures as well as antimicrobial susceptibility testing are also reviewed. Additionally, implementation of MALDI-TOF-MS into routine laboratory workflows is addressed. The second half of this chapter discusses PCR-ESI-MS and its potential applications in the clinical microbiology laboratory in its current state.

MATRIX-ASSISTED LASER DESORPTION IONIZATION TIME-OF-FLIGHT MASS SPECTROMETRY

Bacteria and other microorganisms have an abundance of unique proteins allowing for taxonomic identification based on the mass fingerprint of their proteins. Sample preparation is required before performing matrix-assisted laser desorption ionization time-of-flight mass spectrometry (MALDI-TOF-MS). This can be accomplished by simply applying a colony of bacteria onto a spot on a MALDI-TOF-MS plate (with or without formic acid) or by the more traditional method of using formic acid to disrupt the cells and then extraction of the proteins with acetonitrile. The first method is used for routine identification of bacteria and yeast, whereas the latter has been typically reserved for difficult-to-lyse organisms. Direct colony processing is much simpler, requires less hands-on time, and is less expensive to perform than is protein extraction.

For direct colony processing, a colony is first "picked" from a culture plate and then spotted onto position on a MALDI-TOF-MS plate (Fig. 4.1). These plates can be reusable stainless steel plates in various configurations or disposable 48-well plates in the case of the MALDI Biotyper (Bruker Daltonics, Billerica, Massachusetts), whereas the VITEK MS (bioMérieux, SA, Marcy-l'Etoile, France) uses 48-well disposable plastic slides that fit into a steel plate that is capable of holding four slides. Formic acid can be added to improve the quality of the mass spectra that are generated. Once the formic acid is dried, the spot is overlaid with matrix (typically α-cyano-4-hydroxycinnamic acid dissolved in an acidified mixture of organic solvents and water). MALDI is considered a soft ionization technique, which means that the matrix aids in ionization of proteins without causing them to fragment when the laser is applied to the sample. Once the matrix has dried, the plate is placed into the instrument for measurement.

Once the plate is introduced to the ionization chamber, a pulsed laser is directed onto the matrix/analyte crystals of each spot, causing desorption and ionization of microbial proteins and other molecules in a plume of ionized sample and matrix molecules. The matrix absorbs most of the energy from the laser, leading to an ionized state that transfers charge from matrix to proteins and other molecules through random molecular collisions. The ionized microbial proteins are accelerated in an electromagnetic field and enter a TOF mass analyzer maintained under vacuum in which proteins are separated based on their velocity, with that velocity being inversely proportional to their mass-to-charge (m/z) ratios. In MALDI, the charge is most often +1. Finally, at the end of the flight tube, the ions hit an ion detector, with smaller proteins colliding with the ion detector first, followed by larger ones. The resulting time-resolved spectrum is eventually transformed to a mass spectrum by calibrating the system with peptides of known sizes, relating TOF to m/z. The resulting mass spectrum for a given sample organism can then be compared to reference spectra of known microorganisms (Fig. 4.2).

---

*This chapter expands upon the previous review: Patel R. *Clin Chem* 2015;61:100—111, with permission.

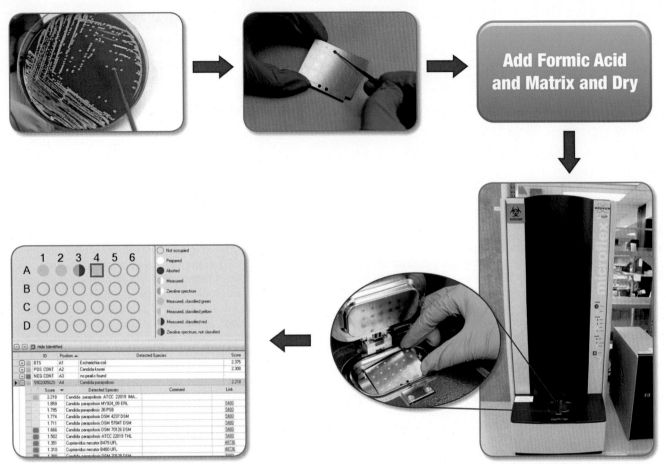

FIGURE 4.1 Matrix-assisted laser desorption ionization time-of-flight mass spectrometry (MALDI-TOF-MS) process. Using a plastic or wooden stick, loop, or pipette tip, a colony is picked from a culture plate to a spot on a MALDI-TOF-MS target plate (a reusable or disposable plate with a number of test spots). One or many isolates may be tested at a time. In this example, cells are treated with formic acid on the target plate and then dried. The spot is then overlaid with matrix and dried. The plate is placed in the ionization chamber of the mass spectrometer (see Fig. 4.2). A mass spectrum is generated and compared against a database of mass spectra by the software, resulting in identification of the organism (*Candida parapsilosis* in position A4 in the example). (Reproduced by permission of Mayo Foundation for Medical Education and Research. All rights reserved.)

History and Development of Commercial Systems

The first proposal that mass spectrometry be used for the identification of bacteria was in 1975, although the process employed an ionization method that fragmented proteins, thereby preventing effective analysis.[1] The technology to study intact proteins by MS was not invented until the 1980s when two groups came up with similar ideas for the detection of intact macromolecules. Koichi Tanaka and coworkers described the use of an ultrafine metal powder mixed with glycerin to detect macromolecules without fragmentation, for which Tanaka was awarded the Nobel Prize in chemistry (shared with John Fenn and Kurt Wüthrich).[2] At the same time that Tanaka published his research, Franz Hillenkamp and Michael Karas published their work on soft desorption/ionization using an organic compound matrix, nicotinic acid, also allowing the analysis of large biomolecules.[3] The term *matrix-assisted laser desorption ionization* (MALDI) was coined from this work. Over the last few decades, computer science, information technology, and aggregation of comprehensive databases of well-characterized reference spectra have led to the adoption of MALDI-TOF-MS as a powerful tool for clinical microbiology laboratories. Most recently, two commercial systems have been approved by the US Food and Drug Administration (FDA) for routine identification of bacteria and fungi in clinical microbiology laboratories. Currently the VITEK MS (bioMérieux SA) and MALDI Biotyper CA System (Bruker Daltonics) have been cleared by the FDA for identification of bacteria and yeast cultured on solid media. A list of claimed organisms as of April 18, 2015 can be found in Box 4.1. Each system consists of a mass spectrometer, software, and a database; however, differences exist in each of the three components, as well as which organisms are cleared by FDA for identification. Differences exist in the size of the instruments; Bruker's mass spectrometer is a bench-top model, and bioMérieux's is a floor model instrument.

Both Bruker's and bioMérieux's FDA-approved and research use only (RUO) systems have been reported on, though the Bruker has been more extensively described in the medical literature. The development of the Bruker system began in 2005 and has led to the relatively recently FDA-approved MALDI Biotyper CA system, although the RUO MALDI

CHAPTER 4 Mass Spectrometry Applications in Infectious Disease

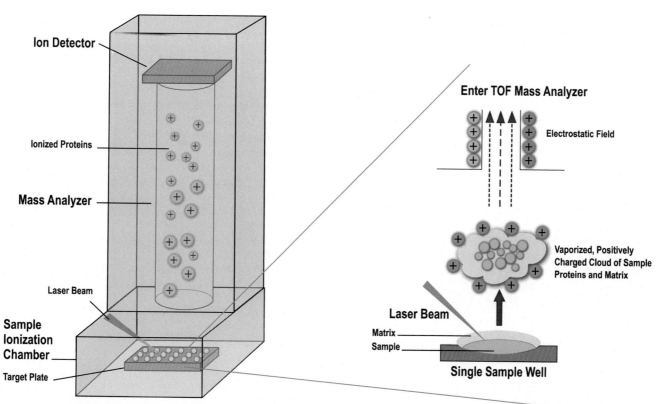

FIGURE 4.2 Matrix-assisted laser desorption ionization time of flight (TOF) mass spectrometer. The target plate is placed into the chamber of the mass spectrometer. Spots to be analyzed are shot by a laser, desorbing and ionizing microbial and matrix molecules from the target plate. The cloud of ionized molecules is accelerated into the TOF mass analyzer, toward a detector. Lighter molecules travel faster, followed by progressively heavier analytes. A mass spectrum is generated, representing the number of ions hitting the detector over time. Separation is by mass-to-charge ratio, but because the charge is typically single for this application, separation is effectively by molecular weight. (Reproduced by permission of Mayo Foundation for Medical Education and Research. All rights reserved.)

BOX 4.1 List of Reportable Organisms for the US Food and Drug Administration–Approved Vitek Mass Spectrometry and Matrix-Assisted Laser Desorption Ionization Biotyper CA Systems

Aerobic Gram-Positive Bacteria
Abiotrophia defectiva[V]
Aerococcus urinae[V]
Aerococcus viridans
Brevibacterium casei[B]
Corynebacterium amycolatum[B]
Corynebacterium bovis[B]
Corynebacterium diptheriae[B]
Corynebacterium glucuronolyticium[B]
Corynebacterium jeikeium[V]
Corynebacterium kroppenstedtii[B]
Corynebacterium macginleyi[B]
Corynebacterium minutissimum[B]
Corynebacterium propinquum[B]
Corynebacterium pseudodiphtheriticum[B]
Corynebacterium riegelii[B]
Corynebacterium tuberculostearicum[B]
Corynebacterium ulcerans[B]
Corynebacterium urealyticum[B]
Corynebacterium xerosis[B]
Corynebacterium aurimucosum group[B]
Corynebacterium striatum group[B]
Dermacoccus nishinomiyaensis[B]

Enterococcus avium
Enterococcus avium group[B]
Enterococcus casseliflavus
Enterococcus durans[V]
Enterococcus faecalis
Enterococcus faecium
Enterococcus gallinarum
Enterococcus hirae[B]
Gardnerella vaginalis
Gemella haemolysans
Gemella morbillorum[V]
Gemella sanguinis[B]
Granulicatella adiacens
Kocuria kristinae[B]
Kytococcus sedentarius[B]
Lactococcus garvieae
Lactococcus lactis ssp *lactis*
Leuconostoc mesenteroides
Leuconostoc pseudomenteroides[V]
Listeria monocytogenes[V]
Macrococcus caseolyticus[B]
Micrococcus luteus[B]
Micrococcus luteus/lylae[V]

Continued

BOX 4.1 List of Reportable Organisms for the US Food and Drug Administration–Approved Vitek Mass Spectrometry and Matrix-Assisted Laser Desorption Ionization Biotyper CA Systems—cont'd

Pediococcus acidilactici[V]
Pediococcus pentosaceus[B]
Rothia aeria[B]
Rothia dentocariosa[B]
Rothia mucilaginosa
Staphylococcus aureus
Staphylococcus auricularis[B]
Staphylococcus capitis
Staphylococcus caprae[B]
Staphylococcus carnosus[B]
Staphylococcus cohnii[B]
Staphylococcus cohnii ssp *cohnii*[V]
Staphylococcus cohnii ssp *urealyticus*
Staphylococcus epidermidis
Staphylococcus equorum[B]
Staphylococcus felis[B]
Staphylococcus haemolyticus
Staphylococcus hominis[B]
Staphylococcus hominis ssp *hominis*[V]
Staphylococcus lugdunensis
Staphylococcus pasteuri[B]
Staphylococcus pettenkoferi[B]
Staphylococcus pseudintermedius[B]
Staphylococcus saccharolyticus[B]
Staphylococcus saprophyticus
Staphylococcus schleiferi
Staphylococcus sciuri[V]
Staphylococcus simulans
Staphylococcus vitulinus[B]
Staphylococcus warneri
Streptococcus agalactiae
Streptococcus anginosus
Streptococcus constellatus
Streptococcus dysgalactiae
Streptococcus gallolyticus ssp *gallolyticus*[V]
Streptococcus gallolyticus[B]
Streptococcus gordonii[B]
Streptococcus infantarius ssp *coli (Str. lutetiensis)*[V]
Streptococcus infantarius ssp *infantarius*[V]
Streptococcus intermedius
Streptococcus lutetiensis[B]
Streptococcus mitis/Streptococcus oralis[V]
Streptococcus mitis oralis group[B]
Streptococcus mutans
Streptococcus pneumoniae
Streptococcus pyogenes
Streptococcus salivarius[B]
Streptococcus salivarius ssp *salivarius*[V]
Streptococcus sanguinis[V]

Gram-Negative Bacteria, Enterobacteriaceae
Citrobacter amalonaticus[V]
Citrobacter amalonaticus complex[B]
Citrobacter braakii[V]
Citrobacter freundii[V]
Citrobacter freundii complex[B]
Citrobacter koseri
Citrobacter youngae[V]
Cronobacter sakazakii[V]
Cronobacter sakazakii group[B]
Edwardsiella hoshinae[V]
Edwardsiella tarda
Enterobacter aerogenes
Enterobacter amingenus[B]
Enterobacter asburiae[V]
Enterobacter cancerogenus[V]
Enterobacter cloacae[V]
Enterobacter cloacae complex[B]
Enterobacter gergoviae[V]
Escherichia coli
Escherichia fergusonii[V]
Escherichia hermannii[V]
Ewingella americana[V]
Hafnia alvei
Klebsiella oxytoca[V]
Klebsiella oxytoca/Raoultella ornithinolytica[B]
Klebsiella pneumoniae
Leclercia adecarboxylata[V]
Morganella morganii
Pantoea agglomerans
Plesiomonas shigelloides[B]
Proteus mirabilis
Proteus penneri[V]
Proteus vulgaris[V]
Proteus vulgaris group[B]
Providencia rettgeri
Providencia stuartii
Raoultella planticola[V]
Salmonella group[V]
Salmonella sp[B]
Serratia fonticola[V]
Serratia liquefaciens
Serratia marcescens
Serratia odorifera[V]
Serratia plymuthica[B]
Serratia rubidaea[B]
Yersinia enterocolitica
Yersinia frederiksenii[V]
Yersinia intermedia[V]
Yersinia kristensenii[V]
Yersinia pseudotuberculosis

Gram-Negative Bacteria, Non-Enterobacteriaceae
Achromobacter denitrificans[V]
Achomobacter xylosoxidans
Acinetobacter baumannii complex
Acinetobacter haemolyticus
Acinetobacter johnsonii[B]
Acinetobacter junii
Acinetobacter lwoffii
Acinetobacter radioresistens[B]
Acinetobacter ursingii[B]
Aeromonas hydrophila/caviae[V]
Aeromonas sobria[V]
Aeromonas ssp.[B]
Alcaligenes faecalis[B]
Alcaligenes faecalis ssp *faecalis*[V]
Bordetella group[B]

BOX 4.1 List of Reportable Organisms for the US Food and Drug Administration–Approved Vitek Mass Spectrometry and Matrix-Assisted Laser Desorption Ionization Biotyper CA Systems—cont'd

Bordetella hinzii[B]
Bordetella parapertussis[V]
Bordetella pertussis[V]
Brevundimonas dimunuta[V]
Brevundimonas dimunuta group[B]
Burkholderia cepacia complex[B]
Burkholderia gladioli[B]
Burkholderia multivorans
Capnocytophaga ochracea[B]
Capnocytophaga sputigena[B]
Chryseobacterium gleum[B]
Chryseobacterium indologenes
Cupriavidus pauculus group[B]
Delftia acidovorans group[B]
Elizabethkingia meningoseptica[V]
Elizabethkingia meningoseptica group[B]
Myroides odoratimimus[B]
Myroides odoratus[B]
Ochrobactrum anthropi[V]
Pasteurella multocida
Pseudomonas aeruginosa
Pseudomonas fluorescens[V]
Pseudomonas fluorescens group[B]
Pseudomonas oryzihabitans[B]
Pseudomonas putida[V]
Pseudomonas putida group[B]
Pseudomonas stutzeri
Ralstonia pickettii[V]
Rhizobium radiobacter
Sphingobacterium multivorum[V]
Sphingobacterium spiritivorum[V]
Sphingomonas paucimobilis[V]
Stenotrophomonas maltophilia
Vibrio cholera[V]
Vibrio parahaemolyticus
Vibrio vulnificus

Fastidious Gram-Negative Bacteria
Aggregatibacter actinomycetemcomitans[V]
Aggregatibacter aphrophilus[V]
Aggregatibacter segnis[V]
Campylobacter coli
Camplylobacter jejuni
Campylobacter ureolyticus[B]
Eikenella corrodens
Haemophilus haemolyticus[B]
Haemophilus influenzae
Haemophilus parahaemolyticus[V]
Haemophilus parahaemolyticus group[B]
Haemophilus parainfluenzae
Kingella denitrificans[V]
Kingella kingae
Legionella pneumophila[V]
Moraxella catarrhalis
Moraxella nonliquefaciens[B]
Moraxella osloensis[B]
Neisseria cinerea[V]
Neisseria gonorrhoeae[V]
Neisseria meningitidis[V]

Neisseria mucosa[V]
Oligella ureolytica
Oligella urethralis

Anaerobic Bacteria
Actinomyces meyeri
Actinomyces neuii
Actinomyces odontolyticus
Actinomyces oris[B]
Anaerococcus vaginalis[B]
Bacteroides caccae[V]
Bacteroides fragilis
Bacteroides ovatus[V]
Bacteroides ovatus group[B]
Bacteroides thetaiotaomicron[V]
Bacteroides thetaiotaomicron group[B]
Bacteroides uniformis[V]
Bacteroides vulgatus
Bacteroides vulgatus group[B]
Clostridium clostridioforme[V]
Clostridium difficile
Clostridium perfringens
Clostridium ramosum[V]
Finegoldia magna
Fusobacterium canifelinum[B]
Fusobacterium necrophorum
Fusobacterium nucleatum
Mobiluncus curtisii[V]
Parabacteroides distasonis[B]
Parvimonas micra[V]
Peptoniphilus asaccharolyticus[V]
Peptoniphilus harei group[B]
Peptostreptococcus anaerobius
Porphyromonas gingivalis[B]
Prevotella bivia
Prevotella buccae
Prevotella denticola
Prevotella intermedia
Prevotella melaninogenica
Propionibacterium acnes

Yeasts
Candida albicans
Candida boidinii[B]
Candida dubliniensis
Candida duobushaemulonii[B]
Candida famata
Candida glabrata
Candida guilliemondii
Candida haemulonii
Candida inconspicua
Candida intermedia[V]
Candida kefyr
Candida krusei
Candida lambica
Candida lipolytica
Candida lusitaniae
Candida metapsilosis[B]
Candida norvegensis

Continued

> **BOX 4.1** List of Reportable Organisms for the US Food and Drug Administration—Approved Vitek Mass Spectrometry and Matrix-Assisted Laser Desorption Ionization Biotyper CA Systems—cont'd
>
> Candida orthopsilosis[B]
> Candida parapsilosis
> Candida pararugosa[B]
> Candida pelliculosa
> Candida rugosa[V]
> Candida tropicalis
> Candida utilis[V]
> Candida valida[B]
> Candida zeylanoides[V]
> Cryptococcus gattii[B]
> Cryptococcus neoformans[V]
> Cryptococcus neoformans var grubii[B]
> Cryptococcus neoformans var neoformans[B]
> Geotrichum candidum[B]
>
> Geotrichum capitatum
> Kloeckera apiculata[V]
> Kodamaea ohmeri[V]
> Malassezia furfur[V]
> Malassezia pachydermatis[V]
> Pichia ohmeri[B]
> Rhodotorula mucilaginosa[V]
> Saccharomyces cerevisiae
> Trichosporon asahii
> Trichosporon inkin[V]
> Trichosporon mucoides[V]
>
> B = Cleared for the MALDI Biotyper CA system.
> V = Cleared for the Vitek MS system.

List of cleared organisms retrieved from http://www.vitekms.com/knowledgebase.html and https://www.bruker.com/products/mass-spectrometry-and-separations/maldi-biotyper-ca-system/overview.html.

Biotyper system (Bruker) has been in use much longer. In 2008, Mellman and associates[4] published a study demonstrating that by constructing their own reference spectrum databases they could accurately identify nonfermenting gram-negative bacterial isolates.[4] Seng and colleagues[5] published a prospective evaluation of Bruker's commercial mass spectrometry database for identification of 1660 bacteria routinely isolated in their laboratory reporting correct identification of 95% to the genus level and 84% to the species level, with a time to identification of 6 minutes per isolate. Over time the analysis software has been upgraded and databases have been better curated, leading to the most recent 510(k) clearance of the MALDI Biotyper CA system (Bruker), which added 170 species and species groups to the latest database, including gram-positive bacteria, fastidious gram-negative bacteria, anaerobic bacteria, Enterobacteriaceae, and yeasts. Before this updated clearance, the system was limited to gram-negative bacteria. RUO databases are available for fungi and mycobacteria.

The VITEK MS system has been developed by bioMérieux, starting in 2008 and expedited by the acquisition of Anagos-Tec and their microbial mass spectral identification system called SARAMIS in 2010.[6] The VITEK MS v2.0 system is FDA-approved for gram-negative and gram-positive bacteria, anaerobes, fastidious bacteria, and yeasts. The VITEK MS Plus configuration allows for use of the FDA-approved database alongside the VITEK MS RUO database on a single mass spectrometer.

Microbial identification using MALDI-TOF-MS has evolved with iterative improvements to the system components, as well as preparation methods that continuously improve its performance. The constant updating of databases, software, and mass spectrometers can make literature difficult to interpret and/or may render it challenging to compare studies because each of these testing components can affect system performance. In addition to the aforementioned components of the system, the cutoff values applied for acceptance criteria, as well as sample preparation, comparators, and organism type studied should be noted when evaluating the literature. Application of user-defined lowered cutoff value acceptance criteria for species or genus identification using the Bruker system, which is not possible for the FDA-approved Bruker Biotyper CA or VITEK MS systems, may enhance the ability of the system to identify certain organisms.

Initially MALDI-TOF-MS systems focused on identification of commonly isolated gram-negative and gram-positive bacteria, and databases were less comprehensive for anaerobic bacteria, mycobacteria, and fungi. Today, however, these organisms are increasingly addressed.[7] Although databases continuously evolve and become more complete, at least for the validly described species, updates resulting from description of new species and revisions in taxonomy will continue to be needed to provide correct microbial identification.

With RUO systems, laboratories have the option to construct their own reference database by adding mass spectra from strains that are of local importance and/or are not well represented in the commercial databases. These databases can be queried alongside commercial databases. Many academic laboratories have undertaken building their own databases to enhance the performance of commercially available systems, although users must be cautious and ensure that entries are carefully curated before adding them to the library. Failure to do so may result in erroneous identification, potentially affecting patient care. The Mayo Clinic Custom MALDI-TOF-MS Library, for example, currently contains 2096 mass spectral entries representing organisms not adequately addressed by commercially available databases.

ANALYSIS OF MATRIX-ASSISTED LASER DESORPTION IONIZATION TIME-OF-FLIGHT MASS SPECTROMETRY DATA

Ribosomal and other abundant "housekeeping" proteins are the main contributors to the mass spectra generated from whole cells of microorganisms. Other molecules are ionized, although ignored because they fall outside the m/z range monitored by systems. Unlike traditional proteomic studies involving MS, individual proteins are not identified and their mass and abundance are merely profiled as an overall fingerprint. Most of these are proteins that make up 70% to 80% of the dry weight of bacteria and have masses that fall between 2 and 20 kDa, with the majority having masses between 4 and 15 kDa.[8] Generally, mass spectra are specific to organism type, with the generated peaks specific to genera, species, and strains. The mass spectrum of the test isolate is compared to a database of reference spectra (main spectra in MALDI Biotyper), to artificial spectra (superspectra in SARAMIS) or to a peak weight matrix (advanced spectrum classifier in VITEK MS) to determine relatedness of the sample's spectrum to spectra in the database. The MALDI Biotyper system uses pattern recognition and generates a score ranging from 0 to 3.000 based on the presence and absence of peaks in the test isolate compared to individual main spectra in the database. According to the manufacturer's criteria for their RUO application, a score of 2.000 or greater is interpreted as species-level identification, whereas 1.700 to 1.999 is genus level, and less than 1.700 represents no identification. System users have found lower cutoff values to be acceptable for identification of certain organism types. Before implementing the approach of applying lowered score cutoffs, an extensive up-front validation by the end user is required. For the FDA-approved MALDI Biotyper CA system, a score of 2.000 or greater is reported as a high confidence identification, (scores of 2.300 to 3.000 represent a highly probable species identification, and scores of 2.000 to 2.299 or less represent a secure genus and probable species identification), whereas a score of 1.999 or less to 1.7 is reported as a low confidence (probable genus) identification and the isolate should be retested with the traditional extraction method or the extended direct transfer (eDT) method in which the bacterial isolate is combined with 70% formic acid on the plate. This extraction method was commonly referred to as on-plate formic acid testing before the latest 510(k) clearance. If the eDT method is employed and fails to yield a high confidence identification, the traditional extraction method can be used. Scores below 1.7 represent no identification. The VITEK MS RUO system's principal identification strategy is based on comparison of an organism's spectrum to supraspectra artificial spectra computed from reference spectra of multiple strains of a given species. In individual supraspectra, peaks are weighted according to their specificity for the targeted microbial taxonomic unit, generally a species, but supraspectra also exist for identification at the genus, subspecies, or family level. The result is a confidence value that is considered significant if 80% or greater. When matches 80% or greater are achieved for supraspectra representing multiple species, for example, in the case of a mixed sample or noisy spectrum, the result is highlighted in red to alert the user. In case the comparison to the supraspectra does yield an ambiguous identification, the sample's spectrum can be compared to all reference spectra in the database. In such cases, no specific interpretive criteria are provided by the manufacturer.

With the VITEK MS system, identification is based on comparison of a sample's spectrum to a bin-weight matrix computed based on a multitude of mass spectra of multiple strains per species (or other targeted taxonomic unit). Reference spectra in the initial database are computationally "binned"; that is, the mass spectra are divided in 1300 mass intervals in which the presence or absence of peaks is recorded. In the bin-weight matrix, each bin is weighted for each species according to the presence or absence of a peak in spectra for the targeted species and in other spectra of the database. For example, bins that are highly specific for individual species receive a higher weight than those that are moderately species-specific. Closely related species with similar patterns and only a small number of distinguishing peaks, such as *Streptococcus pneumoniae* and *Streptococcus mitis* groups may be better (though not perfectly) differentiated using the VITEK MS strategy compared to other strategies.[9,10] For identification, a sample's spectrum is compared to the bin-weight matrix and matches to individual species normalized by the advanced spectrum classifier to yield a probability rank that is reported.

Although these approaches generally perform well to identify species, mass spectra of some organisms are so similar to one another (eg, *Escherichia coli* and *Shigella* species) that their differentiation using available algorithms is unreliable for routine diagnostics with any of the aforementioned strategies. This is a challenge because *E. coli* is one of the most frequently encountered organisms in clinical microbiology laboratories; MALDI-TOF-MS users must use alternative or additional strategies (eg, lactose fermentation, indole production) to distinguish *E. coli* from *Shigella* species. Further problematic groups of organisms that may not be accurately differentiated to the species level are the *Enterobacter cloacae* complex, *Burkholderia cepacia* complex, and *Streptococcus bovis* group species members, although for each group, studies have shown that with database refinement, it may be possible to differentiate the species involved.[11-13] Similarly, identification of all species of *Acinetobacter baumannii* complex may be challenging. *Neisseria polysaccharea* may be misidentified as *Neisseria meningitidis* using the MALDI Biotyper system.[14] Misidentification may occur if such details are not considered during data analysis, leading to the reporting of spurious results. *Bacillus anthracis* could be misidentified as *Bacillus cereus* group organisms because of their close relatedness, although the Bruker Security Relevant database has been used in the identification of clinical isolates of *B. anthracis*.[15] The routine use of the *B. anthracis* spectra contained in the Security Relevant database, however, is not encouraged by the manufacturer because it can lead to misidentification of *B. cereus* group organisms.[15] Other potential bioterrorism agents are not included in the routine Bruker and VITEK databases, thus giving false or no identification of these organisms. The aforementioned limitations aside, most organisms are correctly identified, yield a low score/confidence/probability indicating that reliable identification could not be achieved, or yield a low discrimination result with matches to multiple species exceeding the generally accepted cutoff values. Last, but not least, poor sample preparation can lead to low-quality spectra and failed identification of a sample that is well represented in the databases.

Performance

Identification of Aerobic Bacteria

MALDI-TOF-MS performs as well as, if not better than, automated biochemical identification for commonly encountered bacteria and yeast. In our laboratory, we initially compared the MALDI Biotyper system (v2.0 software and library) and the BD Phoenix Automated Microbiology System (Beckton Dickinson, Franklin Lakes, New Jersey) for their ability to identify 440 common and unusual gram-negative bacilli. The MALDI-TOF-MS system correctly identified 93 and 82%, whereas the biochemical system correctly identified 83 and 75% of these gram-negative bacilli to the genus and species levels, respectively.[16] We further evaluated the MALDI Biotyper system (v2.0 software and library) using 217 clinical isolates of staphylococci, streptococci, and enterococci, 98% of which were correctly identified to the genus level.[17] McElvania Tekippe and associates[18] evaluated the MALDI Biotyper for identification of 239 aerobic gram-positive bacteria using direct on-plate testing with formic acid. Applying the default species cutoff score of 2.000 or greater, genus and correct species identification was achieved for 183 (77%) and 141 (59%) isolates, respectively. When applying a cutoff score of 1.700 or greater for genus- and species-level identification to the data, 92 and 70% of the gram-positive bacteria studied, respectively, were correctly identified.[18] Lau and coworkers[19] evaluated the MALDI Biotyper system (v3.0 software and library v.3.1.2.0) for identification of 67 difficult-to-identify bacteria; 75 and 45% were correctly identified to the genus and species levels, respectively, with 6% (4 of 67) being misidentified. Hsueh and colleagues[20] evaluated the MALDI Biotyper system for identification of 147 isolates of aerobically growing gram-positive bacteria, including *Nocardia* species, *Listeria monocytogenes*, *Kocuria* species, *Rhodococcus* species, *Gordonia* species, and *Tsukamurella* species. All 15 *Kocuria* isolates yielded a top match to the correct species, either *Kocuria kristinae* or *Korcuria marina*, although because of low scores, only 27% of *Kocuria* species were unambiguously identifiable to the species level. Similarly, all 39 *L. monocytogenes* isolates studied yielded a top match to *L. monocytogenes*, although because of low scores only 90% of the isolates were identifiable to the species level. With the exception of *Nocardia nova* and *Nocardia otitidiscaviarum*, *Nocardia* species were not identified, nor were *Tsukamurella* or *Gordonia* species. Schulthess and associates[21] evaluated the MALDI Biotyper system (database v3.1.2.0) for the identification of gram-positive bacilli. Using the manufacturer's interpretative criteria and an initial collection of 190 isolates, 85 and 87% of the gram-positive bacilli studied were identified to the genus level with on-plate formic acid testing and ethanol-formic acid extraction, respectively, compared to 72% with direct colony testing without formic acid. For some isolates, two or three species yielded scores of 2.000 or greater for the same isolate; these included *Corynebacterium aurimucosum* and *Corynebacterium minutissimum*; *Corynebacterium simulans* and *Corynebacterium striatum*; *Lactobacillus gasseri*, *Lactobacillus johnsonii*, and *L. monocytogenes*; and *Listeria ivanovii* and *Listeria innocua*. Reducing the species identification cutoff score to 1.700 or greater increased the percentage of isolates identified. These investigators also compared identification of 215 clinical isolates of gram-positive bacilli by MALDI-TOF-MS using on-plate formic acid testing and a species identification cutoff of 1.700; 87 and 79% of gram-positive bacilli studied, respectively, were identified to the genus and species levels. The authors proposed an algorithm combining MALDI-TOF-MS with nucleic acid sequence analysis for identification of gram-positive bacilli that included score cutoff values lower than those recommended by the manufacturer and covered the most frequently found genera and species. We have found that 87% of 92 non-*diphtheriae Corynebacterium* species could be identified with a cutoff of 1.700 or greater, although *Corynebacterium aurimucosum* was misidentified as *C. minutissimum*.[22] *C. minutissimum* is currently included in the latest cleared database of the MALDI Biotyper CA system. Barberis and associates[23] compared the RUO Bruker Biotyper system to traditional phenotypic identification methods for the identification of 333 clinical gram-positive bacilli. By using cutoffs of 1.500 and 1.700 for genus- and species-level identification they were able to identify 97 and 92% of isolates to the genus and species levels, respectively. Seven misidentifications were reported while using the lower cutoffs, including *Corynebacterium afermentans* subspecies *lipophilum* identified as *Corynebacterium jeikeium*. Conventional methods were able to identify 95 and 86% of isolates to the genus and species level, respectively. The MALDI Biotyper was superior to traditional methods when evaluating aerobic actinomycetes. MALDI-TOF-MS identified 88 and 75% of eight isolates to the genus and species level, respectively, whereas conventional methods yielded genus- and species-level identifications for 75 and 2% of isolates, respectively. In the case of 16 pigmented gram-positive rods, conventional methods failed to identify any isolate to the species level whereas MALDI-TOF-MS identified 75% to the species level.

Multiple evaluations of the VITEK MS system (database v2.0) have been published. Richter and coworkers[24] evaluated 965 routinely encountered Enterobacteriaceae and showed the VITEK MS to correctly identify 97 and 84% to the genus and species level, respectively. Manji and colleagues[25] showed that the VITEK MS correctly identified 91 and 78% of 558 non-*Enterobacteriaceae* gram-negative bacilli to the genus and species levels, respectively. Branda and coworkers[26] evaluated the VITEK MS system for identification of 226 isolates of fastidious gram-negative bacteria; 97 and 96% of the fastidious gram-negative bacteria studied were correctly identified to the genus and species levels, respectively. Rychert and associates[27] evaluated the VITEK MS system for identification of 1146 aerobic gram-positive bacteria and showed that 93% were correctly identified to the species level, with an additional 3% identified to the genus level because of a low discrimination result of two or more species (including *L. monocytogenes* with matches to *L. monocytogenes*, *L. ivanovii*, and *Listeria welshimeri*). A study by Kärpänoja and associates[28] compared the MALDI Biotyper and Vitek MS systems for identification of viridans streptococci using 54 type strains and 97 blood culture isolates. The MALDI Biotyper and VITEK MS systems yielded correct species-level identification for 94 and 69% of the type strains and correctly classified 89 and 93% of the blood culture isolates to the group level, respectively.

Although MALDI-TOF-MS can accurately identify *Francisella tularensis* and *Brucella* species, the general MALDI Biotyper library does not contain these species and so will not identify them; use of Bruker's Security Relevant database enables *Brucella* species and *F. tularensis* identification.[7] Currently the VITEK MS databases do not adequately address these organisms. Working with these organisms on the open laboratory benchtop is hazardous to laboratory personnel; they need to be handled in a biological safety cabinet. Fortunately, once either formic acid or matrix has been placed over the microbial mass and dried, the organisms appear to be rendered nonviable and the remainder of the testing may be performed outside of a biological safety cabinet.[29]

Identification of Anaerobic Bacteria

MALDI-TOF-MS has become the method of choice for identification of anaerobic bacteria, replacing 16S ribosomal RNA gene sequencing and gas-liquid chromatography, techniques that were not available to many clinical microbiology laboratories. MALDI-TOF-MS therefore enables increased routine identification of anaerobic bacteria.

There are several studies examining MALDI-TOF-MS for identification of anaerobic bacteria. Jamal and colleagues[30] tested the MALDI Biotyper system (database v.3.3) with direct on-plate testing and the VITEK MS v1 system/v1.1 database on 274 routinely isolated anaerobic bacteria (with a high proportion of *Bacteroides fragilis*), and reported species-level identification of 89 and 100%, respectively. Hsu and colleagues[31] evaluated the MALDI Biotyper system (v3.0 software) for identification of 101 anaerobic bacteria and showed that using on-plate formic acid testing and a cutoff of 1.700 or greater improved the rate of accurate identification compared to direct on-plate testing without formic acid and the manufacturer-recommended cutoff score. Schmitt and coworkers[32] evaluated a diverse collection of 253 clinical anaerobic isolates using the MALDI Biotyper system (v3.0 software and database v3.3.1.0), on-plate formic acid testing, and a user-supplemented database; 92 and 71% of anaerobic bacteria were correctly identified to the genus and species levels, respectively. Barreau and associates[33] analyzed 1325 anaerobes using the MALDI Biotyper system (v3.0 software) and a score cutoff of 1.900 or greater with direct on-plate testing and showed that 100 and 93% of the isolates were correctly identified to the genus and species levels, respectively. Finally, Garner and colleagues[34] reported the evaluation of the VITEK MS system (database v2.0) in a multicenter study of 651 anaerobic isolates and reported that 93 and 91% of anaerobic bacteria were correctly identified to the genus and species levels, respectively.

Identification of Mycobacteria

As with identification of anaerobic bacteria, identification of mycobacteria has been a challenge for clinical microbiology laboratories; historically it has been done using biochemical testing, DNA probes, high-performance liquid or gas-liquid chromatography, and/or gene sequencing. MALDI-TOF-MS is proving a useful tool for identification of mycobacteria, with a few limitations, such as the inability to differentiate members of the *Mycobacterium tuberculosis* complex and some closely related species such as *Mycobacterium chimaera* and *Mycobacterium intracellulare* or *Mycobacterium mucogenicum* and *Mycobacterium phocaicum*. Although *Mycobacterium abscessus* and *Mycobacterium massiliense* may be challenging to differentiate, Teng and coworkers[35] achieved this using cluster analysis of spectra generated using the MALDI Biotyper system and the genetic algorithm of ClinPro Tools version 3.0.22 ((Bruker Daltonics), a discussion of which is found later in this chapter. This method was able to find six peaks unique for *M. abscessus* (three peaks) and *M. massiliense* (three peaks). MALDI-TOF-MS is easier, less expensive, faster, and more accessible to routine clinical microbiology laboratories than traditional strategies for mycobacterial identification, which will likely make it the method of choice for their identification in the near future although currently neither the VITEK MS nor the MALDI Biotyper databases are FDA approved.

Testing of mycobacteria by MALDI-TOF-MS requires a special workflow to kill suspect mycobacteria and also to disrupt clumped cells and break open the cell envelopes.[36] Biosafety level 3 (BSL-3) procedures should be used in handling cultures of *M. tuberculosis* complex species because of their low infectious dose. For nontuberculous mycobacteria, BSL-2 safety practices can be employed, although because *M. tuberculosis* can resemble a number of slow-growing nontuberculous mycobacteria, BSL-3 practices should be used until *M. tuberculosis* complex has been ruled out. Selection of colonies and spotting of mycobacterial colonies onto the MALDI-TOF-MS plate should be done in a biological safety cabinet. Inactivation of mycobacteria can be accomplished using a number of methods as well as variations thereof. Most employ heat inactivation or mechanical disruption. Heat inactivation involves the suspension of an isolate in a liquid (either water or 70% ethanol) and incubation in a heat bath at 95°C for 30 minutes or longer, causing lysis of the cell. Heat inactivation may require sonication or another method of disruption to free the proteins from the cell, leading to longer procedural times compared to that of mechanical disruption. Mechanical disruption suspends an isolate in 70% ethanol and silica beads followed by vortexing to mechanically disrupt and inactivate mycobacteria. Both methods are effective, although the latter is faster than the former.[36,37] Balada-Llasat and associates[38] evaluated 178 mycobacterial isolates grown on solid and/or broth medium and processed using heat inactivation, treatment with ethanol, and mechanical disruption, as recommended by Bruker, and showed that the MALDI Biotyper system (*Mycobacterium* database v1 and software v3.0) was able to identify 98 and 94% of the *Mycobacterium* species studied to the genus and species levels, respectively. Mather and associates[39] tested two protein extraction protocols, a laboratory-developed protocol that involved heat inactivation, followed by ethanol killing and vortexing with silica beads before pelleting and spotting and a separate bioMérieux extraction protocol, before testing using the MALDI Biotyper and VITEK MS RUO platforms; the MALDI Biotyper database was augmented with mass spectral entries from 123 clinical *Mycobacterium* strains. Of 198 clinical isolates tested, 95% were correctly identified to the species level with the MALDI Biotyper system and augmented database, 79% with the RUO Bruker *Mycobacterium* Library 2.0 database, and 94% with the VITEK MS RUO system.

Identification of Fungi

In addition to identification of bacteria and mycobacteria, MALDI-TOF-MS can identify yeasts and other fungi. Although older studies used elaborate protein extraction procedures for yeasts, many laboratories use the same strategies used for bacteria—direct transfer of a colony with or without on-plate formic acid treatment (see Fig. 4.1).[40] For filamentous fungi, the organisms are inactivated during the extraction procedure (usually 70% ethanol) to safely handle the organisms for downstream processing and identification.

Dhiman and colleagues[41] evaluated the MALDI Biotyper system (database v3.0) using a score cutoff of 1.800 or greater for identification of 138 common and 103 unusual yeast isolates and reported 96 and 85% accurate genus- and species-level identification, respectively, using protein extraction. Lacroix and associates[42] showed that the MALDI Biotyper system, applying the cutoff score recommended by the manufacturer, identified 97% of 1383 routinely isolated *Candida* isolates after protein extraction. Westblade and associates[43] evaluated the VITEK MS system (database v2.0) for identification of 852 yeast isolates in a multicenter clinical trial and found that 97 and 96% were identified to the genus and species levels, respectively. Pence and coworkers[44] compared the VITEK MS (database v.2.0) and MALDI Biotyper (software v3.1) systems for identification of 117 yeast isolates using on-plate formic acid testing. The VITEK MS system correctly identified 95% of the isolates, whereas the Biotyper system correctly identified 83% of the isolates using a species-level cutoff of 1.700 or greater. Hamprecht and colleagues[45] compared the MALDI Biotyper (software v3.0, database v3.0.10.0, species-level cutoff of 2.000 or greater) and VITEK MS (v2.0 knowledge base) systems for identification of 210 yeasts using on-plate formic acid testing.[45] The VITEK MS system identified 96% of the isolates, whereas the MALDI Biotyper system identified 91%. De Carolis and associates[46] created an in-house library using spectra from 156 reference and clinical yeast isolates, generated with a fast sample preparation procedure involving suspending a single colony of yeast in 50 μL of 10% formic acid, vortex mixing, and using 1 μL of the lysate for analysis. Using their database and processing, they identified 96% of 4232 routinely isolated yeasts using a species-level cutoff of 2.000 or greater and the MALDI Biotyper system (software v 3.0). Rosenvinge and colleagues[47] showed that the MALDI Biotyper system identified 88% of 200 yeast isolates to the species level using on-plate formic acid testing and a species cutoff of 1.700 or greater. Mancini and associates[48] compared the Biotyper system (database v3.0) with protein extraction and the VITEK MS system (v1.2.0) with on-plate formic acid testing for the identification of 197 yeast isolates. The rate of correct identification at the species level was comparable using the commercial databases (90 and 84%, respectively) and improved to 100% using the MALDI Biotyper system with an in-house–developed database.

Schulthess and associates[49] evaluated Bruker's Filamentous Fungi Library 1.0. Molds were grown in broth medium for 24 to 48 hours and subjected to protein extraction. First they studied a clinical strain collection of 83 nondermatophyte, nondematiaceous molds and showed that 78 and 54% were identified to the genus and species levels, respectively. Reducing the species identification score cutoff to 1.700 or greater improved identification to the species level to 71%. They then prospectively tested 200 consecutive clinical mold isolates and were able to identify 84 and 79% to the genus and species levels, respectively, using a species cutoff of 1.700 or greater. Lau and colleagues[50] developed an alternative extraction procedure for molds and constructed their own database comprising 294 individual isolates representing 76 genera and 152 species. To extract proteins, a small piece of mold mycelium was excised and subjected to protein extraction with the addition of zirconia-silica bead beating. They then challenged their database with 421 clinical isolates using the MALDI Biotyper software, and demonstrated accurate genus- and species-level identifications in 94 and 90% of isolates, respectively, with no misidentifications.

Theel and coworkers[51] developed a dermatophyte library and showed that when used in conjunction with the MALDI Biotyper database v3.0 and protein extraction, it identified 93 and 60% of 171 isolates to the genus and species levels, respectively, when using phenotypic methods as the reference method and sequencing to resolve discrepancies. Although the species identification for this study may seem low, it is unlikely to affect patient management because literature on species-specific treatment is sparse.[51] De Respinis and colleagues[52] developed an in-house database and performed MALDI-TOF-MS with a VITEK MS RUO system for identification of dermatophytes; of 141 clinical isolates tested, 96% were correctly identified. In a more recent study, De Respinis and colleagues[53] collected more data, which are integrated in the next version of the VITEK MS (v3.0; currently in clinical trials) and validated the extended database with 131 isolates, of which 95% yielded single choice or low-discrimination results, whereas 5% misidentifications occurred. The misidentified isolates primarily resulted from *Trichophyton soudanense* being identified as *Trichophyton violaceum*.[53]

Laboratory Workflow and Cost

In the past, identification of bacteria and fungi has been a challenging, multistep process, with variable procedures for different types of organisms. Students of clinical microbiology have been meticulously trained to interpret colony morphology and Gram stain of bacteria growing on solid media as a prelude to selecting appropriate subsequent test methods such as rapid biochemical tests (eg, catalase and oxidase activity, manual biochemical tests, automated biochemical tests, and specific or broad-range sequencing). With MALDI-TOF-MS, colonies or mycelia are accurately identified in minutes, without prior knowledge of the microorganism type. And because it does not matter whether a bacterium or yeast is tested, with the exception of safe handling of microorganisms that can be hazardous in the laboratory, the complex decision-making process classically surrounding identification of bacteria and fungi on solid media is obviated. Only a small amount of biomass is required; thus testing can be performed from single colonies on primary culture plates, thereby decreasing the amount of media used in the laboratory. Tests for screening for some enteric pathogens may be eliminated. Gram staining of colonies may largely become obsolete because this information is no longer required before testing. For some organisms (eg, *Staphylococcus aureus*), rapid tests performed at the bench may remain the tests of choice or at least an option. DNA sequencing expenses are restricted to esoteric or atypical organisms, and waste disposal is decreased. Importantly, quality control reagents and procedures, as well as the training of technologists required to perform outdated tests, are eliminated.

Seng and associates[54] published over a decade of experience in routine identification of clinical bacterial isolates, including 40 months using MALDI-TOF-MS and 91 months using phenotypic identification. MALDI-TOF-MS and phenotypic identification identified 36 and 19 species in 10,000 isolates, respectively. Additional phenotypic identification was required for 4.5 and 35.2 in 10,000 isolates with MALDI-TOF-MS and primary phenotypic identification, respectively. Additionally, compared with phenotypic identification and sequencing, MALDI-TOF-MS reduced the time for identification by 55- and 169-fold and costs by 5- and 96-fold, respectively.

After implementation of MALDI-TOF-MS into our laboratory for bacterial identification, we have eliminated automated biochemical-based microbial identification and have so far reduced the number of tube-format biochemical set–based identifications from 4668 to 987 per year. To date, we have halved the number of isolates requiring 16S ribosomal RNA gene sequencing; through progressive database improvements, this number continues to decrease. After implementation of MALDI-TOF-MS for yeast identification, the associated supply costs decreased from $30,525 to $5400 per year as a result of the elimination of germ tube and rapid assimilation of trehalose and analytical profile index strip tests (unpublished observation). This was associated with a decrease in turnaround time for identification by 1 to 5 days, depending on the species. Implementation of MALDI-TOF-MS also reduced the number of tests on which our staff must maintain competency and therefore our yearly competency evaluation burden. After implementation of MALDI-TOF-MS for dermatophyte identification, the associated supply costs decreased from $20,020 to $2340 per year, with a turnaround time savings of 1.5 days. Recent adoption of MALDI-TOF-MS in our mycobacteriology laboratory will save an estimated $160,000 per year primarily because of a reduction in sequencing costs. The turnaround time for identification of mycobacteria (after a positive culture result) has been reduced from 24 to 2 hours with the introduction of MALDI-TOF-MS.

Antimicrobial Susceptibility Testing

Antimicrobial susceptibility profiles are not directly determined with the previously described strategy. By rapid organism identification, intrinsic antimicrobial resistance characteristics of particular species (or typical susceptibility of the identified species based on local antibiograms) may guide therapy. Because some resistance factors (eg, β-lactamases) are proteins, and MALDI-TOF-MS detects proteins, it might be intuited that antimicrobial resistance-associated proteins would be detected directly. Unfortunately, at least as currently performed, this has been challenging. For example, although β-lactamases are highly active, they are expressed at low concentrations; further, their molecular weights are similar to those of other bacterial proteins and there are well over 1000 types of β-lactamases, many of which share similar masses. Because MALDI-TOF-MS may provide insight into strain typing, and some strains are more or less likely to be resistant to certain antimicrobial agents, strain typing may infer an association with antimicrobial resistance. MALDI-TOF-MS may be applied as part of a functional assay to measure β-lactam degradation by β-lactamase; in this case, it is not proteins but antimicrobial agents and their chemically modified counterparts that are measured.[55] This approach requires specific system configuration and incubation to allow antibiotic hydrolysis and applies only to resistance mechanisms associated with degradation of antimicrobial agents.[56-58] Another strategy to detect antimicrobial resistance is to grow an organism for a short period (eg, ≤3 hours) in the presence of an antimicrobial agent of interest and isotopically labeled amino acids[59] If the organism is resistant to the antimicrobial agent, it will incorporate isotopically labeled amino acids, increasing protein masses and leading to mass shifts of their corresponding peaks in the profiled spectra.[60] Further details on detection of resistance using MALDI-TOF-MS can be found in a review by Hrabák and associates.[61]

Direct Testing of Clinical Samples

Although direct testing of clinical samples is not generally feasible because of the large number of organisms needed for detection using MALDI-TOF-MS (estimated to be ~10^5 colony-forming units of bacteria[62]), urine may be tested directly owing to the high numbers of bacteria present in clinically infected urine.[63] Urine flow cytometry may be used to screen out negative urine samples, with MALDI-TOF-MS reserved to test positive urine samples.[64,65] The urine specimen must undergo processing before MALDI-TOF-MS testing. Limitations include the inability to reliably identify polymicrobial infections[64] and impairment of database matching caused by urinary proteins, such as α-defensins.[66] Demarco and Burnham[67] described a method using desalting, fractionation, and concentration for preparing urine samples before MALDI-TOF-MS analysis. Desalting fractionation was done by spinning 15 mL of urine in an Amicon centrifuge filter (EMD Millipore, Billerica, Massachusetts) with desalting occurring by adding Milli-Q water to the concentrate and respinning. The resulting concentrate was transferred to a microcentrifuge tube and spun at 14,000 × g for 3 minutes, with the pellet being further desalted and concentrated with another water wash and centrifugation step. The pellet was then used for MALDI-TOF-MS. Other sample types, such as cerebrospinal fluid (CSF),[68,69] also may be tested directly. Treatment of these samples is needed to remove potential interfering ions, proteins, and other impurities that may cause ion suppression, that is, the interference of ionization of the analyte by blocking the ionization process.[70] This could lower confidence scores, preventing direct identification from the previously mentioned sources (as well as blood culture bottles, mentioned later). In addition, if the interfering substance is a protein, peaks may appear that interfere with the correct identification of the microorganism.

Testing of Positive Blood Culture Bottles

MALDI-TOF-MS can be used for rapid identification of microorganisms growing in blood culture bottles. The positive bottles may be subcultured to solid media and growth tested by MALDI-TOF-MS after a short incubation period (eg, 2 to 4 hours[71,72]). Fothergill and colleagues[73] used a lysis filtration method coupled with the VITEK MS RUO system to correctly identify 189 in 259 (73%) positive blood culture bottles to the species level, with 51 (19.7%) having no identification and 6 (2.3%) being misidentified.[73] The remaining 5% consisted of bottles with no growth and no identification, as well as those identified correctly only at the genus level.

Alternatively, blood culture bottles may be tested directly. This application requires preparatory processing because blood culture bottles contain macromolecules from blood and growth media. Processing can be accomplished using differential centrifugation and washings, selective lysis of blood cells, serum separator tubes, or filtration; commercial processing using the Sepsityper (Bruker Daltonics) is available.[74-79] Although results are valid when obtained, yield is generally not as good as direct colony testing; a higher percentage of gram-negative than gram-positive bacteria are typically identified using MALDI-TOF-MS on positive blood culture bottles.[80] As with direct testing of urine, not all organisms present in polymicrobial infections will be identified in all instances.[81] Both the Biotyper and VITEK MS systems have been applied to positive blood culture bottles. Chen and colleagues[82] used the Sepsityper to evaluate the VITEK MS IVD and Biotyper systems for bacterial identification in 181 monomicrobial blood culture bottles. Genus- and species-level identification was provided using the Biotyper and VITEK MS systems in 98% and 82% and 93% and 81% of cases, respectively. Twenty-one blood culture bottles contained two bacterial species. The VITEK MS IVD system identified only the majority species in all 21 instances, whereas the Biotyper system identified both of the two species with greater than 1.6 scores in five mixed cultures in the "top 10 matched pattern choices."[82]

MALDI-TOF-MS performed on positive blood culture bottles can rapidly identify probable contaminants or suggest complementary diagnostic testing in the case of pathogen detection.[83] MALDI-TOF-MS on positive blood culture bottles is rapid (estimated at 30 to 45 minutes),[84,85] albeit not as fast as direct colony testing; such testing is typically batched and performed on a number of occasions throughout the day. By testing positive blood culture bottles using MALDI-TOF-MS, time to organism identification may be reduced by a day or more.[86] A limitation is that antimicrobial susceptibility is not provided. To overcome this limitation, MALDI-TOF-MS has been tied with direct antimicrobial susceptibility testing of positive blood culture bottles,[87] improving time to optimal therapy.[88]

Typing

Because MALDI-TOF-MS can provide accurate species identification, using it for subspecies-level identification, and therefore strain typing, has been proposed. For example, Mencacci and colleagues[89] used MALDI-TOF-MS to type *A. baumannii* and Josten and associates[90] used it to type *S. aureus*. Kuhns and coworkers[91] were able to differentiate *Salmonella enterica* serovar Typhi and nontyphi serovars using isolates from three different areas of Ghana.[91] The study typed the isolates using Biotyper 3.0 system and six biomarker peaks. The advent of bioinformatics and data mining has given rise to software that may aid in typing. ClinPro Tools (Bruker Daltonics) is an integrated set of tools that allows for discovery of biomarkers that are indicative of a specific microorganism. ClinPro Tools performs all the steps necessary for finding distinct biomarker peaks. This includes data pretreatment, visualization, statistics, pattern determination and evaluation, and classification of spectra.[92] This technology has been used to differentiate *E. coli* from *Shigella* species,[93] to differentiate extended-spectrum β-lactamase (ESBL)-producing gram-negative organisms from non-ESBL producers[94] and to type *Mycoplasma pneumoniae* isolates.[95]

Limitations

MALDI-TOF-MS has limitations. Unlike publically available sequence databases such as GenBank, MALDI-TOF-MS databases are proprietary. Low identification percentages for some organisms may be improved by addition of mass spectral entries of underrepresented species or strains (to cover intraspecies variability), but doing so may be beyond the capability of some laboratories. Because of low scores or percentages, repeat testing may be needed. Growth on some media may be associated with low scores or percentages. Tiny or mucoid colonies may fail identification; tiny colonies may be more rapidly identified using 16S ribosomal RNA gene sequencing than MALDI-TOF-MS. Refined criteria are needed to distinguish some closely related species and differentiate them from the next best taxon match. For certain species, genus- or species-specific (including lowered) cutoff values may be appropriate. Newer MS methods may enable separation of microorganisms at deeper taxonomic levels than MALDI-TOF-MS.[96] Laboratory errors may occur as a result of colony inoculation in erroneous target plate locations, testing of impure colonies, smearing between spots, failure to clean target plates, or erroneous data entry into laboratory information systems. There is a learning curve involved in applying ideal colony amounts to target plates. Instrument cost must be considered; laser, software, hardware, or mass spectrometer failure can occur, necessitating an appropriate service plan.

Although MALDI-TOF-MS results are generally reproducible, there are sources of variability, including the mass spectrometer (different types as well as individual instruments, instrument ages, instrument configurations), matrix and solvent composition, preparation methods, technologist training and competence, culture conditions (such as media, colony age, and temperature), and biological variability[18,31,44,97,98]; quality control strategies are being defined. Well-isolated colonies must be tested; if colonies are not well isolated, they may represent more than one organism, and the minority species may be undetected. A Clinical and Laboratory Standards Institute guideline on MALDI-TOF-MS is under preparation.

Because of the ease of use of this technology, technologists may be tempted to work up everything, including clinically nonsignificant isolates, leading to confusion on the part of clinicians receiving such reports and the potential for inappropriate patient treatment. At the same time, because of the ease of use of MALDI-TOF-MS, technologists may lose their skills in visually identifying colonies, although technologists also may build on these existing skills as they gain knowledge on the various morphologies associated with different bacteria because of the immediate feedback that MALDI-TOF-MS provides. Because MALDI-TOF-MS may yield different (generally more accurate) identification than current systems, reporting of organisms with which health care providers are unfamiliar may lead them to ignore potentially clinically significant results or to overtreat clinically insignificant results. Although the latter concerns should be noted, it is entirely plausible and generally believed that MALDI-TOF-MS may have an impact on clinical management of patients. In a retrospective cohort study examining the potential role of MALDI-TOF-MS to diagnose prosthetic joint infection (PJI), Peel and associates[99] demonstrated that the identity of isolates from the site of PJI could inform interpretation of clinical significance.[99] There have been studies demonstrating the benefits of the technology to the laboratory, but few demonstrating an impact on patient outcomes. Perez and associates[100] built on their previous work demonstrating that MALDI-TOF-MS combined with antimicrobial stewardship could decrease hospital costs and conducted an interventional study that examined the effects of MALDI-TOF-MS, rapid antimicrobial susceptibility testing, and antimicrobial stewardship interventions on outcomes for patients who were bacteremic with antibiotic-resistant gram-negative bacteria. The study found that compared to a preintervention phase in which conventional techniques were used for the workup and susceptibility testing of positive blood culture bottles, during their intervention phase, there were decreased intensive care unit and hospital lengths of stays, 30- and 60-day all-cause mortality, and hospital costs.

Future Perspectives

MALDI-TOF-MS will enable adoption of total laboratory automation in clinical microbiology laboratories. Total laboratory automation features automated sample processing, plating, incubation, plate reading using digital imaging, and spotting MALDI-TOF-MS plates. Early growth detection by digital imaging combined with MALDI-TOF-MS may result in faster detection of microorganisms compared to conventional techniques.[101] Laboratory directors and technologists can anticipate new applications of MALDI-TOF-MS, such as using antibodies to capture analytes of interest.[102] Finally, MALDI-TOF-MS will have profound effects on microbiology education of medical technologists, clinical microbiology laboratory directors, and other management staff, medical students, residents, and fellows. For medical students, residents, and fellows who do not practice in the laboratory, it is time to deemphasize conventional biochemical-based microbial identification and to focus instead on interpretation of results of MALDI-TOF-MS—based identification and knowledge of expected corresponding antimicrobial resistance patterns (alongside select quick biochemical reactions, especially those important to microbial pathogenesis).

POINTS TO REMEMBER

MALDI-TOF-MS
- When coupled with an appropriate database, MALDI-TOF-MS is capable of identifying bacteria, mycobacteria, yeast, and fungi.
- MALDI-TOF-MS performs as well as or better than automated biochemical platforms for identification of common bacteria and yeasts.
- Most proteins analyzed by MALDI-TOF are 4 to 15 kDa.
- Rare isolates or strains may be identified using databases generated by the end user.
- MALDI-TOF-MS is less expensive, with faster turnaround times than most standard identification methods.

ELECTROSPRAY IONIZATION MASS SPECTROMETRY

Electrospray ionization mass spectrometry (ESI-MS) has been applied in the research of macromolecules since 1984, when the ESI technique was described by Masamichi Yamashita and John Bennett Fenn.[103] The technique was applied to biological macromolecules in 1989, and for this work Fenn won the 2002 Nobel Prize in chemistry.[104] ESI is a soft ionization technique much like MALDI; however, it differs in that it produces multiply charged ions and requires an extended mass range of the analyzer.[105]

Polymerase chain reaction followed by electrospray ionization mass spectrometry (PCR-ESI-MS) has undergone much iteration as improvements have been made to the platform to ensure optimal performance. The latest version of the platform, known as IRIDICA (Abbott Laboratories, Chicago, Illinois), improves on the technology present in the previous platforms, PLEX-ID (Abbott) and the original IBIS T-5000 (Abbott). PCR-ESI-MS for pathogen identification, in the form of the T-5000, was originally developed by Ibis Biosciences (Carlsbad, California) for use in biodefense, public health, and safety to allow for rapid detection and identification of pathogens in clinical and environmental samples.[106-109] In 2009, Abbott Molecular (Des Plaines, Illinois) bought the IBIS technology and renamed the platform PLEX-ID. PLEX-ID has been the focus of the majority of studies in the clinical microbiology setting.[110-115] The most recent version of the technology, known as IRIDICA, is now available for clinical diagnostic use (CE-IVD) in Europe and the Middle East. It is not yet available in the United States, where it is undergoing clinical trials. This latest rendition of the platform has assays for detection of bacteria in blood, sterile fluids, soft tissues, bronchoalveolar lavage (BAL) fluids, and endotracheal aspirates (ETAs). The BAC LRT assay for BAL fluids and ETAs includes a semi-quantitative component. Other assays include a fungal assay capable of detecting more than 200 fungi directly from BAL samples or cultured isolates and the viral IC assay capable of detecting over 130 viruses in plasma. A potential advantage of PCR-ESI-MS is its ability to detect bacteria, fungi, viruses, and other microorganisms directly from the source, which may provide actionable results within hours of collecting the sample compared to the days and potentially weeks required for identification and antimicrobial susceptibility testing using current strategies. The turnaround time from the time of nucleic acid extraction until a final report is generated is approximately 6 hours. In the sections that follow, an overview of how PCR-ESI-MS works and studies that have evaluated this technology in the clinical microbiology setting are discussed.

The Polymerase Chain Reaction and Electrospray Ionization–Mass Spectrometry Method

ESI-MS is also a soft ionization technology like MALDI-TOF-MS, although instead of a solid matrix, ESI-MS requires the analytes to be present in an aqueous solution to which a high voltage is applied creating an aerosol. The workflow requires extraction of nucleic acids from either patient samples or cultured isolates followed by PCR amplification with multiple primer pairs that provide broad-range amplification as well as detection of specific species and/or characteristics (eg,

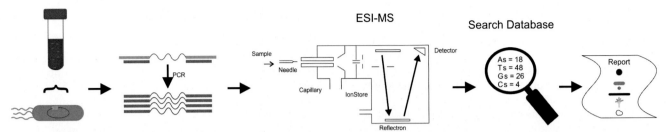

FIGURE 4.3 Sample workflow for polymerase chain reaction followed by electrospray ionization mass spectrometry (PCR-ESI-MS). DNA isolated from patient samples directly (or an isolate) is amplified using multiple PCR assays, followed by a desalting step. Once the amplified DNA is free from salt and other impurities, the newly purified amplified DNA is subjected to electrospray ionization time-of-flight mass spectrometry. Computer software uses the precise calculated mass of each piece of amplified DNA to generate a fingerprint that corresponds to the numbers of each base present. The data generated, including the PCR assays that yield a product as well as the base composition of the amplified DNA, is then searched against a database of reference organisms and resistance genes. A report is generated identifying the microorganism(s) present in the patient sample, if any, and their associated antimicrobial susceptibility, where interrogated.

antimicrobial resistance genes).[116] The broad-range primers are homologous to conserved regions across broad groups of organisms. The conserved region primers are designed to have a variable region between them allowing mass spectrometric differentiation of amplified DNA. Following PCR amplification, the amplified DNA is desalted to prevent interference with the mass spectrometer and subjected to ESI-MS. At this point, the amplified DNA is passed through heated capillaries and hit with a high-voltage laser leading to a plume of ions and ionized nucleic acids in the form of single strands of double-stranded amplified DNA that is analyzed based on TOF and collision with an ion detector, with the smaller pieces of amplified DNA reaching the detector sooner than larger ones (Fig. 4.3). The mass spectrometer and associated software determine the exact molecular weight, and the software deconvolutes the data and determines the base composition (ie, number of As, Gs, Ts, and Cs) of the amplified DNA, but not the order of the individual nucleic acids. Once the masses are established, a unique fingerprint is provided that is compared to a reference database to aid in identification of the organism or resistance gene(s), if applicable. The specificity of this method is increased by using multiple loci that are amplified by the initial PCR step to triangulate information. The end result is a report that yields pathogen identification (and in some cases limited susceptibility information) from the primary specimen along with two different metrics, the Q-score and level. The Q-score is a rating between 0 (low) and 1 (high) of the confidence in the identification of the organism. The level is a reflection of signal abundance relative to a set of competitive PCR standards of known input quantity and thus serves as an indirect estimate of how much specific template was amplified. This is calculated with reference to an internal calibrant construct (the amplification control). For a more complete review of PCR-ESI-MS in clinical microbiology, see the review by Wolk and associates.[117]

Bacterial Diagnostics

PCR-ESI-MS can be applied to amplified specimens, such as blood culture bottles, or to patient specimens directly; studies using both approaches have been published. Using PCR-ESI-MS, we demonstrated that PCR-ESI-MS was more sensitive albeit less specific than culture for detecting bacteria in sonicate fluids from explanted orthopedic prostheses for diagnosis of PJI, with sensitivities of 78 and 70% and specificities of 94 and 99%, respectively.[118] One advantage of using PCR as a component of testing is that viable organism is not needed to make a diagnosis; this was evident because the sensitivity of PCR-ESI-MS compared to culture in patients receiving antibiotics in the 14 days before surgery was 86 and 66%, respectively. For those patients with PJI receiving antibiotics in the 28 days before surgery, sensitivities of PCR-ESI-MS and culture were 86 and 74%, respectively. In a similar study that looked at using synovial fluid in place of sonicate fluid for 103 failed knee arthroplasties (21 PJI, 82 aseptic failure), PCR-ESI-MS and culture of synovial fluid were found to have sensitivities of 81 and 86%, and specificities of 95 and 100%, respectively.[119] There was a high level of concordance between the two methods when comparing detection of antimicrobial resistance genes and phenotypic antimicrobial susceptibility, with 10 of 11 staphylococci that were oxacillin resistant having been isolated from synovial fluids that tested positive for *mecA* by PCR-ESI-MS.

Sepsis is a leading cause of mortality; survival is directly linked to the amount of time it takes to initiate specific, tailored antimicrobial therapy. To examine the potential role of PCR-ESI-MS in accelerating diagnosis in bacteremic patients, Kaleta and colleagues[116] evaluated the RUO Ibis T5000's BAC assay's ability to detect microorganism(s) present in positive blood culture bottles compared to clinical reference standards. Concordance was 99 and 97% at the genus and species level, respectively. Also highlighted in this article is that *Saccharomyces cerevisiae* was detected in 39 of 45 negative blood culture bottles, a finding attributed to contamination as a result of yeast extracts present in the broth in the bottles. Also of note was the inability of PCR-ESI-MS to detect a second organism in 7 of 29 culture bottles positive with multiple organisms. One potential explanation for the failure of the assay to detect a second organism by the assay is that the nucleic acids present from the first organism may have been present at such a high concentration that they saturated the primers, effectively suppressing the amplification of the nucleic acids of the second organism. When comparing this platform to MALDI-TOF-MS, the same group found that the two methods were equivalent when compared to standard blood culture procedure.[115]

A more recent study by Bacconi and associates[120] evaluated the current IRIDICA platform's ability to detect bacterial and yeast infections in blood. To improve sensitivity of the assay compared to the PLEX-ID, 5 mL of whole blood was used instead of the previously used 1 mL. The new extraction technique employed by the new method used high-impact percussive beating with zirconium-yttrium beads followed by automated extraction and purification of total nucleic acids. The aim of the study was to evaluate this new method by comparing its sensitivity and specificity to culture. The PCR step of the new assay (BAC) was optimized to perform well in the presence of high levels of human DNA (up to 12 µg per PCR). To test the accuracy of identification of the new system, 331 prospectively collected, deidentified blood samples were compared to the results of standard clinical microbiology cultures. Of the 331 subjects, 35 (11%) were positive by PCR-ESI-MS compared to 18 positive samples by culture (5%). Fifteen samples were positive by both methods. Interestingly, one specimen was found to be infected with more than one organism, with two organisms detected by culture and three by PCR-ESI-MS. The concordance for the 16 cases for which culture identified an organism was 94% (15 of 16). The single mismatch came from a case in which culture failed to identify the organism in the primary culture used for comparison, although a subsequent blood draw yielded the organism (*E. coli*). The sensitivity and specificity of PCR-ESI-MS were 83% and 94%, respectively, when using culture as the comparator method. The sensitivity and specificity increased to 91 and 99%, respectively, if the definition of a true positive was amended to include samples that were positive by PCR-ESI-MS twice while culture remained negative. This method increases the sensitivity of the assay compared to previous versions of the technology that missed up to half of culture-positive cases, making it useful in cases in which a culture-negative infection is suspected.

Endocarditis is a serious infection. Although microbiology is an important component of the diagnosis, cultures are negative in up to 31% of cases,[121] meaning that diagnosis must rest on serologic and/or histopathologic findings without antimicrobial susceptibility information on the infecting organism being available. It is in this population of endocarditis cases that PCR-ESI-MS may find a niche in testing resected heart valve vegetations. In our evaluation of 83 formalin-fixed and paraffin-embedded (FFPE) heart valves from valve or blood culture–positive cases of endocarditis using the BAC 2.0 assay, there was 55% detection of an organism that had also been detected in culture, whereas 11% had discordant results between PCR-ESI-MS and culture and 34% had no detection.[122] Detection of antimicrobial resistance genes by PCR-ESI-MS was 100% concordant with results of phenotypic antimicrobial susceptibility testing on isolated bacteria. In one instance, *Tropheryma whipplei* was detected when *Staphylococcus lugdunensis* had been considered the infectious agent, based on a single positive blood culture. *T. whipplei* detection was confirmed by histopathologic findings and a *T. whipplei*-specific PCR, and the *S. lugdunensis* on review was considered a contaminant. Such a finding suggests a benefit to a PCR-ESI-MS platform, although it should be noted that the assay may produce false-positive results. In one instance, for example, *Candida tropicalis* was found throughout multiple simultaneously tested specimens from different patients, suggesting contamination. This particular study evaluated the use of FFPE tissue, which is not sterile; results of such testing using sensitive molecular techniques require an extra degree of scrutiny considering how FFPE blocks are handled and stored. The use of tissue blocks also may account for the infective endocarditis cases missed by the PCR-ESI-MS system because sensitivity of PCR may be compromised in formalin-fixed tissue because nucleic acid degradation and cross-linking to histones.[123,124] A separate study by Wallet and colleagues[125] tested 13 fresh, frozen cardiac valves from 12 patients by PCR-ESI-MS using the BAC Spectrum SF assay compared to conventional valve culture and blood cultures collected at the time of surgery.[125] Blood culture at the time of surgery was positive in only 2 of 13 valves, and valve culture was positive in 4 of 13 valves. It should be noted though that 10 of 12 patients were receiving antimicrobial therapy at the time of blood culture and valve removal. PCR-ESI-MS was positive in 10 of 13 valves. The results from valves that did not have corresponding positive conventional cultures were evaluated using Duke endocarditis criteria, C-reactive protein, leukocyte count, and results of blood cultures collected before surgery. Using this information, PCR-ESI-MS had a sensitivity and specificity of 91 and 100%, respectively. The difference in sensitivities between the two studies may be due to the use of FFPE tissue in our study versus fresh, frozen tissue that was processed almost immediately after collection. Although the differences in tissue types likely account for the differences in sensitivity, the use of a different kit as well as nucleic acid purification methods cannot be totally discounted.

Fungal Diagnostics

Identification of fungi grown in culture typically relies on examination of macroscopic and microscopic colony characteristics, alongside MALDI-TOF-MS. When these techniques fail, the isolates may be sequenced, which delays results by a day or more. The broad-range fungal kit is a single-plex assay that contains nine broad-range primer sets, six more specific primer sets, and one control primer set. Simner and coworkers[126] evaluated the PLEX-ID system's broad fungal assay's ability to identify 91 fungal isolates consisting of 64 manufacturer verified organisms and 27 non-verified organisms and found that PLEX-ID was able to correctly identify 96 and 81% of isolates to the genus and species level, respectively. For verified species alone, the PLEX-ID broad fungal assay had 100% (64 of 64) and 92% agreement at the genus and species levels, respectively. For the 4% of isolates incorrectly identified at the species level, D2 sequencing confirmed a misidentification by PLEX-ID. Performance was poorest for the nonverified species with 85 and 56% of isolates identified to the genus and species level, respectively.

PCR-ESI-MS can be performed directly on patient samples, potentially providing rapid, accurate, actionable results. The study by Simner and coworkers[126] examined 395 respiratory specimens and evaluated the concordance between PLEX-ID results and culture. Of the 395 specimens tested, only 223 had growth from the same specimen tested by

PLEX-ID. Direct analysis of specimens by PLEX-ID resulted in 68 and 67% agreement with cultures at the genus (267 of 395) and species levels (263 of 395), respectively. Sixty-four (16%) sample results from direct testing by PLEX-ID were discordant with culture, and PLEX-ID failed to detect an organism in another 64 (16%) specimens. More than 95% of negative samples by PLEX-ID did ultimately have growth in culture. The potential for the increased sensitivity of the assay was evident by its ability to detect pathogenic organisms where culture methods had no growth (35 of 172). Four of the specimens were positive for *Pneumocystis jiroveci*, which cannot be cultured, demonstrating the ability of PLEX-ID to detect nonculturable organisms. In the remaining discrepant samples in which PLEX-ID detected an organism and culture did not, it was not possible to determine whether the findings were attributable to contaminating DNA, colonization, or detection of actual infecting organism.

Of particular importance in the culturing of fungal and mycobacterial organisms is the prevention of overgrowth of bacteria that may contaminate or prevent the culture of fungi or mycobacteria. This is typically accomplished using selective media that contain antibiotics to prevent the overgrowth of bacteria. Shin and colleagues evaluated 691 BAL fluid specimens collected over 6 years and frozen at −80°C before testing with the broad-range fungal assay and looked at the ability of PCR-ESI-MS to detect fungi compared to routine fungal culture.[127] One hundred sixty-four specimens (24%) were positive by culture, and 377 (55%) specimens were positive by PCR-ESI-MS. Those samples that were positive by both methods demonstrated a concordance of 63 and 81% at the genus and species levels, respectively. PLEX-ID was able to detect *Pneumocystis* species, similar to the study by Simner and coworkers. Another major advantage of PLEX-ID over fungal culture is that a pure isolate is not needed. Of the 53 specimens that were overgrown with bacteria, PLEX-ID identified a fungal organism in 26. One hundred fifty-four samples were positive for having fungus detected, but no identification was given. This likely could be attributed to the dilute nature of BAL fluid, colonization, or nonviable fungal DNA. PLEX-ID also had higher rates of detection of single and multiple fungal organisms from BAL fluid compared to culture, although the authors had no clinical history and could not distinguish whether the results would have been clinically relevant. Huttner and associates[128] also looked at the utility of PCR-ESI-MS of BAL fluids compared to standard techniques and found a poor concordance of 45%, which was increased to 66% if discordance for commensal flora was excluded, underscoring a need for more studies on PCR-ESI-MS in complex nonsterile sources.

Application to Mycobacterial Testing

Similar to fungal cultures, growth of mycobacteria can take a significant amount of time, and although molecular methods for diagnosis do exist, aside from *M. tuberculosis*, specific detection methods rely on growth of the organisms in culture, with subsequent identification using MALDI-TOF-MS or traditional techniques.

Simner and coworkers[129] examined the ability of PLEX-ID multidrug-resistant (MDR) tuberculosis (TB) assay to correctly identify *M. tuberculosis* complex as well as nontuberculous mycobacteria. The MDR TB assay is designed to detect *M. tuberculosis* complex and nontuberculous mycobacteria as well as drug resistance genes for *M. tuberculosis* complex organisms in sputum, culture, nasal and throat swabs, tissue and cells, and body fluids. In addition to *M. tuberculosis* complex ($n = 68$) and nontuberculous mycobacteria ($n = 97$) isolates, 57 positive and 50 negative mycobacteria growth indicator tube (MGIT) bottles were examined by PLEX-ID. PLEX-ID correctly identified all *M. tuberculosis* complex isolates. PLEX-ID identified 98% of nontuberculous mycobacteria from agar cultures and 96% from MGIT broth cultures. The one misidentification occurred when *Mycobacterium rhodesiae* was identified as *Mycobacterium aicheiense*. In three instances, the isolates tested by PLEX-ID identified the correct species as well as a second nontuberculous *Mycobacterium* species. These included *M. austroafricanum* identified as *M. austroafricanum/M. vanbaalenii* and *M. murale* and *M. tokaiense* that were both identified as *M. murale/M. tokaiense*. These results were counted as correct as a result of the assay's inability to distinguish the two species from one another. From MGIT broth culture the PLEX ID MDR-TB assay correctly detected 100% of *M. tuberculosis* complex when compared to AccuProbe (Hologic, Marlborough, Massachusetts) identification. There was also 100% concordance with regard to susceptibility testing as the VersaTREK system (ThermoFisher Scientific, Waltham, Massachusetts) and PLEX-ID found the organisms to be susceptible to all first-line agents active against *M. tuberculosis* complex. PLEX-ID performed similarly with regard to MGIT positive for nontuberculous mycobacteria, showing a concordance of 100 and 96% at the genus and species levels, respectively. Two *Mycobacterium gordonae* isolates were identified as *Mycobacterium* species, and an *M. avium* by AccuProbe had identifications of *M. avium* and *M. gordonae* by PLEX-ID. With regard to the negative tubes, PLEX-ID was in agreement in all but one case in which *M. gordonae* was identified from a culture-negative sample. Sequencing of the MGIT lysate was negative, confirming the negative culture results. PLEX-ID performed well in determining drug resistance of 48 well-characterized isolates (isoniazid resistant $n = 16$, rifampin resistant $n = 9$, ethambutol resistant $n = 12$, fluoroquinolone resistant $n = 2$) as sensitivity and specificity for isoniazid resistance in *M. tuberculosis* complex was 100 and 94%, respectively. Two major errors occurred in which a *katG* mutation (S315T) was detected when the isolates were found to be susceptible by the agar proportion method. Detection of rifampin resistance in characterized *M. tuberculosis* complex isolates by the MDR-TB assay had a sensitivity and specificity of 100 and 92%, respectively. Three discordant findings resulted in major errors as the PLEX-ID found the mutations L511P, D516G, and D516F in *rpoB* of the isolates deemed susceptible by agar proportion. Sensitivity and specificity for detection of ethambutol resistance in *M. tuberculosis* complex were 92 and 94%, respectively. Two major errors occurred in which *embB* mutations M306V and M306I were detected in two isolates that were susceptible by the agar proportion method. In addition, a very major error occurred in which no resistance was detected when agar proportion found the isolate to be resistant. PLEX-ID had 100% sensitivity and specificity for fluoroquinolone susceptibility.

Performance for Viral Diagnostics

Although all areas of diagnostic testing in the clinical microbiology laboratory present specific challenges, viral diagnostics has been the most complicated and labor intensive. Because of this, molecular methods have largely usurped culture as the primary method of viral diagnostics in clinical laboratories. Viral cytopathic effect (CPE) often can be ambiguous and at other times can take weeks for conclusive CPE to develop, should it develop at all. These issues were somewhat abrogated with the onset of shell vials; however, a limitation of shell vials is that there must be some idea of the virus suspected so that the proper antibodies are selected for direct fluorescent antibody (DFA) testing. A second issue is that the ability to perform a DFA assay relies on antibodies being available for the suspected virus. Although culture and shell vials have given way to the newer, more sensitive multiplex PCRs, these assays are limited in the viruses they can detect as they tend to be grouped by symptoms or systems affected. A PCR-ESI-MS approach could expand the number of targets tested at a single time without placing limitations on the viruses that can be detected. The current IRIDICA platform is currently able to detect 132 viruses broken into 13 reporting groups using the Viral IC (immunocompromised) testing kit though its use is limited to plasma at this time.

LeGoff and associates[130] compared standard methods to PCR-ESI-MS to determine the ability of the latter to detect adenovirus and herpesvirus infections in hematopoietic stem cell transplant recipients. A total of 100 stool and 92 plasma samples were tested. The authors reported that 97% of patient samples with a plasma viral load of 4.47 $\log_{10}$ copies/mL were detected by PCR-ESI-MS. A median viral load of 3.1 $\log_{10}$ copies/mL was noted for samples that were negative by PCR-ESI-MS. PCR-ESI-MS showed 100% concordance at the species level for positive samples with single infections. PCR-ESI-MS had a lower sensitivity in stool samples in which only 78% were positive with a median viral load of 5.25 $\log_{10}$ copies/mL with a median viral load of 3.42 $\log_{10}$ for negative PCR-ESI-MS samples. All negative samples tested negative by PCR-ESI-MS, and in the 92 positive plasma samples, 67% were found to have coinfection with BK virus ($n = 41$), cytomegalovirus (CMV) ($n = 36$), Epstein-Barr virus (EBV) ($n = 26$), JC virus ($n = 9$), and herpes simplex virus ($n = 6$). Sixty samples had corresponding whole blood samples that were tested for CMV and EBV, of which 27 were positive for CMV and 28 were positive for EBV. PLEX-ID of the corresponding plasma samples detected 23 of the positive CMV samples (85%) and 17 of the positive EBV samples (61%). Specificities for the assay were 100% for CMV and 94% for EBV. In 11 plasma ($n = 5$) and stool ($n = 6$) samples from three patients infected with two adenoviruses, PLEX-ID was able to detect coinfection only in one sample and the remaining 10 had only the predominant adenovirus detected.

Although the utility of influenza testing with PCR-ESI-MS is questionable with the availability of faster, less expensive influenza testing, there have been studies looking at the performance of the PCR-ESI-MS flu detection kit on the PLEX-ID. Mengelle and colleagues[111] tested 293 respiratory samples and found a concordance of 93% with the multiplex RespiFinder kit (PathoFinder, Maastricht, the Netherlands).[111] The sensitivity, specificity, and positive and negative predictive values of the PCR-ESI-MS system were 87%, 97%, 92% and 94%, respectively. A comparison of PCR-ESI-MS and the Center for Disease Control and Prevention RT-PCR showed an overall agreement of 89% for 75 nasopharyngeal samples.[131]

Limitations and Future Perspectives

PCR-ESI-MS remains an emerging technology for the clinical microbiology laboratory and, although CE marked for in vitro diagnostic use in Europe, the technology is not currently available in the United States. The platform can be used for detecting bacteria in blood, sterile fluids, BAL fluids, ETA, and soft tissues as well as fungi in BAL fluids and from isolates, and for detecting viruses in plasma. The studies mentioned in this chapter have shown the potential utility of this platform because it may provide direct-from-specimen results with a faster turnaround time than culture combined with subsequent MALDI-TOF-MS. However, it remains to be determined whether the improvements in the platform overcome the problems with ease-of-use and throughput that have been observed with previous renditions of the technology. The PLEX-ID can hold 15 plates, limiting the number of patient specimens that can be tested in a batch. Although PCR-ESI-MS can detect polymicrobial infections, this remains an area where improvement is needed. Also, although the sensitivity of the platform is a strength, it remains an open system, so strict adherence to a unidirectional workflow is imperative to prevent amplified DNA contamination. Finally, this technology will most likely be limited to large laboratories because of the high costs of reagents and instrumentation, although the latest redesign of the platform may lower the costs. Laboratories that adopt this technology must decide which patients will benefit most from this testing. PCR-ESI-MS may have the potential to affect patient care by reducing lengths of stay, and/or time to initiate targeted therapy, thus improving antimicrobial stewardship; however, outcome-based studies are needed.

POINTS TO REMEMBER

PCR-ESI-MS
- PCR-ESI-MS couples PCR to TOF-MS to detect a wide range of pathogens and some antimicrobial resistance genes.
- PCR-ESI-MS allows testing directly from patient samples.
- PCR-ESI-MS is capable of providing semi-quantitative data.
- PCR-ESI-MS is not reliant on viable organisms being present.
- PCR-ESI-MS is an open system requiring strict adherence to unidirectional workflow.

CONCLUSIONS

Clinical microbiology laboratories are undergoing several changes in diagnostic processes and offerings, with MS front and center. The ease of use and widespread adoption of MALDI-TOF-MS have reduced costs and turnaround times for microbial identification from culture. PCR-ESI-MS has the potential to offer same-day direct-from-patient specimen testing for bacteria, fungi, and viruses, which may be helpful when a rapid diagnosis is needed or traditional methods have failed to provide a diagnosis.

REFERENCES

1. Anhalt JP, Fenselau C. Identification of bacteria using mass spectrometry. *Anal Chem* 1975;**47**:219—25.
2. Tanaka K. The origin of macromolecule ionization by laser irradiation (Nobel lecture). *Angew Chem* 2003;**42**:3860—70.
3. Karas M, Hillenkamp F. Laser desorption ionization of proteins with molecular masses exceeding 10,000 daltons. *Anal Chem* 1988;**60**:2299—301.
4. Mellmann A, Cloud J, Maier T, et al. Evaluation of matrix-assisted laser desorption ionization-time-of-flight mass spectrometry in comparison to 16S RNA gene sequencing for species identification of nonfermenting bacteria. *J Clin Microbiol* 2008;**46**:1946—54.
5. Seng P, Drancourt M, Gouriet F, et al. Ongoing revolution in bacteriology: routine identification of bacteria by matrix-assisted laser desorption ionization time-of-flight mass spectrometry. *Clin Infect Dis* 2009;**49**:543—51.
6. Kallow W, Erhard M, Shah HN, et al. MALDI-TOF MS for microbial identification: years of experimental development to an established protocol. In: *Mass spectrometry for microbial proteomics*. New York, NY: John Wiley & Sons; 2010. p. 255—76.
7. Cunningham SA, Patel R. Importance of using bruker's security-relevant library for biotyper identification of *Burkholderia pseudomallei*, *Brucella* species, and *Francisella tularensis*. *J Clin Microbiol* 2013;**51**:1639—40.
8. Murray PR. What is new in clinical microbiology: microbial identification by MALDI-TOF mass spectrometry. A paper from the 2011 William Beaumont Hospital Symposium on Molecular Pathology. *J Mol Diagn* 2012;**14**:419—23.
9. Branda JA, Markham RP, Garner CD, et al. Performance of the VITEK MS v2.0 system in distinguishing Streptococcus pneumoniae from nonpneumococcal species of the Streptococcus mitis group. *J Clin Microbiol* 2013;**51**:3079—82.
10. Dubois D, Segonds C, Prere M-F, et al. Identification of clinical *Streptococcus pneumoniae* isolates among other alpha and nonhemolytic streptococci by use of the VITEK MS matrix-assisted laser desorption ionization—time of flight mass spectrometry system. *J Clin Microbiol* 2013;**51**:1861—7.
11. Pavlovic M, Konrad R, Iwobi AN, et al. A dual approach employing MALDI-TOF MS and real-time PCR for fast species identification within the *Enterobacter cloacae* complex. *FEMS Microbiol Lett* 2012;**329**:46—53.
12. Vanlaere E, Sergeant K, Dawyndt P, et al. Matrix-assisted laser desorption ionisation-time-of of-flight mass spectrometry of intact cells allows rapid identification of *Burkholderia cepacia* complex. *J Microbiol Methods* 2008;**75**:279—86.
13. Hinse D, Vollmer T, Erhard M, et al. Differentiation of species of the *Streptococcus bovis/equinus*-complex by MALDI-TOF mass spectrometry in comparison to soda sequence analyses. *Syst Appl Microbiol* 2011;**34**:52—7.
14. Cunningham SA, Mainella JM, Patel R. Misidentification of *Neisseria polysaccharea* as *Neisseria meningitidis* with the use of matrix-assisted laser desorption ionization—time of flight spectrometry. *J Clin Microbiol* 2014;**52**:2270—1.
15. Holzmann T, Frangoulidis D, Simon M, et al. Fatal anthrax infection in a heroin user from southern Germany. *Euro Surveill* June 2012;**2012**(17):26.
16. Saffert RT, Cunningham SA, Ihde SM, et al. Comparison of Bruker biotyper matrix-assisted laser desorption ionization—time of flight mass spectrometer to BD Phoenix automated microbiology system for identification of gram-negative bacilli. *J Clin Microbiol* 2011;**49**:887—92.
17. Alatoom AA, Cunningham SA, Ihde SM, et al. Comparison of direct colony method versus extraction method for identification of gram-positive cocci by use of Bruker biotyper matrix-assisted laser desorption ionization—time of flight mass spectrometry. *J Clin Microbiol* 2011;**49**:2868—73.
18. McElvania TeKippe E, Shuey S, Winkler DW, et al. Optimizing identification of clinically relevant gram-positive organisms by use of the Bruker biotyper matrix-assisted laser desorption ionization—time of flight mass spectrometry system. *J Clin Microbiol* 2013;**51**:1421—7.
19. Lau SKP, Tang BSF, Teng JLL, et al. Matrix-assisted laser desorption ionisation time-of-flight mass spectrometry for identification of clinically significant bacteria that are difficult to identify in clinical laboratories. *J Clin Pathol* 2014;**67**: 361—6.
20. Hsueh P-R, Lee T-F, Du S-H, et al. Bruker biotyper matrix-assisted laser desorption ionization—time of flight mass spectrometry system for identification of *Nocardia, Rhodococcus, Kocuria, Gordonia, Tsukamurella*, and *Listeria* species. *J Clin Microbiol* 2014;**52**:2371—9.
21. Schulthess B, Bloemberg GV, Zbinden R, et al. Evaluation of the Bruker MALDI biotyper for identification of gram-positive rods: development of a diagnostic algorithm for the clinical laboratory. *J Clin Microbiol* 2014;**52**:1089—97.
22. Alatoom AA, Cazanave CJ, Cunningham SA, et al. Identification of non-*diphtheriae Corynebacterium* by use of matrix-assisted laser desorption ionization—time of flight mass spectrometry. *J Clin Microbiol* 2012;**50**:160—3.
23. Barberis C, Almuzara M, Join-Lambert O, et al. Comparison of the Bruker MALDI-TOF mass spectrometry system and conventional phenotypic methods for identification of gram-positive rods. *PLoS ONE* 2014;**9**:e106303.
24. Richter SS, Sercia L, Branda JA, et al. Identification of *Enterobacteriaceae* by matrix-assisted laser desorption/ionization time-of-flight mass spectrometry using the VITEK MS system. *Eur J Clin Microbiol Infect Dis* 2013;**32**:1571—8.
25. Manji R, Bythrow M, Branda JA, et al. Multi-center evaluation of the VITEK® MS system for mass spectrometric identification of non-*Enterobacteriaceae* gram-negative bacilli. *Eur J Clin Microbiol Infect Dis* 2014;**33**:337—46.
26. Branda JA, Rychert J, Burnham C-AD, et al. Multicenter validation of the VITEK MS v2.0 MALDI TOF mass spectrometry system for the identification of fastidious gram-negative bacteria. *Diagn Microbiol Infect Dis* 2014;**78**:129—31.
27. Rychert J, Burnham C-AD, Bythrow M, et al. Multicenter evaluation of the VITEK MS matrix-assisted laser desorption ionization—time of flight mass spectrometry system for identification of gram-positive aerobic bacteria. *J Clin Microbiol* 2013;**51**:2225—31.
28. Kärpänoja P, Harju I, Rantakokko-Jalava K, et al. Evaluation of two matrix-assisted laser desorption ionization—time of flight mass spectrometry systems for identification of viridans group streptococci. *Eur J Clin Microbiol Infect Dis* 2014;**33**: 779—88.
29. Cunningham SA, Patel R. Standard matrix-assisted laser desorption ionization—time of flight mass spectrometry reagents may inactivate potentially hazardous bacteria. *J Clin Microbiol* 2015;**53**:2788—9.
30. Jamal WY, Shahin M, Rotimi VO. Comparison of two matrix-assisted laser desorption/ionization-time of flight (MALDI-TOF) mass spectrometry methods and API 20AN for identification of clinically relevant anaerobic bacteria. *J Med Microbiol* 2013;**62**:540—4.

31. Hsu Y-MS, Burnham C-AD. MALDI-TOF MS identification of anaerobic bacteria: assessment of pre-analytical variables and specimen preparation techniques. *Diagn Microbiol Infect Dis* 2014;**79**:144–8.
32. Schmitt BH, Cunningham SA, Dailey AL, et al. Identification of anaerobic bacteria by Bruker biotyper matrix-assisted laser desorption ionization–time of flight mass spectrometry with on-plate formic acid preparation. *J Clin Microbiol* 2013;**51**:782–6.
33. Barreau M, Pagnier I, La Scola B. Improving the identification of anaerobes in the clinical microbiology laboratory through MALDI-TOF mass spectrometry. *Anaerobe* 2013;**22**:123–5.
34. Garner O, Mochon A, Branda J, et al. Multi-centre evaluation of mass spectrometric identification of anaerobic bacteria using the VITEK® MS system. *Clin Microbiol Infect* 2014;**20**:335–9.
35. Teng S-H, Chen C-M, Lee M-R, et al. Matrix-assisted laser desorption ionization–time of flight mass spectrometry can accurately differentiate between *Mycobacterium masilliense* (*M. abscessus* subspecies *bolletti*) and *M. abscessus* (sensu stricto). *J Clin Microbiol* 2013;**51**:3113–6.
36. Dunne WM, Doing K, Miller E, et al. Rapid inactivation of *Mycobacterium* and *Nocardia* species before identification using matrix-assisted laser desorption ionization–time of flight mass spectrometry. *J Clin Microbiol* 2014;**52**:3654–9.
37. Machen A, Kobayashi M, Connelly MR, et al. Comparison of heat inactivation and cell disruption protocols for identification of mycobacteria from solid culture media by use of VITEK matrix-assisted laser desorption ionization–time of flight mass spectrometry. *J Clin Microbiol* 2013;**51**:4226–9.
38. Balada-Llasat JM, Kamboj K, Pancholi P. Identification of mycobacteria from solid and liquid media by matrix-assisted laser desorption ionization–time of flight mass spectrometry in the clinical laboratory. *J Clin Microbiol* 2013;**51**:2875–9.
39. Mather CA, Rivera SF, Butler-Wu SM. Comparison of the Bruker biotyper and VITEK MS matrix-assisted laser desorption ionization–time of flight mass spectrometry systems for identification of mycobacteria using simplified protein extraction protocols. *J Clin Microbiol* 2014;**52**:130–8.
40. Theel ES, Schmitt BH, Hall L, et al. Formic acid-based direct, on-plate testing of yeast and *Corynebacterium* species by Bruker biotyper matrix-assisted laser desorption ionization–time of flight mass spectrometry. *J Clin Microbiol* 2012;**50**:3093–5.
41. Dhiman N, Hall L, Wohlfiel SL, et al. Performance and cost analysis of matrix-assisted laser desorption ionization–time of flight mass spectrometry for routine identification of yeast. *J Clin Microbiol* 2011;**49**:1614–6.
42. Lacroix C, Gicquel A, Sendid B, et al. Evaluation of two matrix-assisted laser desorption ionization-time of flight mass spectrometry (MALDI-TOF MS) systems for the identification of *Candida* species. *Clin Microbiol Infect* 2014;**20**:153–8.
43. Westblade LF, Jennemann R, Branda JA, et al. Multicenter study evaluating the VITEK MS system for identification of medically important yeasts. *J Clin Microbiol* 2013;**51**:2267–72.
44. Pence MA, McElvania TeKippe E, Wallace MA, et al. Comparison and optimization of two MALDI-TOF MS platforms for the identification of medically relevant yeast species. *Eur J Clin Microbiol Infect Dis* 2014;**33**:1703–12.
45. Hamprecht A, Christ S, Oestreicher T, et al. Performance of two MALDI-TOF MS systems for the identification of yeasts isolated from bloodstream infections and cerebrospinal fluids using a time-saving direct transfer protocol. *Med Microbiol Immunol* 2014;**203**:93–9.
46. De Carolis E, Vella A, Vaccaro L, et al. Development and validation of an in-house database for matrix-assisted laser desorption ionization time of flight mass spectrometry-based yeast identification using a fast protein extraction procedure. *J Clin Microbiol* 2014;**52**:1453–8.
47. Rosenvinge FS, Dzajic E, Knudsen E, et al. Performance of matrix-assisted laser desorption-time of flight mass spectrometry for identification of clinical yeast isolates. *Mycoses* 2013;**56**:229–35.
48. Mancini N, De Carolis E, Infurnari L, et al. Comparative evaluation of the Bruker biotyper and VITEK MS matrix-assisted laser desorption ionization–time of flight (MALDI-TOF) mass spectrometry systems for identification of yeasts of medical importance. *J Clin Microbiol* 2013;**51**:2453–7.
49. Schulthess B, Ledermann R, Mouttet F, et al. Use of the Bruker MALDI biotyper for identification of molds in the clinical mycology laboratory. *J Clin Microbiol* 2014;**52**:2797–803.
50. Lau AF, Drake SK, Calhoun LB, et al. Development of a clinically comprehensive database and a simple procedure for identification of molds from solid media by matrix-assisted laser desorption ionization–time of flight mass spectrometry. *J Clin Microbiol* 2013;**51**:828–34.
51. Theel ES, Hall L, Mandrekar J, et al. Dermatophyte identification using matrix-assisted laser desorption ionization–time of flight mass spectrometry. *J Clin Microbiol* 2011;**49**:4067–71.
52. de Respinis S, Tonolla M, Pranghofer S, et al. Identification of dermatophytes by matrix-assisted laser desorption/ionization time-of-flight mass spectrometry. *Med Mycology* 2013;**51**:514–21.
53. De Respinis S, Monnin V, Girard V, et al. Matrix-assisted laser desorption ionization–time of flight (MALDI-TOF) mass spectrometry using the VITEK MS system for rapid and accurate identification of dermatophytes on solid cultures. *J Clin Microbiol* 2014;**52**:4286–92.
54. Seng P, Abat C, Rolain JM, et al. Identification of rare pathogenic bacteria in a clinical microbiology laboratory: impact of matrix-assisted laser desorption ionization–time of flight mass spectrometry. *J Clin Microbiol* 2013;**51**:2182–94.
55. Wang L, Han C, Sui W, et al. MALDI-TOF MS applied to indirect carbapenemase detection: a validated procedure to clearly distinguish between carbapenemase-positive and carbapenemase-negative bacterial strains. *Anal Bioanal Chem* 2013;**405**:5259–66.
56. Sparbier K, Schubert S, Weller U, et al. Matrix-assisted laser desorption ionization–time of flight mass spectrometry-based functional assay for rapid detection of resistance against β-lactam antibiotics. *J Clin Microbiol* 2012;**50**:927–37.
57. Hoyos-Mallecot Y, Cabrera-Alvargonzalez JJ, Miranda-Casas C, et al. MALDI-TOF MS, a useful instrument for differentiating metallo-β-lactamases in *Enterobacteriaceae* and *Pseudomonas* spp. *Lett Appl Microbiol* 2014;**58**:325–9.
58. Álvarez-Buylla A, Picazo JJ, Culebras E. Optimized method for *Acinetobacter* species carbapenemase detection and identification by matrix-assisted laser desorption ionization–time of flight mass spectrometry. *J Clin Microbiol* 2013;**51**:1589–92.
59. Jung JS, Eberl T, Sparbier K, et al. Rapid detection of antibiotic resistance based on mass spectrometry and stable isotopes. *Eur J Clin Microbiol Infect Dis* 2014;**33**:949–55.
60. Sparbier K, Lange C, Jung J, et al. MALDI biotyper-based rapid resistance detection by stable-isotope labeling. *J Clin Microbiol* 2013;**51**:3741–8.

61. Hrabák J, Chudáčková E, Walková R. Matrix-assisted laser desorption ionization—time of flight (MALDI-TOF) mass spectrometry for detection of antibiotic resistance mechanisms: from research to routine diagnosis. *Clin Microbiol Rev* 2013;**26**:103—14.
62. Croxatto A, Prod'hom G, Greub G. Applications of MALDI-TOF mass spectrometry in clinical diagnostic microbiology. *FEMS Microbiol Rev* 2012:380—407.
63. Sánchez-Juanes F, Siller Ruiz M, Moreno Obregón F, et al. Pretreatment of urine samples with SDS improves direct identification of urinary tract pathogens with matrix-assisted laser desorption ionization—time of flight mass spectrometry. *J Clin Microbiol* 2014;**52**:335—8.
64. Wang XH, Zhang G, Fan YY, et al. Direct identification of bacteria causing urinary tract infections by combining matrix-assisted laser desorption ionization-time of flight mass spectrometry with UF-1000i urine flow cytometry. *J Microbiol Methods* 2013;**92**:231—5.
65. March Rosselló GA, Gutiérrez Rodríguez MP, de Lejarazu Leonardo RO, et al. Procedure for microbial identification based on matrix-assisted laser desorption/ionization-time of flight mass spectrometry from screening-positive urine samples. *APMIS* 2014;**122**:790—5.
66. Köhling HL, Bittner A, Müller K-D, et al. Direct identification of bacteria in urine samples by matrix-assisted laser desorption/ionization time-of-flight mass spectrometry and relevance of defensins as interfering factors. *J Med Microbiol* 2012;**61**:339—44.
67. DeMarco ML, Burnham C-AD. Diafiltration MALDI-TOF mass spectrometry method for culture-independent detection and identification of pathogens directly from urine specimens. *Am J Clin Pathol* 2014;**141**:204—12.
68. Nyvang Hartmeyer G, Kvistholm Jensen A, Böcher S, et al. Mass spectrometry: pneumococcal meningitis verified and *Brucella* species identified in less than half an hour. *Scand J Infect Dis* 2010;**42**:716—8.
69. Segawa S, Sawai S, Murata S, et al. Direct application of MALDI-TOF mass spectrometry to cerebrospinal fluid for rapid pathogen identification in a patient with bacterial meningitis. *Clin Chim Acta* 2014;**435**:59—61.
70. Annesley TM. Ion suppression in mass spectrometry. *Clin Chem* 2003;**49**:1041—4.
71. Bhatti MM, Boonlayangoor S, Beavis KG, et al. Rapid identification of positive blood cultures by matrix-assisted laser desorption ionization—time of flight mass spectrometry using prewarmed agar plates. *J Clin Microbiol* 2014;**52**:4334—8.
72. Idelevich EA, Schüle I, Grünastel B, et al. Rapid identification of microorganisms from positive blood cultures by MALDI-TOF mass spectrometry subsequent to very short-term incubation on solid medium. *Clin Microbiol Infect* 2014;**20**:1001—6.
73. Fothergill A, Kasinathan V, Hyman J, et al. Rapid identification of bacteria and yeasts from positive-blood-culture bottles by using a lysis-filtration method and matrix-assisted laser desorption ionization—time of flight mass spectrum analysis with the SARAMIS database. *J Clin Microbiol* 2013;**51**:805—9.
74. March-Rosselló GA, Muñoz-Moreno MF, García-Loygorri—Jordán de Urriés MC, et al. A differential centrifugation protocol and validation criterion for enhancing mass spectrometry (MALDI-TOF) results in microbial identification using blood culture growth bottles. *Eur J Clin Microbiol Infect Dis* 2013;**32**:699—704.
75. Leli C, Cenci E, Cardaccia A, et al. Rapid identification of bacterial and fungal pathogens from positive blood cultures by MALDI-TOF MS. *Int J Med Microbiol* 2013;**303**:205—9.
76. Tadros M, Petrich A. Evaluation of MALDI-TOF mass spectrometry and Sepsityper kit™ for the direct identification of organisms from sterile body fluids in a Canadian pediatric hospital. *Can J Infect Dis Med Microbiol* 2013;**24**:191—4.
77. Gray TJ, Thomas L, Olma T, et al. Rapid identification of gram-negative organisms from blood culture bottles using a modified extraction method and MALDI-TOF mass spectrometry. *Diagn Microbiol Infect Dis* 2013;**77**:110—2.
78. Riederer K, Cruz K, Shemes S, et al. MALDI-TOF identification of gram-negative bacteria directly from blood culture bottles containing charcoal: Sepsityper® kits versus centrifugation—filtration method. *Diagn Microbiol Infect Dis* 2015;**82**:105—8.
79. Bidart M, Bonnet I, Hennebique A, et al. An in-house assay is superior to Sepsityper® for the direct MALDI-TOF mass spectrometry identification of yeast species in blood culture. *J Clin Microbiol* 2015;**53**:1761—4.
80. Nonnemann B, Tvede M, Bjarnsholt T. Identification of pathogenic microorganisms directly from positive blood vials by matrix-assisted laser desorption/ionization time of flight mass spectrometry. *APMIS* 2013;**121**:871—7.
81. Rodríguez-Sánchez B, Sánchez-Carrillo C, Ruiz A, et al. Direct identification of pathogens from positive blood cultures using matrix-assisted laser desorption-ionization time-of-flight mass spectrometry. *Clin Microbiol Infect* 2014;**20**:O421—7.
82. Chen JHK, Ho P-L, Kwan GSW, et al. Direct bacterial identification in positive blood cultures by use of two commercial matrix-assisted laser desorption ionization—time of flight mass spectrometry systems. *J Clin Microbiol* 2013;**51**:1733—9.
83. Martiny D, Debaugnies F, Gateff D, et al. Impact of rapid microbial identification directly from positive blood cultures using matrix-assisted laser desorption/ionization time-of-flight mass spectrometry on patient management. *Clin Microbiol Infect* 2013;**19**:E568—81.
84. Foster AGW. Rapid identification of microbes in positive blood cultures by use of the VITEK MS matrix-assisted laser desorption ionization—time of flight mass spectrometry system. *J Clin Microbiol* 2013;**51**:3717—9.
85. Jamal W, Saleem R, Rotimi VO. Rapid identification of pathogens directly from blood culture bottles by Bruker matrix-assisted laser desorption laser ionization-time of flight mass spectrometry versus routine methods. *Diagn Microbiol Infect Dis* 2013;**76**:404—8.
86. Clerc O, Prod'hom G, Vogne C, et al. Impact of matrix-assisted laser desorption ionization time-of-flight mass spectrometry on the clinical management of patients with gram-negative bacteremia: a prospective observational study. *Clin Infect Dis* 2013;**56**:1101—7.
87. Machen A, Drake T, Wang YF. Same day identification and full panel antimicrobial susceptibility testing of bacteria from positive blood culture bottles made possible by a combined lysis-filtration method with MALDI-TOF VITEK mass spectrometry and the VITEK2 system. *PLoS ONE* 2014;**9**:e87870.
88. Perez KK, Olsen RJ, Musick WL, et al. Integrating rapid pathogen identification and antimicrobial stewardship significantly decreases hospital costs. *Arch Pathol Lab Med* 2013;**137**:1247—54.

89. Mencacci A, Monari C, Leli C, et al. Typing of nosocomial outbreaks of *Acinetobacter baumannii* by use of matrix-assisted laser desorption ionization–time of flight mass spectrometry. *J Clin Microbiol* 2013;51:603–6.
90. Josten M, Reif M, Szekat C, et al. Analysis of the matrix-assisted laser desorption ionization–time of flight mass spectrum of *Staphylococcus aureus* identifies mutations that allow differentiation of the main clonal lineages. *J Clin Microbiol* 2013;51:1809–17.
91. Kuhns M, Zautner AE, Rabsch W, et al. Rapid discrimination of *Salmonella enterica* serovar *typhi* from other serovars by MALDI-TOF mass spectrometry. *PLoS ONE* 2012;7:e40004.
92. Ketterlinus R, Hsieh S-Y, Teng S-H, et al. Fishing for biomarkers: analyzing mass spectrometry data with the new ClinProTools software. *Biotechniques* 2005;38:37–40.
93. Khot PD, Fisher MA. Novel approach for differentiating *Shigella* species and *Escherichia coli* by matrix-assisted laser desorption ionization–time of flight mass spectrometry. *J Clin Microbiol* 2013;51:3711–6.
94. Li B, Guo T, Qu F, et al. Matrix-assisted laser desorption ionization: time of flight mass spectrometry-identified models for detection of ES BL-producing bacterial strains. *Med Sci Monit Basic Res* 2014;20:176–83.
95. Xiao D, Zhao F, Zhang H, et al. Novel strategy for typing *Mycoplasma pneumoniae* isolates by use of matrix-assisted laser desorption ionization–time of flight mass spectrometry coupled with ClinProtools. *J Clin Microbiol* 2014;52:3038–43.
96. Gekenidis M-T, Studer P, Wüthrich S, et al. Beyond the matrix-assisted laser desorption ionization (MALDI) biotyping workflow: in search of microorganism-specific tryptic peptides enabling discrimination of subspecies. *Appl Environ Microbiol* 2014;80:4234–41.
97. Anderson NW, Buchan BW, Riebe KM, et al. Effects of solid-medium type on routine identification of bacterial isolates by use of matrix-assisted laser desorption ionization–time of flight mass spectrometry. *J Clin Microbiol* 2012;50:1008–13.
98. Ford BA, Burnham C-AD. Optimization of routine identification of clinically relevant gram-negative bacteria by use of matrix-assisted laser desorption ionization–time of flight mass spectrometry and the bruker biotyper. *J Clin Microbiol* 2013;51:1412–20.
99. Peel TN, Cole NC, Dylla BL, et al. Matrix-assisted laser desorption ionization time of flight mass spectrometry and diagnostic testing for prosthetic joint infection in the clinical microbiology laboratory. *Diagn Microbiol Infect Dis* 2015;81:163–8.
100. Perez KK, Olsen RJ, Musick WL, et al. Integrating rapid diagnostics and antimicrobial stewardship improves outcomes in patients with antibiotic-resistant gram-negative bacteremia. *J Infect* 2014;69:216–25.
101. Mutters NT, Hodiamont CJ, de Jong MD, et al. Performance of Kiestra total laboratory automation combined with MS in clinical microbiology practice. *Ann Lab Med* 2014;34:111–7.
102. Razavi M, Johnson LDS, Lum JJ, et al. Quantification of a proteotypic peptide from protein c inhibitor by liquid chromatography–free SISCAPA-MALDI mass spectrometry: application to identification of recurrence of prostate cancer. *Clin Chem* 2013;59:1514–22.
103. Yamashita M, Fenn JB. Electrospray ion source: another variation on the free-jet theme. *J Phys Chem* 1984;88:4451–9.
104. Cho A, Normile D. Mastering macromolecules. *Science* 2002;298:527–8.
105. Ho CS, Lam CWK, Chan MHM, et al. Electrospray ionisation mass spectrometry: principles and clinical applications. *Clin Biochem Rev* 2003;24:3–12.
106. Ecker DJ, Sampath R, Massire C, et al. Ibis T5000: a universal biosensor approach for microbiology. *Nat Rev Micro* 2008;6:553–8.
107. Ecker DJ, Sampath R, Li H, et al. New technology for rapid molecular diagnosis of bloodstream infections. *Exp Rev Mol Diagn* 2010;10:399–415.
108. Ecker DJ, Drader JJ, Gutierrez J, et al. The Ibis T5000 universal biosensor: an automated platform for pathogen identification and strain typing. *JALA* 2006;11:341–51.
109. Hofstadler SA, Sampath R, Blyn LB, et al. Tiger: the universal biosensor. *Int J Mass Spectrom* 2005;242:23–41.
110. Romero R, Miranda J, Chaiworapongsa T, et al. A novel molecular microbiologic technique for the rapid diagnosis of microbial invasion of the amniotic cavity and intra-amniotic infection in preterm labor with intact membranes. *Am J Reprod Immunol* 2014;71:330–58.
111. Mengelle C, Mansuy J-M, Da Silva I, et al. Evaluation of a polymerase chain reaction–electrospray ionization time-of-flight mass spectrometry for the detection and subtyping of influenza viruses in respiratory specimens. *J Clin Virol* 2013;57:222–6.
112. Farrell JJ, Sampath R, Ecker DJ, et al. "Salvage microbiology": detection of bacteria directly from clinical specimens following initiation of antimicrobial treatment. *PLoS ONE* 2013;8:e66349.
113. Jacovides CL, Kreft R, Adeli B, et al. Successful identification of pathogens by polymerase chain reaction (PCR)-based electron spray ionization time-of-flight mass spectrometry (ESi-TOF-MS) in culture-negative periprosthetic joint infection. *J Bone Joint Surg Am* 2012;94:2247–54.
114. Chen K-F, Rothman RE, Ramachandran P, et al. Rapid identification viruses from nasal pharyngeal aspirates in acute viral respiratory infections by RT-PCR and electrospray ionization mass spectrometry. *J Virol Methods* 2011;173:60–6.
115. Kaleta EJ, Clark AE, Cherkaoui A, et al. Comparative analysis of PCR–electrospray ionization/mass spectrometry (MS) and MALDI-TOF/MS for the identification of bacteria and yeast from positive blood culture bottles. *Clin Chem* 2011;57:1057–67.
116. Kaleta EJ, Clark AE, Johnson DR, et al. Use of PCR coupled with electrospray ionization mass spectrometry for rapid identification of bacterial and yeast bloodstream pathogens from blood culture bottles. *J Clin Microbiol* 2011;49:345–53.
117. Wolk DM, Kaleta EJ, Wysocki VH. PCR–electrospray ionization mass spectrometry: the potential to change infectious disease diagnostics in clinical and public health laboratories. *J Mol Diagn* 2012;14:295–304.
118. Greenwood-Quaintance KE, Uhl JR, Hanssen AD, et al. Diagnosis of prosthetic joint infection by use of PCR-electrospray ionization mass spectrometry. *J Clin Microbiol* 2014;52:642–9.
119. Melendez DP, Uhl JR, Greenwood-Quaintance KE, et al. Detection of prosthetic joint infection by use of PCR-electrospray ionization mass spectrometry applied to synovial fluid. *J Clin Microbiol* 2014;52:2202–5.

120. Bacconi A, Richmond GS, Baroldi MA, et al. Improved sensitivity for molecular detection of bacterial and *Candida* infections in blood. *J Clin Microbiol* 2014;**52**:3164−74.
121. Brouqui P, Raoult D. Endocarditis due to rare and fastidious bacteria. *Clin Microbiol Rev* 2001;**14**:177−207.
122. Brinkman CL, Vergidis P, Uhl JR, et al. PCR-electrospray ionization mass spectrometry for direct detection of pathogens and antimicrobial resistance from heart valves in patients with infective endocarditis. *J Clin Microbiol* 2013;**51**:2040−6.
123. Srinivasan M, Sedmak D, Jewell S. Effect of fixatives and tissue processing on the content and integrity of nucleic acids. *Am J Pathol* 2002;**161**:1961−71.
124. Miething F, Hering S, Hanschke B, et al. Effect of fixation to the degradation of nuclear and mitochondrial DNA in different tissues. *J Histochem Cytochem* 2006;**54**:371−4.
125. Wallet F, Herwegh S, Decoene C, et al. PCR−electrospray ionization time-of-flight mass spectrometry: a new tool for the diagnosis of infective endocarditis from heart valves. *Diagn Microbiol Infect Dis* 2013;**76**:125−8.
126. Simner PJ, Uhl JR, Hall L, et al. Broad-range direct detection and identification of fungi by use of the PLEX-ID PCR-electrospray ionization mass spectrometry (ESI-MS) system. *J Clin Microbiol* 2013;**51**:1699−706.
127. Shin JH, Ranken R, Sefers SE, et al. Detection, identification, and distribution of fungi in bronchoalveolar lavage specimens by use of multilocus PCR coupled with electrospray ionization/mass spectrometry. *J Clin Microbiol* 2013;**51**:136−41.
128. Huttner A, Emonet S, Harbarth S, et al. Polymerase-chain reaction/electrospray ionization-mass spectrometry for the detection of bacteria and fungi in bronchoalveolar lavage fluids: a prospective observational study. *Clin Microbiol Infect* 2014;**20**:O1059−66.
129. Simner PJ, Buckwalter SP, Uhl JR, et al. Identification of *Mycobacterium* species and *Mycobacterium tuberculosis* complex resistance determinants by use of PCR-electrospray ionization mass spectrometry. *J Clin Microbiol* 2013;**51**: 3492−8.
130. LeGoff J, Feghoul L, Mercier-Delarue S, et al. Broad-range PCR-electrospray ionization mass spectrometry for detection and typing of adenovirus and other opportunistic viruses in stem cell transplant patients. *J Clin Microbiol* 2013;**51**: 4186−92.
131. A Murillo L, Hardick J, Jeng K, et al. Evaluation of the pan influenza detection kit utilizing the PLEX-ID and influenza samples from the 2011 respiratory season. *J Virol Methods* 2013;**193**:173−6.

5

Development and Validation of Small Molecule Analytes by Liquid Chromatography-Tandem Mass Spectrometry

Russell P. Grant and Brian A. Rappold

ABSTRACT

Background
The application of liquid chromatography coupled to tandem mass spectrometry (LC-MS/MS) represents one of the most compelling opportunities for advancements in human health through the combination of reference measurement procedure capabilities, broad chemical coverage, and a rich history in support of drug development from the 1990s onward. Clinical application of these technologies has begun to gather pace in many laboratories, with diverse applications ranging from expanded newborn screening to identification of emerging toxicants. The promise of these technologies is vast and the need is palpable; however, the journey can be exacting. Perhaps somewhat unique among analytical techniques, LC-MS/MS assay development and validation requires significant knowledge of a number of specialties: a mastery of chemistry (sample preparation, chromatography, ionization), physics (ion manipulation), engineering principles (automation, order of experiments, programming), and mathematics (data reduction and interpretation) applied to questions of a biological origin (normal, disease, metabolism).

Content
This chapter provides a stepwise roadmap for systematically developing and validating an LC-MS/MS assay for small molecule analytes. Starting from first principles (salt correction in gravimetric weighing), each component of the LC-MS/MS assay (mass spectrometer tuning, ionization enhancements, chromatography, extraction) is detailed with best-practice experiments for development and data reduction techniques to fulfill performance goals. After refinement of each component of the assay, prevalidation experiments are described to enable efficient execution of validation. Finally, an array of validation guidance documents is reduced to a coherent process for burden of proof.

The analysis of small molecule analytes (atomic mass <1000 amu) using mass spectrometric techniques represents the oldest and most diverse application of the technology within the clinical field. Forensic and clinical toxicology assays were published and deployed in the 1970s using gas chromatography–mass spectrometry (GC-MS)[1,2]; notably, many of the same analytes are still measured by GC-MS today.

The development of atmospheric pressure ionization interfaces (see Chapter 2) enabled the genesis of metabolomics workflows and the application of newborn screening using blood spots, which represents the next real explosion in clinical chemistry from the early 1990s.[3] Continuing advances in mass spectrometer sensitivity enabled the application of liquid chromatography–tandem mass spectrometry (LC-MS/MS) for targeted metabolomics such as the measurement of steroid hormones and biogenic amines in the early 2000s.[4-6] During the same time frame, therapeutic drug monitoring applications became de rigeur, with analysis of immunosuppressants driving uptake of the technology in the hospital setting to enable preoperative and postoperative or treatment serial monitoring.[7]

Advancements in high-performance liquid chromatography (HPLC) hardware (late 2000s) and mandates for measurement of drugs of abuse has led to highly multiplexed measurement of drug classes such as benzodiazepines and opiates within a single injection from a common sample preparation protocol in urine.[8,9] Combinations of HPLC, GC, and ionization mode, often with accurate mass time-of-flight technology, are currently being extensively used in untargeted metabolomics studies to generate new combinations of metabolites that may uncover disease phenotypes with the promise of precision medicine.[10-12] This avenue of potential new clinical assays bodes well for the advancement of MS technologies in small molecule analysis, and the industry is poised to translate these into clinical practice. Finally, the real promise of MS technologies in providing confidence in measurement is being realized through a number of international standardization and harmonization efforts involving the generation of certified reference materials and reference method procedures.[13,14]

The use of isotope dilution and external calibration curves enables exceedingly low error and imprecision in the accurate measurement of analytes within complex mixtures and provides an opportunity for cross-platform harmonization and improved patient care.[15]

As noted earlier, many potential technologies can be applied to small molecule analysis in clinical practice; however, the most contemporary application is highlighted in Fig. 5.1 and will be the predominant foundation for discussion within this chapter. The goals of this chapter are to provide guidance for systematically developing and validating an LC-MS/MS

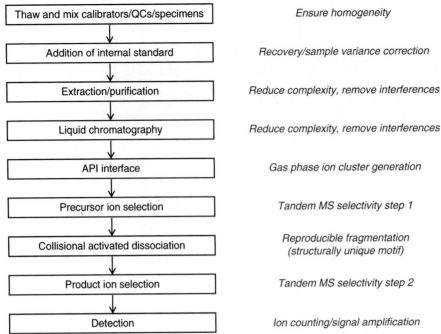

FIGURE 5.1 The quantitative liquid chromatography–tandem mass spectrometry (MS) experiment. The individual components of the method are described, together with the function of each step within the experiment. *API,* Atmospheric pressure ionization; *QCs,* quality controls.

assay from first principles to enable clinically relevant measurement in the support of patient care. There have been a number of well-informed examples of when MS assays fail.[16] The tools and considerations for determining and addressing such challenges are covered in this chapter. We have taken the liberty of describing, to our knowledge, the best practices extant in the literature and from our own experiences to underpin these processes and concepts here.

The LC-MS/MS development process is often a very personal journey with bias toward certain known and mastered technologies. In our experience, it is often beneficial to allow the analytes to dictate their own optimal conditions, determined through highly parallel experiments in selection of mobile phase, stationary phase, column dimensions, and sample preparation. For completely new analytes or classes, it is recommended that development flows from the mass spectrometer back toward the sample preparation in a systematic manner to evaluate optimal conditions and a compendium of solutions as go-to next steps in the event of developmental failure. Burden of proof mandates a thorough evaluation of method performance through prescriptive validation experiments with predefined acceptance criteria. Currently there is limited guidance that describes in detail the "what to do and how to do it" for clinical assay validation.[17,18] This chapter synthesizes a number of validation guidance documents into a coherent process for consideration. Notably, these processes are adhered to for each assay in our laboratories to provide high-quality results.

PRACTICAL CONSIDERATIONS FOR LIQUID CHROMATOGRAPHY—TANDEM MASS SPECTROMETRY ASSAYS

Calibration Standard and Matrix

The foundation of LC-MS/MS assays relies on well-characterized standard materials for generation of a calibration curve of a known analyte quantity. The definition of well-characterized is as follows: the structure and molecular composition of a standard material should be clearly listed, with any salt of crystallization described in detail. When dissolved into solution, the standard will dissociate into the free acid-base form; thus the salt content must be accounted for in concentration assignment. Further, the composition of all other impurities must be considered and subtracted from the final assignment of stock materials concentration.

Practically, the numbers of reference standards are limited relative to the clinical measurement landscape. Thus particular attention should be paid to the certificate of analysis provided with the standard materials. At a minimum, additional details, including the determination of purity of analysis by HPLC, should be known and requested as a chromatogram, together with water content through Karl-Fischer analysis. In the absence of these materials, it is recommended that two individual sources of standard materials be acquired for developmental and validation efforts. Note, determination that the alternative source of standard is of a different provenance as many alternative vendors relabel standards from a common preparation source is critical. Finally, storage conditions should be clearly listed (and followed) in standard operating procedures (SOPs), together with expiry date (proved and adhered to) under that condition. Reassessment of standard materials purity is recommended where materials are repeatedly accessed, because of the hygroscopic nature of most standards, unless particular care is taken to control for this eventuality (ie, storage under argon).

These standard materials are added using gravimetric techniques[19] to an analyte-free matrix representative of the specimen type being assayed. Best manufacturing practice of calibrators requires the overall organic composition to be less than 5% by volume.[20] Further, the hierarchy of calibration matrices[21] clearly stipulates that the calibration matrix should be identical to the specimen being measured; one could argue that this is true even to the choice of counter-ion (Li vs. Na heparin). Superficially this would appear to be facile, certainly for exogenous drug assays in blood-based matrices; however, two particular types of clinical assay are confounding. Firstly, endogenous analyte assays (eg, steroid hormones) cannot be prepared in a true blank matrix. Although various quantitative approaches may be used in small molecule analysis, such as standard addition or internal standard (IS) signal matching,[22] more commonly a surrogate matrix is used. Commonly, this matrix is charcoal stripped or manufactured from standard chemicals to be physiologically similar to the patient sample (isotonic and containing protein for cerebrospinal fluid [CSF] or synthetic urine). Second, the current trend toward high-density LC-MS/MS assays for drugs of abuse spanning several orders of magnitude in measurement interval (analytical measurement range [AMR]) precludes the ability to combine stocks without failing the aforementioned 5% solvent rule. Thus solvent-based calibration systems are often used, although solvent-based calibration is the lowest desirable calibrator matrix defined in guidance.[21] As will be described later, defined validation experiments demonstrating the comparable nature of calibrators and patient specimens is required for confidence in measurement, particularly in these situations.

Internal Standard Selection and Utility

The "killer-app" of LC-MS/MS assays is undoubtedly the intrinsic ability to control measurement error and imprecision through the use of isotopically labeled ISs. Optimal ISs are a stable (nonradioactive) labeled version of the analytes and thus a physicochemical mimic, usually enriched with deuterons ^{2}H, nitrogen (^{15}N), or carbon ^{13}C. Alternatively, an analog IS structurally dissimilar to the analyte of interest may be used. For quantitative diagnostic applications of MS, analog ISs are strongly discouraged. The selection of an IS is often constrained by availability and cost; however, if at all possible, a stable labeled version of the analytes should be used in clinical assays, with preference given to ^{13}C- and ^{15}N-labeled forms.

The nature of enrichment should follow four basic rules (Table 5.1); first, ensure incorporation of at least three heavy atoms, unless the analyte contains chlorine or bromine. Ideal ISs for these analytes contain ^{13}C$_5$ or ^{2}H$_5$ as a minimum as the +2 isotopes of chlorine and bromine in combination with naturally occurring ^{13}C in the analyte may contribute to the measurement of the IS. Second, the IS should not contain excess deuterons (more than ^{2}H$_6$, often generated from historical GC-MS workflows) because this leads to chromatographic resolution of analyte and IS. Use of these ISs requires additional sample preparation constraints and validation studies to confirm that the IS controls for matrix effects.[23] Third, the IS material should not contain excessive unlabeled analyte; this reduces the amount of IS that can be added without contribution to the analyte, leading to assay imprecision and inaccuracy. Finally, interrogation of the position of deuterons in the IS is necessary. De-deuteration can occur in sample preparation (acid hydrolysis, solid-phase extraction [SPE]) and the LC-MS interface, resulting in excessive variability in measurement.[24]

Practical use of an IS is both quantitative (isotope dilution) and qualitative (recovery, matrix effects, chromatographic variance, and mass spectrometer sensitivity drift). For quantitative use of the IS (see Fig. 5.1, step 2), the IS is added to all samples except double-blank samples (no analyte or IS) at the same concentration and at a reproducible volume greater

TABLE 5.1 Internal Standard Considerations

Isotopic Internal Standard Labeling Guidelines	Reasoning
Minimum of +3 (^{2}H$_3$, ^{15}N$_3$, ^{13}C$_3$)	Sufficient mass resolution from naturally occurring ^{13}C isotopes of the analyte (requires greater labels if analyte has Cl$_{35}$/Cl$_{37}$ or Br$_{79}$/Br$_{81}$)
If deuterium is used, less than ^{2}H$_6$	Prevent chromatographic dissimilarities between the analyte and internal standard caused by the deuterium isotope effect
Internal standard labeling purity ensures no observable analyte	Contribution to the analyte shifts the intercept of the calibration curve, obfuscating low-level measurement
If deuterium is used, locations of the ^{2}Hs should be at least β to active moieties (COOH, NH$_3$, etc.)	In solution or in-source hydrogen-deuterium exchange can vary between samples, as well as provide false-positive response for the analyte

than 25 µL unless high-quality robotics are used. The IS should account for extraction, injection, and ionization variance throughout the LC-MS/MS experiment (analog IS or chemically similar/nonisotopically labeled compounds used as the IS usually fail these criteria) and must be mass resolved as a discrete transition.

The calibration curve is generated by determining the concentration ratio of analyte to IS on the x-axis, albeit the IS value is usually set to 1, so the x-axis describes the concentration of analyte. Notably, only nonzero calibrators are used. The variable y-axis is defined as the integrated peak area ratio of analyte divided by IS. Assignment of unknown samples is thus extrapolated by measuring the peak area ratio (y) of analyte to IS and solving for the analyte concentration using a generic calibration curve equation ($x = y$/slope − intercept).

Although often underappreciated, the qualitative use of ISs is highly informative. Analysis of IS peak area trends across a batch is used to monitor underrecovery, pipetting errors in preparation, or matrix effects (IS response between 50% and 150% of calibrator/quality control [QC] mean, predicating a decision for repeat sample analysis) or determining deleterious instrument drift resulting in corrective action (instrument clean). An appropriate IS may be used as a surrogate for the analyte in recovery studies in sample preparation development or, perhaps more importantly, to define the absolute retention properties of the analyte, informing the data reviewer of analyte peak shape and retention time.

When considering the nature and number of ISs to use in a high-density multianalyte LC-MS/MS experiment (>15 compounds), the answer is rather trite—as many as possible and isotopically labeled appropriately. In appraising the value of the IS and the almost unique ability of MS to incorporate external calibration and internal standardization, one should aspire to leverage these strengths in LC-MS/MS assays deployed in the care of patients.

Mass Spectrometry Source Setup and Tuning

The first step in the practical development process involves interrogation of the analyte of measure, specifically the potential to ionize functional groups in solution. Consideration should be given to the upstream process constraints that the interface modality engenders (Table 5.2; Chapter 2). Analytes containing amines or amide bond(s) will form positive pseudomolecular protonated ions (biogenic amines), whereas analytes containing carboxylic acids (eg, free fatty acids) or an electronegative ion (eg, thyroxine and carboxylic acid moieties) will form negative pseudomolecular deprotonated ions. Analytes containing these motifs are candidates for electrospray ionization (ESI). Analytes that are neutral in solution require the inclusion of adduct ions for electrospray[25] or require gas phase ionization,[26] as provided by atmospheric pressure chemical ionization (APCI) interfaces. Small molecules almost exclusively form singly charged ions; thus knowledge of the molecular weight of the analyte and ionization mode provides a good starting point to predict the expected precursor ion mass-to-charge ratio (m/z).

Two instrument configurations are shown in Fig. 5.2, postcolumn infusion (PCI; A) and flow injection analysis (FIA; B). The PCI setup will be used for other developmental efforts, most notably matrix effect determination,[27,28] and is a valuable technique for any method development process. The PCI setup involves delivery of a mobile phase to a mixing tee (T), where a concentrated solution of analyte(s) at approximately 1 μg/mL in mobile phase (or methanol) is added at 5 to 10 μL/min.

The mixed eluent flow is introduced into the interface for rapid determination of ionization mode (ESI/APCI, positive or negative ion mode) and source parameters under conditions of solvent load. Initial determination of precursor ion(s) may be performed without the makeup flow from the chromatographic pumps for ESI to reduce chemical noise. ACPI requires solvent flow (>200 μL/min) to stabilize discharge current (Chapter 2 and Table 5.2). As the analyte is predetermined, the expected molecular ion's m/z ratio can be quickly calculated. Pseudomolecular ions will have an expected m/z of the analytes of molecular weight +1 for positive ionization mode and molecular weight −1 for negative ion mode.

The FIA setup differs by the removal of the HPLC column and T, incorporating injection of analyte solution prepared in a mobile phase to enable signal to noise (S/N)-based optimization or to remove the influence of solvent ions from the determination of precursor ions. PCI evaluation using ESI in positive ion mode for $^2H_4$-hydroxymidazolam is shown in Fig. 5.3; notably, the proposed precursor ion ($m/z = 346$) is difficult to discern in this example. Modification of the default interface parameters (heat, orifice voltage, gas flow, etc.) is often required to generate appropriate selection of the precursor ion. Alternatively, shutting off the HPLC solvent flow (infusion pump only) or performing a secondary infusion of solvent only may enable precursor ions to be identified and differentiated from solvent ions by comparison of mass spectra.

Interface ion chemistry may be confounding (Table 5.3). Scanning the first quadrupole at ±80 amu of the expected intact precursor ion ($[M+H]^+$ or $[M-H]^-$ for positive or negative ionization, respectively) to observe the formation of adducts (ie, ammonium for immunosuppressants) or thermal losses (water losses for steroids) is recommended, as is scanning slowly (1 second per 100 amu scan range). As a last resort, increase the concentration of the infusion solution or, alternatively, use FIA, which facilitates background subtraction of interfering ions, a functionality for all MS software systems, in order to ascertain the appropriate precursor ion characteristics. The influence of interface temperature on ion population can be performed in PCI mode, allowing 1 minute of equilibration for each interface temperature step of 100°C before determining precursor ions formed.

After confirmation of precursor ion(s), determination of product ions from each individual precursor is performed in PCI mode unless confounded by solvent ions (FIA mode). A precursor ion is selected to traverse the first quadrupole and accelerated into a collision gas in the second quadrupole with a collision energy (CE) to induce collisional activated dissociation, generating product ions. The product ion population is then determined by scanning across a mass range with the third quadrupole. As noted earlier, scan is performed slowly from m/z of 50 to precursor + 50 amu (1 s/100 amu) and the CE and sum spectra are stepped or, preferably, the CE is ramped from 0 to 100 eV, noting the product ions formed as the CE changes. Fig. 5.4 shows product ions for $^2H_4$-hydroxymidazolam at 20, 40, and 60 eV,

TABLE 5.2 Considerations for Liquid Chromatography–Tandem Mass Spectrometry Interface Selection Between the Two Most Common Interfaces*

Differentiator	ESI	APCI
Charged analytes	Good	Average
Neutral analytes	Poor (adduct)	Good
Volatile analytes	No	Yes
Thermally labile analytes	Good	Poor
Mass range (amu)	Broad	<2000
Dynamic range (calibrations)	Good	Better
Buffer concentration	Poor	Better
LC gradient	Response affected	Less affected
Smaller ID columns	Good (concentration dependent)	Average (mass sensitive)
Rugged	Good	Better
LC flow (mL/min)	0.00001–2	0.1–2
Matrix effects	Poor	Better
Utility	**ESI**	**APCI**
Eluent pH	Charged analyte in solution	Less affected
Solvent	Greater organic desired	Less affected
Smaller ID column	Yes	≥2.1 mm
Flow rates (2.1 mm ID)	High as possible	High as possible
Flow rates (<1 mm ID)	Low/flow split	>200 μL/min
Buffer concentration	<10 mM	<50 mM
Use divert valve	Yes	Yes

APCI, Atmospheric pressure chemical ionization; *ESI*, electrospray ionization; *ID*, internal diameter of the LC column; *LC*, liquid chromatography.

*Considerations take into account molecular features and hardware limitations.

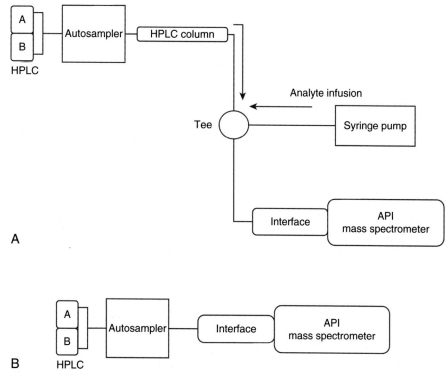

FIGURE 5.2 Instrument configuration for postcolumn infusion *(PCI)* **(A)** and flow injection analysis **(B)**. Postcolumn infusion is used for both evaluation of mobile phase-based ionization efficiency as well as determination of regions of ion suppression. Note that the flow-injection setup does not have a column between the autosampler and the mass spectrometry interface. *API,* Atmospheric pressure ionization; *HPLC,* high-performance liquid chromatography.

together with an extracted ion chromatogram for four precursor-to-product ion pairs (transitions).

Two things should be considered when selecting transitions. First, facile neutral losses of water, ammonia, and carbon dioxide (see Fig. 5.4A, low CE) should be avoided if at all possible. Transitions incorporating these losses are ubiquitous and thus not selective. Using these losses places a significant burden on chromatography and sample preparation to ensure measurement specificity of the LC-MS/MS method. Secondly, low-mass product ions generated at high CE (see Fig. 5.4C, high CE >60 eV) retain limited structural content relative to the precursor ion and may have limited structural specificity (putative product ions of different analytes). Product ions incorporating 30% to 70% of the precursor ion mass are desirable to balance these two factors if observed (see Fig. 5.4B).

Performing product ion CE ramp scans to generate a total ion current (TIC) plot and extracting product ions facilitates the generation of CE versus yield plots (CE voltage vs. intensity as counts per scan; see Fig. 5.4D). Optimal CEs are readily interpolated. In the absence of multiple product ions, maintaining the same neutral loss but with a discrete CE is recommended for transition ratio analysis. The CE offset should be sufficient to generate alternative transitions at 50 to 75% yield (ideally $n = 4$ transitions). A number of clinically relevant analyte measures follow this concept, notably tramadol and cortisol,[29] by offsetting the precursor or product ions by 0.01 to 0.1 amu (to generate a distinct transition) and selecting a CE for a readily observable difference. In the example shown, a second transition for $^2H_4$-hydroxymidazolam at m/z 346.1 to 203.1 and CE approximately 25 eV would yield a 1:2 response ratio versus optimal CE ($\sim$36 eV). Initially, more than four transitions per analyte should be taken forward to further development with the final goal of two or more transitions per analyte in clinical sample analysis (transition ratio monitoring) to enable purity assignment and therefore confidence in the quality of reported results.

Analyzing as many transitions as possible is a good practice; redundant or unacceptable transitions will be systematically removed during HPLC and sample preparation development. Finally, the product ions generated for the unlabeled analyte ideally should be structurally identical to those selected for the matched stable isotope-labeled IS with the same CE voltage for each transition. As long as the precursor ion is differentiated between analyte and labeled IS, product ions can be identical because most modern mass spectrometers have eradicated cross-talk (the detection of precursor ions retained in the collision cell from a previously analyzed molecule, signal from analyte in the IS transition, or vice versa). An example would be the measurement of testosterone and its $^3C_{13}$ IS. The position of labeling of C_{13} atoms is on the A ring and thus is retained in the predominant product ions formed.[30,31] Selected transitions are generally m/z 289 to 97 or 289 to 109 for testosterone and m/z 292 to 100 or 292 to 112 for the IS with CEs of approximately 35 and approximately 38 eV for each matched pair of transitions.

An initial determination of acceptable interface conditions should be briefly determined. This can be performed in either PCI mode or by FIA. Fig. 5.5A shows the modification of declustering potential, gas flows, and ion spray voltages modified during PCI. Alternatively, methods can be built and the provisional source settings assessed by FIA (see Fig. 5.5B). These are not final settings; without a sufficient chromatographic separation, noise or artificial signal cannot be discriminated from the signal generated solely by the analyte of interest.

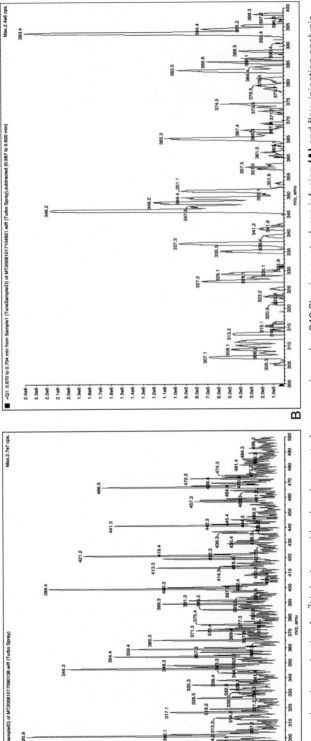

FIGURE 5.3 Precursor ion determination for $^2H_4$-hydroxymidazolam (protonated precursor ion $m/z = 346.2$) using postcolumn infusion **(A)** and flow injection analysis (FIA) **(B)** with background subtraction of FIA peaks. The relative cleanliness of the FIA spectra allows for review of ions created by the analyte exclusive of chemical noise present in the mass spectrometer's source.

TABLE 5.3 Development Phase 1: Preliminary Mass Spectrometry Conditions

Experiment	Outcome
Precursor ion screening, $m/z \pm 80$ amu from expected	Determination of all possible precursors, including adducts and in-source decompositions
Product ion screening, m/z 50 − precursor ion + 50	Determination of all possible product ions with optimal collisionally activated dissociation energies. In the absence of multiple possible transitions, determination of CE offset for transition ratio analysis
PCI infusion or flow-injection analysis for optimal mobile phase determination	One to four mobile phases that exhibit the highest sensitivity for the ion generation
PCI infusion or flow-injection analysis for provisional source conditions	Initial values for source temperatures, declustering potential, gas flow rates, etc.

CE, Collision energy; *PCI*, postcolumn infusion.

Solvent Chemistry Optimization

After selection of preliminary interface conditions and relevant transitions, the next step in the development process involves determination of the appropriate volatile solvent chemistry to introduce the analytes into the interface.[32] Using a four-solvent (quaternary) pump and the PCI modality from Fig. 5.2, a programmed solvent sequence can be used to modify solvent chemistries and evaluate their influence on sensitivity in selected reaction monitoring mode (to reduce the observed influence of chemical noise). Fig. 5.6A demonstrates simple programming of solvent changes involving acetonitrile, water, methanol, and 100 mmol/L ammonium formate for the measurement of reverse triiodothyronine (rT$_3$) in negative ion ESI mode. Simple observations of response changes enable selection of potential HPLC solvent chemistries. In the example shown, rT$_3$ demonstrates improved response for methanol over acetonitrile and either no or minimal counter ion (formate). Any variety of solvents, buffers, or other mobile phase additives can be used in this screening mode.

Alternatively, Fig. 5.6B and C demonstrate FIA-based injections of atenolol in positive ESI mode whereby atenolol is spiked at the same amount into individual wells containing mixtures of solvents, buffers, acids, or bases, with the same final liquid volume in each independent well and thus the same concentration of atenolol in each well.[33] Each well contains a unique combination of solvent chemistries and the influence of solvent chemistry on ionization is determined as the peak height of each injection (there is no column and thus limited mixing). As shown in Fig. 5.6, a 10-fold difference in response is observed when atenolol is injected in methanol and trifluoroacetic acid versus acetonitrile and formic acid when both solvents have the same composition of ammonium formate. This experimental setup enables faster screening of a broader array of potential solvent chemistries than repeatedly changing and priming new solvents in LC-PCI mode. In this manner, 96 individual samples are prepared and assayed and the data are reduced in approximately 3 hours in the previous reference. Finally, assessment of optimal chemistries in FIA mode can be used to determine which analyte has the lowest response (ionization and MS/MS transmission efficiency) relative to the desired lower limit of measurement interval (LLMI, otherwise called the lower limit of quantification [LLOQ]). When developing methods with multiple analytes, it is best to focus on optimizing sensitivity for the least sensitive analyte in terms of on-column detection requirements (analytical sensitivity) and circulating levels required to measure for clinical care, if known.

Liquid Chromatography Development

The true complexity of clinical sample analysis with LC-MS/MS technologies arises in the interplay between selectivity and sensitivity. The next challenge is to evaluate and optimize the chemical purification methods of HPLC and sample preparation independently and then in combination, together with further streamlining of MS/MS transitions for validation. The scale and combinations of technologies that could be used is staggering; however, the techniques to optimize are predictable.

There is no singular aspect of the LC-MS/MS method that has more myth and mystery associated with it than the HPLC step (Table 5.4). Certain myths are that HPLC separations are robust (HPLC systems leak, usually just after you have left the laboratory), columns should last longer than 1000 injections (after 500 the column has paid its way, it is a consumable, albeit an expensive one), smaller particles lead to better efficiency (they also lead to higher operating pressure), isocratic separations are faster (theoretically, although at the possible costs of retention time variations and strongly bound materials eluting a number of samples later, inducing false-positive results or ion suppression). Chapter 1 incorporates significant detail on the principles of HPLC; the practicality of development is covered here.

The first step in the HPLC experiment is interrogation of the molecular features of the analytes. Table 5.5 provides details for consideration of the modality of HPLC separation based on analyte polarity (charge) and knowledge from the previous solvent chemistry screen. Many highly charged small molecule analytes are amenable to reverse-phase HPLC (RPLC), the separation technique that is most understood with the broadest range of commercially available stationary phase chemistries. The inclusion of ion pairing reagents to effect retention control[34] or the retentive capacity (k′) of the earliest eluting analytes can be helpful for the most polar of small molecules chromatographed in reverse-phase mode. Although significantly less understood, hydrophilic interaction liquid chromatography (HILIC) is an excellent separation modality for clinically relevant metabolites, and a paper by the authors provides further guidance on the empirical nature of the HILIC development process.[35] For analytes for which the optimal solvent for ionization is acetonitrile and the analytes are highly polar (amino acids, nicotine and its metabolite, cotinine, etc.), an exploration of HILIC may be desirable. For the sake of brevity, this chapter shall focus on RPLC developmental processes and its functional use, although many experimental considerations hold true for HILIC.

The primary goals for HPLC separation are to resolve analytes from interferences and ion suppression–inducing compounds (simplifying the composition introduced into

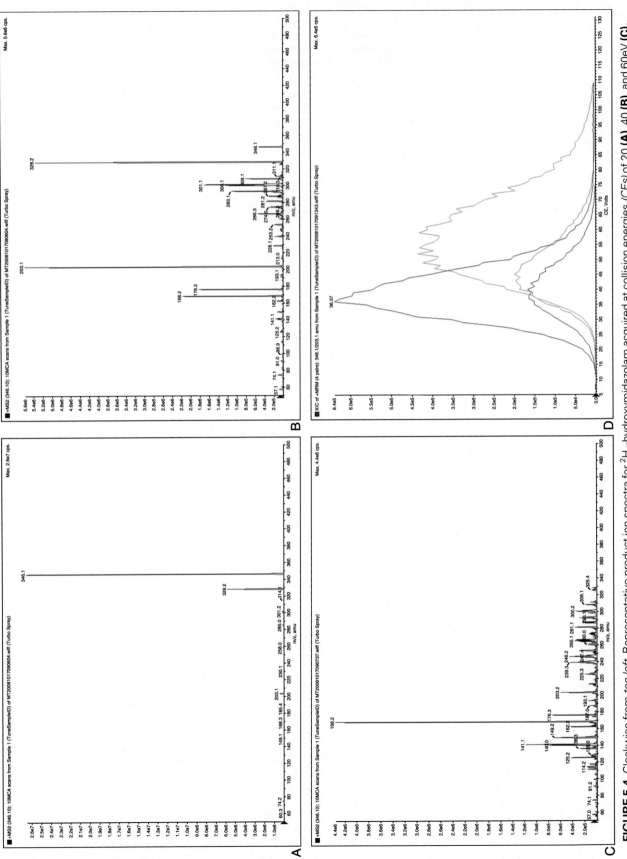

FIGURE 5.4 Clockwise from *top left*. Representative product ion spectra for $^2H_4$-hydroxymidazolam acquired at collision energies *(CEs)* of 20 **(A)**, 40 **(B)**, and 60eV **(C)**, together with extracted ion chromatograms **(D)** displaying optimal *CEs* for selected transitions *346.2 → 301.1, 346.2 → 203.1, 346.2 → 176.2, and 346.2 → 268.2.*

CHAPTER 5 Development and Validation of Small Molecule Analytes

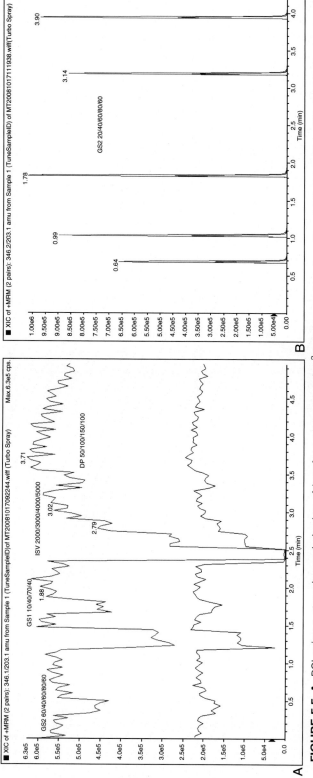

FIGURE 5.5 A, PCI using a stepwise optimization of interface parameters for $^2H_4$-hydroxymidazolam. Desolvation gas flow *(GS 1)*, electrospray voltage *(ISV)*, and declustering potential *(DP,* orifice voltage). This experiment is performed with the user manually adjusting values in the acquisition software during infusion with real-time data review. **B,** Flow injection analysis (FIA) using a stepwise optimization of desolvation gas *(GS 2)*. FIA experiments can be preprogrammed for automated analysis and postanalysis data review.

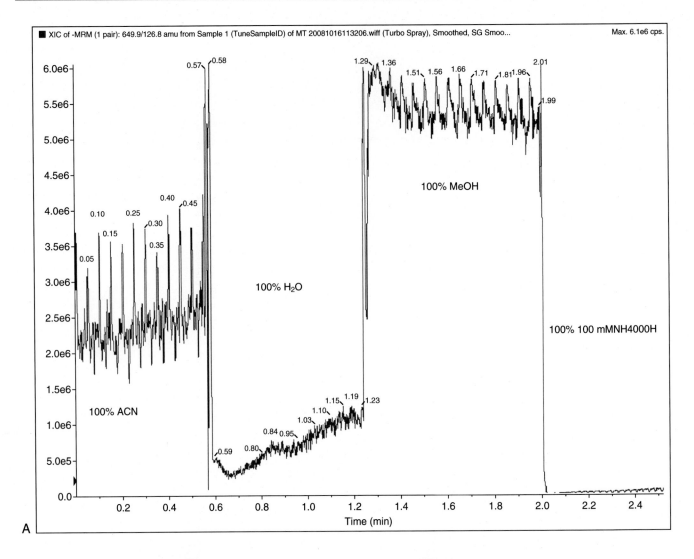

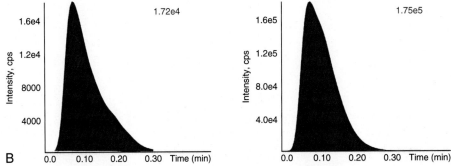

FIGURE 5.6 A, Liquid chromatography–postcolumn infusion (LC-PCI) solvent screening with a quaternary pump for reverse triiodothyronine. The noise in the data, particularly the 100% acetonitrile and 100% methanol steps is caused by the flow variance of the LC pump and the infusion pump. Flow infusion analysis–based injections of in-well solvent chemistries for atenolol in CH_3CN:1 mmol/L NH_4COO + 0.1% HCOOH (**B,** *Left*) and CH_3COOH:1 mmol/L NH_4COO + 0.1% CF_3COOH (**B,** *Right*).

the interface) and facilitate reproducible retention properties (retention time, baseline noise, and peak shape) among calibrators, QCs, and specimens. The first step in the developmental process is to explore classical chromatographic parameters (selectivity, retentive capacity, and peak shape) by screening combinations of stationary phases and the preferential solvent compositions determined previously. The injection solution should be compatible with RPLC (50% to 100% methanol, substitute with acetonitrile for HILIC) and provide an S/N ratio greater than 100, usually generated at 100× the assay LLMI with a 10-μL injection. The use of high amounts of methanol in the injection solvent may be counterintuitive. However, the reasoning is that adsorptive loss of compounds to the walls of the injection vessel often

TABLE 5.4 Common Liquid Chromatography Myths

Myth	Reality
Liquid chromatographs are rugged and robust	The number of fittings, tubing and failure points in an LC system far outnumber those in a mass spectrometer
LC columns should last for >2000 injections	Possible, but not a requirement. LC columns are consumables
Smaller particles = greater efficiency	At the expense of higher pressure (needs higher class LC system), reduced flow rate (slower inject-to-inject time), smaller interparticle spaces (propensity for fouling with biological material)
The flow of liquid through a column can only be in one direction	Current day columns have frits at both ends to prevent the loss of stationary phase in bidirectional flow
Injection solvent must be the same as the mobile phase	Determination of solvent and volume of injection is empirical. High variety of options in gradient separations
Autosampler injection loops should be overfilled	Stable labeled internal standards provide normalization for injection variance
A maximum of 50 mmol/L buffer can be introduced to the mass spectrometer source	Determination of buffer influence on source is empirical. Present day sources are more rugged than those of 20 years ago

LC, Liquid chromatography.

TABLE 5.5 Considerations for Liquid Chromatography Selection Based on Molecular Features, Ionization Mode Preference, and Chromatographic Determinants*

Differentiator	RPLC	NP/HILIC
Charged analytes	Yes	Yes
Neutral analytes	Yes	No
Highly polar	No	Yes
ACN preference	Yes	Yes
MeOH preference	Yes	No
Elevated flow rate	Good	Better
ESI sensitivity	Good	Better
APCI sensitivity	Good	Good
Urine dilute/inject	Good	Average
Protein precipitate injection	Good	Good
LLE (NP solvent) injection	No	Yes
Isomer resolution (selectivity)	Good	Better
Stationary phase coverage	Good	Average
Column ruggedness	Better	Good
Generic LC protocol	Yes	Limited
Matrix effects control	Yes	Better
Understood mechanistically	Yes	No
<2 μm particles stationary phase coverage	Yes	Limited

ACN, Acetonitrile; *APCI*, atmospheric pressure chemical ionization; *ESI*, electrospray ionization; *HILIC*, hydrophilic interaction liquid; *LC*, liquid chromatography; *LLE*, liquid-liquid extraction; *MeOH*, methanol; *NP*, normal phase; *RPLC*, reverse-phase high-performance liquid chromatography.

*These are generalizations and exclude the use of ion-pairing reagents for polar compounds in reverse-phase LC.

can be mediated with organic solvent and a column that can properly retain the analytes of interest in a high organic environment typically denotes good retentive capacity. Two additional samples are injected, a blank diluent sample (always run a negative control in all experiments) and, if the assay is to be performed in a blood-based matrix, protein precipitated (PPT) matrix at a minimum 3:1 solvent-to-sample ratio[36] with IS included in the precipitating solvent to evaluate matrix-induced retention variance and response changes. The inclusion of the blank sample ensures chromatographic system cleanliness (absence of analytes when screening new columns); the precipitated sample is used to monitor the retention properties of known interfering species, namely phospholipids, and to assess the variation in retention time of the analyte in a complex sample.

Phospholipids are well known for the ability to modify ionization cross-section of the analyte in ESI.[37] To identify the retention properties of phospholipids, additional transitions are added in positive ion mode for two lysoglycerophosphatidyl cholines (1-P-2-OH: 1-palmitoyl-2-OH-sn-glycero-phosphocholine [18:2], m/z 496 to 184 and 1-S-2-OH: 1-stearoyl-2-OH-sn-glycero-phosphocholine [18:0], m/z 524 to 184) and two glycerophosphatidyl cholines (1-La-2-Li: 1-lauroyl-2-linoleoyl-sn-glycero-phosphocholine [12:0, 18:2], m/z 702 to 184 and 1-Li-2-A: 1-linoleoyl-2-arachidonoyl-sn-glycero-phosphocholine [18:2, 20:4] m/z 806 to 184), all using CEs of approximately 10eV. These phospholipids comprise early and late eluting forms of the abundant matrix effect—causing species for both classes and will require resolution through either chromatographic separation or sample preparation.[27,37] When analytes are being assayed in negative ion mode, a separate acquisition of the precipitated sample in positive ion mode (monitoring phospholipids) is recommended to determine the elution properties of phospholipids relative to the analytes (or use polarity switching) as additional information during the screen. The measurement of phospholipids is unnecessary for urine-based assays; salts and metabolites are the primary concern for urine.

If research has indicated the probability of circulating isomers of the analyte of interest, that material should be sourced if available and included in the test solutions described herein. To discriminate the interfering peak from the analyte of interest, inclusion of the IS provides the expected retention time of the analyte. In the presence of a single isobar, three peaks should be observed for an appropriate HPLC separation. Two peaks that coelute represent the

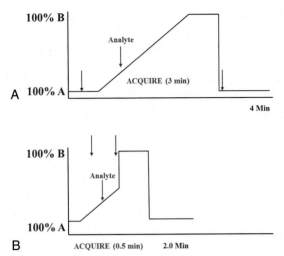

FIGURE 5.7 A, Reversed-phase liquid chromatography (RPLC) scouting gradient from 100% aqueous (100% A) to 100% organic solvent (100% B). **B,** Truncated RPLC gradient minimizing cycle-time and recondition composition, retaining scouting gradient pitch, in which the starting percent organic content is increased and the gradient is shortened. Empirical determination of efficient wash and reequilibration times is required.

analyte and its isotopically labeled IS; the third (isomer) is fully resolved from the former two.

It is recommended to screen different stationary phases using 50- × 2-mm columns with 5-μm particles because of the many different phases available. The goal is to screen the chromatographic characteristics of the phase and not change particle size or column dimensions (additional variables), which could confound conclusions. Fig. 5.7A indicates an example screening gradient; note that HPLC system dead-volume, maximum HPLC operating pressure (<80% pressure maximum recommended), and maximum interface eluent flow compatibility should be considered when setting the time or flow rate for the four steps in the gradient screen. The scouting gradient at 1 mL/min comprises injection focus at 100% A (30 seconds, 1.5 system volumes), gradient separation to 100% B (100 seconds), column clean at 100% B (50 seconds, 2.5 system volumes), and column reconditioning to 100% A (60 seconds, 3 system volumes) over 4 minutes with data acquisition for 3 minutes initiated when the gradient starts (30 seconds).

Modern HPLC systems have significantly reduced dead volumes (<150 μL); however, large injector loop volumes (>20 μL) and poorly made tubing fittings will add unwanted gradient delay and should be avoided and corrected. Further, it is good practice to divert solvent away from the ion source before and after the acquisition window to maintain interface and MS ion optics cleanliness; the first and last 30 seconds are diverted to waste during the scouting gradient at the flow rates listed. Screening multiple columns requires variety for serendipity; recommended RPLC screening functionalities are C18 (two manufacturers), C8, pentafluoro phenyl (PFP), phenyl hexyl, biphenyl, cyano, and amide, among others.

Further consideration of the gradient is warranted; the 1% per second change in elution solvent over 100 seconds provides an approximation to the elution composition when analytes elute. When a peak elutes at 50 seconds, the pump is delivering 50% B; however, analyte retention time is affected by gradient delay and the HPLC system dead volume. With the conditions listed using standard HPLC hardware, the gradient delay is approximately 20 seconds, and thus the analyte is eluting from the HPLC column at approximately 30% organic composition. Understanding the interplay of observed retention time and dwell volume can enable the reduction of HPLC method cycle time by truncation of unnecessary steps. Fig. 5.7B, indicates an accelerated gradient profile (0 to 100% over 100 seconds). This is the intermediate goal, but column screening must be performed before optimization occurs.

The previous solvent chemistry screen (development phase 1) should define the aqueous and organic solvent compositions applied to the RPLC screen. Because of the variety of aqueous-soluble buffers, three aqueous and one organic solvents are routinely used. The three aqueous compositions should be materially different to ensure retention of some degree of variety. Note the two solvents in Fig. 5.6 (B and C) with trifluoroacetic acid versus formic acid. The experiment is performed (on a binary mixing pump) as follows: inject blank solvent twice (ignore the first such so the system is dynamically equilibrated), then neat analyte in 50% to 100% organic, and then the 3:1 organic-to-matrix precipitated sample. This precipitated sample is used only for phospholipid retention determination and not for the analyte itself. This sample should be centrifuged and the supernatant transferred and injected to prevent the loading of precipitated proteins onto the head of the column, inducing increased pressure. After data acquisition, remove the first column, replace with the second column, and repeat the sample sequence. After all columns have been screened with the first set of solvents, the experiment is repeated with the second set of optimal solvents. It is more efficient to replace an HPLC column than to swap or purge mobile phases in screening HPLC modalities.

If the chosen eluting solvent is acetonitrile, 2 to 4 column volumes (1 mL for 50- × 2-mm column) of methanol should be applied to the column after the precipitated sample to remove phospholipids for the next time the column is used.[37] Use of a quaternary pump allows three aqueous solvents to be screened between switching columns (12 injections sequentially) or the automatic addition of methanol after injection of the PPT sample to clean an RPLC column off after addition of phospholipid-rich precipitated matrix.

Fig. 5.8A describes data generated using a methanol-to-water gradient screening for a 10-steroid profile. Because of knowledge of isobaric interferences (structural isomers, no specific transitions for analytes are generated) the gradient pitch was reduced from 1% to 0.25% B per second (0% to 100% over 6.67 minutes). This example is included as a reminder that the HPLC screening stage should incorporate resolution from known interferents spiked into the analyte sample for determination of appropriate stationary phase and solvent chemistries that enable selective MS/MS measurement of the target analyte. Reduction of the data is shown in Table 5.6. Amino, Cyanopropyl, BetaMax Base, and Hypercarb columns are qualitatively poor and excluded. Phenyl-Hexyl, PFP, Fluophase RP, and particularly C8 are excluded because of a lack of resolution of known isobars (C8 only

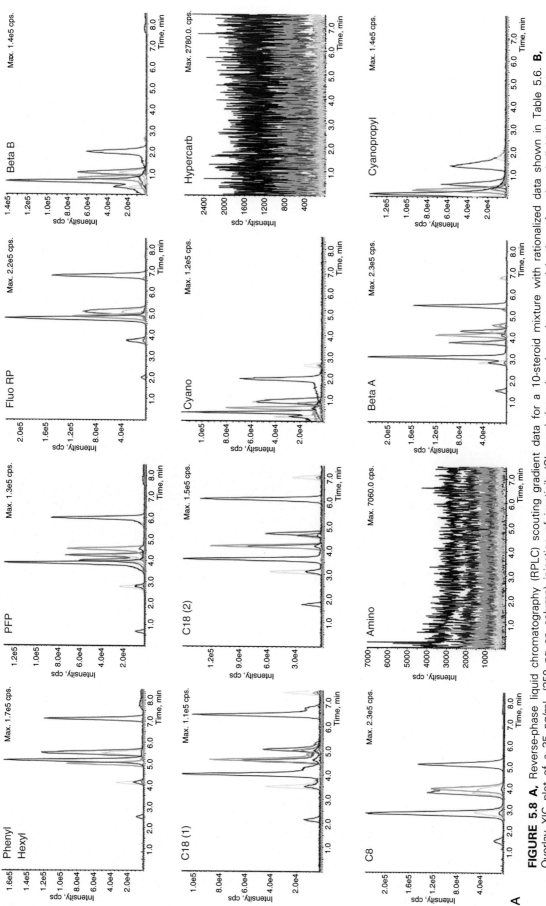

FIGURE 5.8 A, Reverse-phase liquid chromatography (RPLC) scouting gradient data for a 10-steroid mixture with rationalized data shown in Table 5.6. **B,** Overlay XIC plot of a 35 ng/mL (350 pg on column) injection of imatinib (Gleevec) *(arrow)* and a subsequent injection of a phospholipid-rich precipitated sample from RPLC screening. The phospholipids monitored include lysophosphatidyl cholines 1-palmitoyl-2-OH (16:0 acyl chain, m/z 496.3 to 186.1), 1-steroyl-2-OH (18:0 acyl chain, m/z 524.3 to 186.1) and select glycerophospholipids 1-lauric-2-linoleic (12:0 and 18:2 acyl chains, m/z 780.4 to 186.1) 1-linoleic-2-arachidonic (18:2 and 20:4 acyl chains, m/z 804.4 to 186.1).

FIGURE 5.8 cont'd.

TABLE 5.6 Qualitative Evaluation of Liquid Chromatography Conditions for the Steroid Profile From FIGURE 5.8

Column	Acceptable Retention	Isomeric Separation	Peak Asymmetry
Phenyl hexyl	Yes	No	Acceptable
PFP	Yes	Partial	Fronting
Fluophase RP	Yes	No	Acceptable
BetaMax Base	No	No	Fronting
C18 (1)	Yes	No	Tailing
C18 (2)	Yes	Yes	Acceptable
Cyano	No	No	Fronting
Hypercarb	No	NA	NA
C8	No	No	Acceptable
Amino	No	NA	NA
BetaMax Acid	Yes	Yes	Acceptable
Cyanopropyl	No	No	Tailing

NA, Not applicable; *PFP*, pentafluorophenyl.

has five peaks, peak counting is an additional tool for qualitative assessment of isobar resolution, as noted earlier) or peak fronting (PFP). Of the two C18 columns, the second demonstrates superior resolution of isobars over the first brand; however, when comparing to the BetaMax Acid separation, it is clear that two isomers (11-desoxycortisol and 17-hydroxyprogesterone, pink) are resolved but coelute on C18 (1). Thus Betamax Acid and C18 (2) would be good HPLC column functionalities for further development. This particular example demonstrates a complicated data set reduction and highlights some of the concepts of qualitative assessment of chromatographic data.

A secondary qualitative assessment is shown in Fig. 5.8B whereby imatinib (Gleevec; 35 ng/mL, 350 pg on column) was injected, followed by a precipitated sample to monitor the presence of phospholipids; chromatograms were overlaid to assess for selectivity. The monitoring of phospholipids was performed by analysis of the expected $[M+H]^+$ of highly abundant phospholipid species, including both lysophospholipids (1-acyl-2-OH glycerophosphocholines) and diacyl species (1-acyl-2-acyl glycerophospholipids). Owing to the additional fatty acid chain in the diacyl species, two distinct retention regions may be observed in reverse-phase separations. The ability to effectively resolve phospholipids chromatographically provides flexibility in sample preparation approaches.

Quantitative assessment of chromatographic performance (Chapter 1) is routinely performed after peak integration. Commonly reviewed features include analyte peak height, retention time, peak width, peak asymmetry, and selectivity (between isobars), notably, the S/N ratio is less relevant here. Often, a first-pass screen of columns and solvents (see Fig. 5.8) is such that initially successful columns lead to a second-tier screen of a particular functional modality as similar stationary phases from the same or different manufacturers may lead to dramatic changes in performance because of proprietary functionality and secondary chromatographic retention characteristics related to composition of the silica bed. For example, not all C18 columns are the same because various endcapping techniques, carbon load, formation, and abundance of residual silanols can induce the subtle retention variations a separation may require. Stationary phases exhibiting poor retentive or elution characteristics are abandoned, and additional energy is focused on the class of stationary phase that demonstrates preliminary success.

Fig. 5.9A indicates data reduction for temazepam from a benzodiazepine mixture; note that temazepam demonstrated the lowest transmission efficiency with the lowest desired LLMI, and thus was the key analyte of focus in optimization of sensitivity. First-tier HPLC screening indicated C18 was marginally better overall; however, Phenyl-Hexyl and fluorinated PFP phases also demonstrated acceptable performance and so were carried forward into the second tier. Review of the tabulated data indicates a number of key results that will enable appropriate selection of primary, secondary, and tertiary column (and solvent chemistry) choices. The choice of columns is rationalized by reduction of data for peak height (a measure of response), retention time (measure of acceptable retention), peak width (to assess broad peaks; compressed peaks generate greater signal above noise), and peak asymmetry (a measure of fronting or tailing).

The data shown in Table 5.7 are derived from the values shown in Fig. 5.9 and are rank-ordered by the characteristics detailed previously. Firstly, peak height was demonstrably better on the Hypersil Gold with Luna Phenyl-Hexyl, SB C18, and Extend C18 following in order. Peak height relates to sensitivity, theoretically enabling improved S/N ratio at the LLMI. The retention time of 1.7 minute likely accounts for some of the observed increase in sensitivity as the organic eluent composition when temazepam elutes is greater using this media than the other columns screened. The organic composition during elution of this analyte (as calculated using retention time and dwell volume of the system) is approximately 60%. For effective chromatographic control (or the ability to manipulate retention time by gradient pitch/shape modifications), analyte elution between 40% and 80% organic facilitates injection focusing and gradient shaping to resolve interferences. Thus analyte retention time in the panel should fall within this solvent composition window. Here again, the Hypersil Gold demonstrates the strongest retention, followed by the Atlantis T3 and Luna Phenyl-Hexyl (tied at third most retentive).

Evaluation of the peak width and, more importantly, peak asymmetry are key variables for consideration. Review of the data indicates the peak width is approximately 0.13 minutes (7.8 seconds) and the Fluophase PFP peak width is 0.077 minutes (4.6 seconds). These differences are relevant when complex separation is required (resolution from isobars); at this stage the detail is good information when selecting a secondary (backup column) to move forward. More importantly, peak asymmetry is a final deciding factor, calculated as (retention time − peak start)/(peak end − retention time). Perfectly Gaussian peaks have peak asymmetry (As) of 1. Peaks exhibiting asymmetry values between 0.8 and 1.5 are considered acceptable for further evaluation. The consideration of peak asymmetry is important for downstream data reduction because automated peak integration programs integrate Gaussian peaks more reproducibly than fronting (As >0.8) or excessively tailing (As >2) peaks. The rank ordering for As in Table 5.7 is presented as the absolute

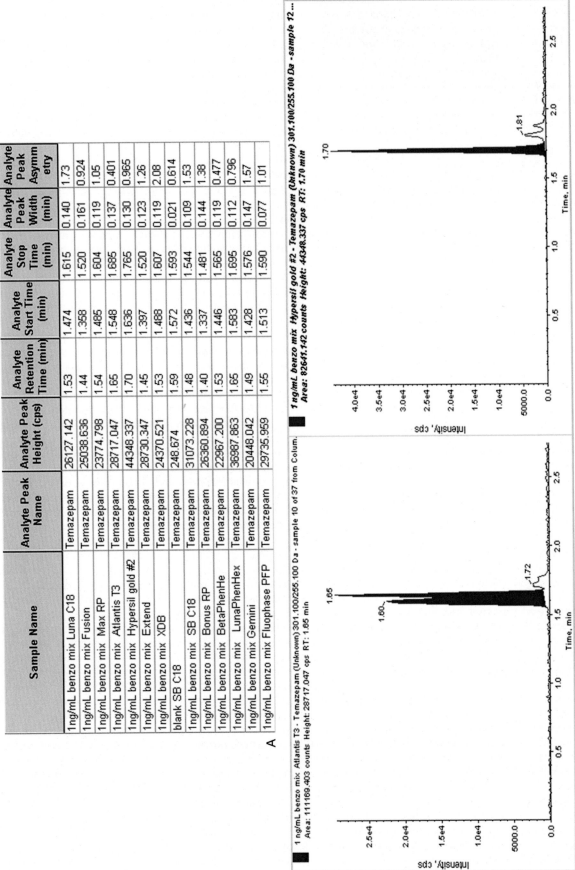

FIGURE 5.9 A, Reverse-phase liquid chromatography second tier screening data reduction for temazepam. **B,** Exemplar chromatograms for temazepam demonstrating screening failure (*left*) and success (*right*) based on peak asymmetry and peak height. Rationalized data are shown in Table 5.7.

TABLE 5.7 Second-Tier Column Screening Results in FIGURE 5.9 (1 = Favorable)

Column	Peak Height	Retention Time	Peak Width	Peak Asymmetry
Luna C18	8	8	10	12
Fusion	9	12	13	4
Max RP	11	5	6	3
Atlantis T3	6	3	9	11
Hypersil Gold	1	1	8	2
Extend	5	11	7	6
XDB-C18	10	8	6	13
SB-C18	3	10	2	9
Bonus RP	7	13	11	7
BetaSil Phenyl/Hexyl	12	7	6	8
Luna Phenyl Hexyl	2	3	3	5
Gemini C18	13	9	12	10
Fluophase PFP	4	4	1	1

value of the difference between the measured As and the value of 1 (perfectly symmetrical) peak. The Fluophase RP achieves the highest ranking, followed by the Hypersil Gold and Max RP. The poorest asymmetry performers were the XDB-C18 and Luna C18 for substantial fronting and the Atlantis T3 for substantial tailing.

It is at this point that a single column is optimized, and the additional column ranking will be used again later when development dictates the need for a ready-made backup column in the event of manufacturing failure of lots or obsolescence of a particular stationary phase (clinical methods often survive for >10 years). Notably, the particular column characteristics were not chosen before initiation of this screening process. In these cases, it is the molecule's preferential mobile phases (examined in solvent screening) and that molecule's preference for a stationary phase (as deduced from empirical data) that selects the column chemistry of choice.

Additional High-Performance Liquid Chromatography Optimization

After selection of hardware configuration (solvents, stationary phases), the chromatographic program should remove superfluous steps to increase efficiency. The next stage in the development process involves reducing the cycle time of the HPLC method, as shown in Fig. 5.7B. This involves four steps:
- Increase percent of organic at injection
- Remove excess gradient time
- Reduce acquisition window/increase mass spectrometer bypass time
- Reduce or increase reequilibration times

In Fig. 5.7, the scouting gradient initiated at 0% organic. However, acceptable retention of the analyte at the head of the column may be achieved at a higher percent organic, facilitating the pass-through of poorly retained molecules in the injection slug (hopefully to waste by way of a bypass valve). Thus the starting solvent composition (percent solvent B) was increased to approximately 10% less than the solvent composition of the earliest eluting analyte. This also can be performed empirically by increasing the starting percent organic in 5% steps until breakthrough is observed (peak fronting).

Second, the gradient continues far after the peak of interest elutes; this additional gradient time does not provide any more useful data. It is thus removed by truncating the gradient to end at a solvent composition that is approximately 10% higher organic than the analyte elution composition (percent solvent B). The HPLC method is then immediately stepped to the washing conditions to remove strongly bound materials from the column. Here, the mass spectrometer acquisition window should be reduced to include just the analyte of interest (with 6 to 10 seconds either side) and specifically exclude the materials eluting during the washing phase of the separation. A bypass valve should be used to divert the waste (ie, unmeasured materials) away from the source of the mass spectrometer, preserving the cleanliness of the ion optics.

Finally, empirical evidence can be generated to shorten the reequilibration time of the system back to the original starting conditions. Often, 3 to 5 column volumes are recommended (5 to 10 for HILIC), but this is not a steadfast rule. Reducing the amount of reequilibration time at the end of the gradient and injecting the analyte of interest ($n = 3$ per time reduction) should generate identical peak shape and retention time compared to longer equilibration steps in HPLC methods.

Adsorptive Loss Evaluation

After removal of unnecessary HPLC cycle time and reduction of the eluent window directed to the mass spectrometer, the possibility of analyte adsorption to containers and pipettes must be addressed. This is an exploratory experiment consisting of stressing the solubility of the analyte in various neat solutions against possible loss of the analyte to the walls of vessels and pipette tips. Equimolar solutions of the analyte (at some measureable concentration in the LC-MS/MS assay) are prepared in methanol, acetonitrile, and water, preferably in glass tubes. A moderate volume of these solutions is then transferred to three individual autosampler vials or wells of a 96-well plate (150 μL for a 200-μL well plate, 500 μL for a 1.0-mL well plate, etc.) in which the final sample will be stored for injection. One well is unmodified and serves as the control. Using a fresh pipette tip, perform a single aspiration and dispense event, then dispose of the pipette tip. This is then repeated 10 times, inducing the possibility of loss of the analyte to the walls of the pipette tip. The third sample is transferred to 10 unused vials or wells using the same pipette tip (do not dispose after the aspirate/dispense cycle).

Step 1: Prepare equimolar solution in methanol, acetonitrile, and water (all LC-MS—grade reagents), with three vials or wells per solution.

Step 2: Aliquot a fixed volume of each solution in triplicate to an autosampler vial or well of a 96-well plate. Note that the volume depends on maximum volume of container. Use appropriate judgment.

Step 3: Aspirate and dispense one sample of each solution using 10 different pipette tips, dispensing the liquid back to the origin vial or well.

Step 4: Transfer entire contents of a separate vial or well for each solution to an unused vial or well using the same pipette tip. Repeat for a total of 10 transfers.

Step 5: Aliquot the same fixed volume of an unspiked solution (neat water, methanol, and acetonitrile) to an autosampler vial or well of a 96-well plate as a negative control (contamination).

Four different specimens have been created from each of the three solutions ($n = 12$ total). One set is the experimental baseline (interaction with a single pipette tip and single vessel), one set is completely blank solvent (contamination), one set repeatedly had interactions with fresh pipette tips, and one set has undergone repeated access to container surfaces. Injection of blank samples (first so there is no possibility of carryover) provides details on selectivity of containers—that is, do pipet tips and container walls generate interferences. Interferences generated from vessels require the sourcing of a different material or manufacturer and replication of the experiment. The comparison of adsorption in the different solvents across the different conditions to the single interaction (baseline) provides insight into the loss of analyte after sample preparation and can guide in the required final solution after sample preparation. Absolute losses greater than 10% should be eradicated through materials or chemistry changes (buffers, carrier solutions, etc.). This can also be information regarding the preparation and storage of calibrators and QCs.

After determination of the appropriate solution to prevent adsorptive loss, perform increasing injections from 10 to 100 μL in 10-μL increments (after placing a 100-μL loop on the autosampler) and monitor for the degradation in peak shape. Fig. 5.10 indicates increasing injection of neat methanol solutions of cortisol onto an RPLC separation where cortisol elutes at 25% acetonitrile in the gradient. A subtle retention time shift and substantial chromatographic peak fronting are observed at 20% and 30% acetonitrile, respectively. The previously mentioned experiment was then replicated to facilitate a balance between the adsorptive loss of cortisol to a polypropylene 96-deep well plate and the appropriate retention of the molecule on the stationary phase at higher injection volumes. In the particular case of cortisol, a 1:1 mixture of methanol to water enabled an 80-μL injection to be performed without losses to either tips or wells in the 96-well plate. Note, however, that 80 μL is not the final injection volume for the assay but the maximum injectable volume at which peak shape is unaffected. Maximum injection volumes are used to facilitate robust operations only when the mass spectrometer performance has degraded. Ideally, the typical assay injection volume should not exceed 80% of the maximum injectable amount to allow flexibility (source contaminated) and should be used as a prompt for corrective action (instrument maintenance).

On-Column Detection Limits

An estimation of the on-column detection limit that can be achieved is required. Stock solutions at 100× the assay expected LLMI are serially diluted in 10-fold steps through 6 cycles, generating final solutions from 100 to 1/1000th of the assay LLMI, together with the blank diluent solution. For example, an assay has an expected LLMI of 5 ng/mL. Solutions of that analyte are prepared in the injection solvent (determined previously) at 500, 50, 5, 0.5, 0.05, and 0.005 ng/mL. These concentrations are then converted from nanograms per milliliter to picograms per microliter, such that the final concentrations are 500, 50, 5, 0.5, 0.05, and 0.005 pg/μL. These solutions are injected at a fixed volume (microliters), allowing for the calculation of amount of material on the column. For a 10-μL injection of each of the solutions, the on-column load would be 5000, 500, 50, 5, 0.5, and 0.05 pg on column.

After triplicate injection of these materials at a volume equal to 50% of the maximum injectable volume determined previously, data are reduced by peak response and reproducibility of peak characteristics (peak width, peak area). The lowest concentration observed in the on-column detection study (S/N ratio >20:1 and consistent peak width and peak area determined as coefficient of variation [CV] <10%) can be extrapolated to deduce the on-column detection limit. Note that the value for the S/N ratio can be misleading because the region of noise selected for all peaks may exhibit differences between injections due to the scanning nature of the mass spectrometer. This determination of on-column detection limits via peak area reproducibility is critical to decision making in the sample preparation process.

In an ideal world, the on-column detection limit would enable analysis of a reasonable sample volume (<100 μL for blood-based matrices) allowing for a 20% loss during sample preparation (desirable, to be discussed), yet providing an extract that can be injected twice. Following on from the previous example with a 5-ng/mL LLMI, desirable on-column detection limit = maximum sample volume × concentration of LLMI.

$$100 \text{ μL} \times 5 \text{ pg}/\text{μL} = 500 \text{ pg}$$

Assuming for 80% extraction efficiency:

$$500 \text{ pg} \times 80\% \text{ recovery} = 400 \text{ pg}$$

Allowing for duplicate injection as well as residual volume after the injection:

$$400 \text{ pg}/3 = 133 \text{ pg on column detection limit for } 100-\text{μL sample}$$

The determined on-column detection limit guides the sample extraction strategy. Using the earlier example, having a lower on-column detection limit of 1 pg on column would allow for a 100-fold dilution before sample injection. A protein precipitation (PPT) extraction or dilute-and-shoot workflow may well suffice. Should the on-column detection limit be only 500 pg, however, the assay would require some form of concentration (approximately fivefold) and quite likely a larger sample volume (>100 μL), particularly when considering the necessity of volume for repeat injections.

Final Development Stage 2 Considerations

Before moving into sample preparation evaluation, ensure that the HPLC system is set up correctly, with the shortest tubing lengths installed to reduce systemic dwell volume, incorporating stainless steel tubing where appropriate (where

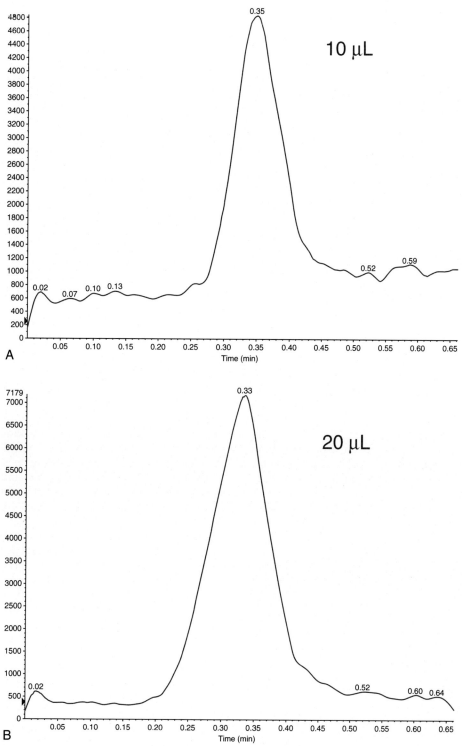

FIGURE 5.10 Increasing injections volume of cortisol in neat methanol. Appropriate peak shape is observed for 10-μL injections with fronting (breakthrough) observed at 20 and 30 μL. To allow for increased injection volume to facilitate higher on-column concentrations, modification of the injection volume to 1:1 H_2O to methanol facilitated up to 80-μL injections.

pressures exceed 200 bar or 3000 psi). Internal diameters of 0.005 inches are preferred for most HPLC tubing. Polyetheretherketone tubing must be cut flush (all fittings checked), and stainless steel tubing should be laser cut and polished by the manufacturer because hand cutting in the laboratory can create burrs at outlets. An appropriate sample loop should be installed based on maximum injection volume allowable. Use of an inline filter frit between the autosampler and HPLC column (0.5 to 2 μm frit filter) can help reduce particulates clogging the interstitial spaces of the HPLC

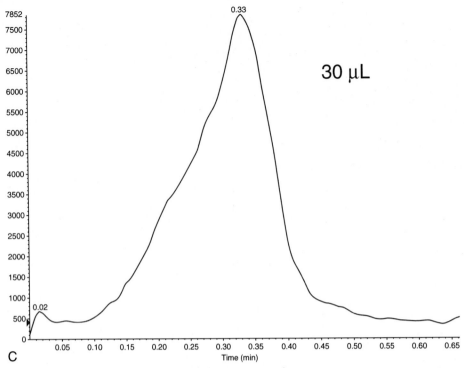

FIGURE 5.10 cont'd.

column, which can be responsible for increased back pressures and necessitate column replacement. Ensure there are multiple lots of the optimal columns on hand, as well as multiple lots of high-purity solvents; analytical-grade chemicals and HPLC- or MS-grade solvents should be used exclusively (Table 5.8).

Finally, an evaluation of transition dwell times to enable adequate points across a peak (peak sampling) should be performed. When determining the appropriate dwell time per transition, the following details should be considered. At the very minimum, 10 points should be used to generate the peak shape, with 15 to 30 being far more reproducible. This must be balanced against the individual transition dwell time. The detection event is an averaging of the ion counts across a certain time span (counts per second [cps]). Very fast scanning (<5 ms dwell time) can incorporate random variation because of electronic or chemical noise in detection, resulting in jagged peaks that may be difficult to integrate. To assess appropriate dwell times, use the peak with the smallest width at its lowest abundance (the LLMI) and calculate the peak width, but exclude any fronting or tailing in seconds and multiply by 1000 to convert to milliseconds. Determine the total intertransition delay (number of transitions × fixed intertransition delay time) in milliseconds. The delta of these values is the available scan time. An example using 20 transitions, a 3-second wide peak, and a 5-ms intertransition delay is shown in the following equations:

$$3 \text{ seconds} \times 1000 = 3000 \text{ ms}$$

$$20 \text{ transitions} \times 5 \text{ ms} = 100 \text{ ms}$$

Available scan time is thus 2900 ms available to scan across the 3-second-wide peak. Dividing this again by the total number of transitions (20) equals the cycle time per transition:

$$2900 \frac{\text{ms}}{20} \text{ transitions} = 145 \frac{\text{ms}}{\text{transition}}$$

TABLE 5.8 Development Phase 2: Chromatography Conditions

Experiment	Outcome
Provisional determination of acceptable LC mode	Reverse-phase or HILIC separation motif selected
Generation of phospholipid screening method in MS for blood-based matrix analysis	MS/MS Method for ion-suppression inducing phosphatidyl cholines
Multiple column screening using optimal solvents from phase 1	One to three columns, which demonstrate acceptable retentive characteristics, response, and peak shape
LC efficiency and adsorptive loss evaluation	Optimize LC cycle time. Ensure that analyte is not being sequestered to containers or pipettes
On-column detection limits	Establish minimum observable amount necessary for detection in the mass spectrometer.

HILIC, Hydrophilic interaction liquid chromatography; *LC*, liquid chromatography; *MS*, mass spectrometry; *MS/MS*, tandem MS.

TABLE 5.9 Transition Dwell Time Versus Peak Sampling Considerations

Peak Width (ms)	No. Transitions	Dwell Time (ms)	Intertransition Delay (ms)	Total Cycle Time (ms)	Points Across Peak
3000	20	5	5	200	15
3000	20	10	5	300	10
3000	20	25	5	600	5

To maintain a minimum of 10 scans across each peak, the time per transition is divided by 10 to generate the necessary scan time per transition:

$$145 \frac{\text{ms per transition}}{10} \text{ scans per transition} = 14.5 \frac{\text{ms}}{\text{scan}}$$

Note here that the value of 10 scans per transition is the absolute minimum recommended. A more preferable scan rate would include a higher number of points (scanning faster) but may come at the cost of noisier data. A number of theoretical examples are described in Table 5.9 using 3-second-wide peaks (3000 ms) and a fixed intertransition delay to evacuate the collision cell (Q2) between transitions. For most modern instruments, the interchannel delay is approximately 1 to 5 ms.

SAMPLE PREPARATION

Tools and Considerations

All LC-MS/MS assays have a requisite sample preparation component and because of the variety of measures performed, the greatest diversity of technologic solutions (see Chapter 3). Sample preparation provides the following attributes: (1) reduction of the sample complexity (selectivity for the analytes), (2) addition of ISs for improved control in measurement, and (3) concentration of analyte into an injectable volume. Sample preparation is ideally simple, reproducible, and orthogonal to the HPLC method (eg, ion-exchange SPE with RPLC). The uncontrolled nature of clinical specimens places a significant burden on sample preparation to work in concert with HPLC separation and the MS transitions selected to add confidence in measurement. Therefore clinical methods should not focus on quantitative recovery (>95%). Instead, selective recovery is the end goal.

Before performing sample preparation development, a frame of reference is needed to ensure system performance and comparator—a system suitability test (SST) injection. Following on from adsorption studies, two solutions are prepared in bulk (store the remainder in a freezer or refrigerator) in large aliquots and placed in vials or wells in the autosampler for repeated injections. The first is the blank sample diluent and should be injected to ensure the analytical system is free of interferences (<20% analyte LLMI response and <5% IS response). The second is a neat solution of analytes at 100× LLMI, and the IS at 10× LLMI of the analytes. Where samples are expected to be concentrated or diluted by more than 10-fold, adjust the concentrations of the analyte and IS in the SST accordingly.

Before injecting samples, ensure the LC-MS/MS system is free of analyte or IS interferences at the expected retention times (two blank injections), inject the analyte SST solution in triplicate, then inject blanks again (2×) ensuring no appreciable system carryover is observed in the first blank (response <20% analyte LLMI and <5% IS). When performing provisional sample preparation optimization, the injected sample content can deleteriously affect instrument response. Repeated SST injections are monitored within-run to assess for instrument measurement drift (increases or losses for analyte/IS) that can confound assay recovery conclusions.

Three key experimental approaches are used in evaluation of sample preparation include classic spike and recovery, determination of zones of suppression using PCI,[27] and preextract and postextract spiking.[38] The latter method is recommended and involves the individual determination of recovery, matrix effects, and total efficiency using three distinct samples. The first is a neat sample (A) that does not undergo extraction or contain sample matrix. This sample represents 100% recovery and zero matrix effects (the SST solution serves this purpose). The A sample should be fortified or diluted appropriately to correct for the expected extracted concentration for the sample preparation scheme. The second sample is spiked at 100× LLMI with analytes and 10× LLMI IS after extraction (B). The final concentration of this material should be equivalent to that of the A sample, although the detected response may be different. This specimen contains extracted matrix but no recovery losses for analyte or IS. The third sample is spiked at 100× LLMI for analytes before extraction and 10× LLMI IS after extraction (C). This specimen includes extracted matrix content and recovery losses with the opportunity to create a peak area ratio (IS added afterward) when onization effects are confounding. The three calculations made from this sample set are:

$$\text{Matrix effects (\%)} = \text{Peak area response (B)}/\text{Peak area response (A)} \times 100$$

$$\text{Recovery efficiency (\%)} = \text{Peak area response (C)}/\text{Peak area response (B)} \times 100$$

$$\text{Total efficency (\%)} = \text{Peak area response (C)}/\text{Peak area response (A)} \times 100$$

This approach is much simpler to perform for exogenous drug analytes because true blank matrix is readily attainable. For endogenous biomarkers, a starting pool (as low as possible) that is overspiked to 100× LLMI for analytes is used or the ISs are used to evaluate because they are unique to the testing specimen (if selected and chromatographed properly) and serve as a surrogate for the analyte where residual analyte concentrations are confounding. Consideration that stripped matrix may not contain the same binding proteins as intact matrix is important because sample preparation may be acceptable for charcoal-stripped but not for authentic matrices. When performing these studies, spiking of the low-level pool preextraction should be performed in bulk where the spike represents less than 5% of the total volume. Binding equilibration must be ensured before use to provide recovery calculations that are representative of the true nature of analytes in a sample before extraction. Absolute equilibration in binding is difficult to determine without a validated assay, and thus some assumptions concerning the equilibration must be made. Generally, extracting spiked samples immediately is discouraged; mixing for an hour or allowing a spiked sample to equilibrate in refrigerated conditions over a weekend is a good starting point.

The majority of sample preparation schemes used in clinical chemistry fall into one of five approaches (Table 5.10): PPT, liquid-liquid extraction (LLE), supported liquid extraction (SLE), SPE, and online-direct injection (online). Several considerations are important in selection of the appropriate technique to start with; structural features (more specifically the uniqueness of structural features), the need to concentrate or dilute, and process efficiency.

The nature of the analytes is the first differentiator. For analytes that are charged in solution (carboxylic acids, tertiary amines, etc.) or highly polar, ion-exchange SPE can provide effective purification because of ionic or multimodal retention and the ability to use aggressive washing protocols to elute unwanted materials before elution of analytes. By contrast, LLE and SLE are preferred for neutral analytes to enable partition into the predominantly apolar solvent phase. Where analytes are thermally labile (acyl glucuronides), evaporation and reconstitution should be avoided; thus a "lossless" online approach or PPT injection is preferred over SLE, LLE, or SPE. The latter often creates diluted extracts relative to the starting sample volume, requiring concentration and usually a solvent exchange (evaporation/reconstitution) before injection. PPT protocols are generic, requiring little optimization but producing extracts with a significant amount of interferences, most notably phospholipids and salts, which lead to ionization effects. LLE protocols are often fine-tuned with just enough polar composition added to the extraction solvent to effect necessary recovery, are thus less generic, but provide the opportunity for cleaner extracts for subsequent analysis.[27] In Table 5.10, SPE is listed as selective (*) for ion-exchange modality only and hydrophobic SPE is predominantly a desalting and deproteinizing technique that affords limited orthogonal selectivity to the method.

High sample volume tests require considerations to the process efficiency. Dilute and direct inject represent a two-step addition process (sample + IS in diluent) and is readily automated in a 96-well plate—simple and fast. PPT is similar, usually including an additional postprecipitation transfer of supernatant to a fresh plate for injection. Multistep SPE and SLE can be performed with 96-well plate–based formats. LLE in tubes is inherently a poor candidate for liquid handlers, and in-plate LLE often suffers from well-to-well contamination (wicking). Multistep SPE and LLE are much lower throughput than SLE, PPT, or online, and preparative materials costs are worthy of consideration. Developmental experiment will be considered individually for each technique, although certain approaches will apply to multiple techniques.

Sample preparation development represents a two-step process. The first uses overspiked samples and comparison of SST with IS spiked before or after extraction to assess early success or failure of a probable extraction procedure. The second step involves generation of calibration curves in an appropriate matrix, together with analysis of patient samples for evaluation of interferences, spike/recovery, and dilutional mixing. Each preparative technique has step 2 as a common process. Discussion of first-pass developmental principles will be described for each technique first. The step 2 process will be described; however, if there are known interferents, such as isomers to the analyte of interest, that could be removed during sample preparation evaluation, consider adding to the screening protocol designed (phospholipids are just one example).

For final consideration, the MS/MS transitions that are selective have not yet been chosen; thus evaluation of baseline

TABLE 5.10 Considerations for Sample Preparation Selection, Rank Order Preferred (1) to Suboptimal (4)

Differentiator	PPT	LLE	SLE	SPE	Online
Charged analytes recovery	2	4	3	1	3
Neutral analytes recovery	4	1	1	2	2
Highly polar analytes recovery	2	4	4	1	3
Thermally labile analytes recovery	2	4	4	3	1
Generic protocol	1	4	3	2	2
Assay ruggedness	4	1	1	3	2
Matrix effects	4	1	1	2	2
Selectivity	4	1	3	2*	3
Sample concentration	4	1	1	2	3
96-Well plate preparation	3	4	2	2	1
Automatable	3	4	2	2	1
Simplicity	1	2	2	3	4
Method development speed	1	3	3	4	2
Cost (cheapest)	1	2	3	3	4
Sample type variance	2	1	3	3	4
Preparation time (fastest)	2	3	3	4	1
Sample volume (limited)	2	3	3	3	1
Extract direct injection	2	4	4	3	1

LLE, Liquid-liquid extraction; PPT, protein precipitation; SLE, supported liquid extraction; SPE, solid-phase extraction.
*Selectivity for SPE is contingent on the use of a stationary phase orthogonal to the modality of chromatographic separation. SPE is a low-resolution technique; thus the use of hydrophobic SPE with hydrophobic LC yields concentration rather than cleanup of the extracts.

noise (S/N ratio) and the potential of interferences to extract and coelute in a transition as the sample preparation motif is modified should be evaluated. Only after real human samples have been evaluated can the determination be made of which transitions (within the linear measureable range of the mass spectrometer) are of acceptable specificity.

Protein Precipitation

PPT represents the simplest and most generic protocol and is applicable to many analytes. A number of variables require optimization, as shown in Table 5.11: the precipitating solvent (chemistry), ratio of solvent to sample (minimum 3:1 for organic precipitating reagents),[36] IS equilibration for equivalent recovery between analyte and IS, mixing time during precipitation, temperature of the precipitating reagent or precipitant, centrifugation time and force, and supernatant injection versus transfer to a fresh container. When considering the nature of the precipitating chemistry, it should be noted that different PPT solvents (methanol, acetonitrile, and isopropanol as organic-based or tricholoroacetic acid and sulfosalicylic acid as acidic precipitants) generate extracts with different recoveries and composition.[36,39]

For initial development of PPT, spiked (100× analyte LLMI only, sample B) and unspiked samples are aliquoted into a vial or well. An appropriate amount of precipitating reagent is added, the samples are then mixed vigorously for 10 minutes, centrifuged to separate, and the clear supernatant is transferred. The naïve sample supernatant is spiked with equivalent to 100× LLMI (sample C) for analyte (accounting for the volume differences in the total experiment) and both sample extracts spiked with 10× LLMI IS solution. Dilution should be less than 25% overall. These samples are bracketed with SST injections (sample A) before, during, and after the sequence to assess drift (<10% acceptable). Comparison of postextract spiked sample IS response (B) to SST IS response (A) provides insight on matrix effects. Monitoring the retention time of phospholipids (if not previously performed in HPLC method development) is important to understand matrix effects observations from this comparison. The optimal goal for matrix effects is less than 10% ionization suppression or enhancement. Comparison of prespiked (C) to postspiked (B) analyte peak area response provides quantitative measure of recovery efficiency, with a goal of more than 80% absolute recovery. Where matrix effects (ionization suppression) are observed, the calculation of recovery is performed as the peak area ratio of analyte to IS for each matched pair of samples. This is performed because the IS was not extracted and should coelute with analytes and undergo the same relative ionization effect difference, if selected appropriately in the initial development phase.

Fig. 5.11A indicates the differences in extracted ion-suppressing species for gonadotropin-releasing hormone (GnRH). The PCI (see Fig. 5.2A) setup was deployed, and stable labeled IS was infused after column. The IS is preferred for PCI here because the extracts are of true human matrix and expected to have some endogenous concentration of GnRH. Injection of water is used as a negative control for the expected response change during the gradient in the absence of ionization effects. Subsequent injections of methanol- and acetonitrile-precipitated serum samples demonstrate deviation (suppression) relative to the water injection; most notably, the degree of ion suppression observed at the retention time of GnRH is markedly reduced for acetonitrile versus methanol and thus is preferred between the two potential precipitating solutions. The use of PCI in this instance is to inform regarding the relative degree and chromatographic location of interference between samples and not as an absolute measurement of matrix effects (use sample B/sample A approach previously).

Additional consideration is warranted when performing PPT for endogenous analyte measurement because of the milieu of potential binding partners within the specimen. The IS must be in the same state as the analyte (equilibrated) before removal of supernatant. Fig. 5.11B indicates recovery measurement for serotonin in serum and whole blood (C/B sample ratios). Serum and whole blood were overspiked to reduce the bias related to residual serotonin and allowed to incubate for 24 hours before these studies. Material differences were observed between serotonin recoveries in serum versus whole blood after aggressive mixing for 10 minutes. This underrecovery required correction because of the need to assay both matrices in clinical practice. The study was repeated using class A pipetting (to rule out spiking errors) and modifications to incubation time before extraction, and results were the same. At this point, differential equilibrium between analyte and IS (preextract vs. postextract spiked) is

TABLE 5.11 Protein Precipitation Variables

Variable	Notes
Precipitating solvent	Methanol, acetonitrile, ethanol, isopropanol, acetone, trichloroacetic acid (TCA), sulfosalicyclic acid (SSA), flocculation reagents
Ratio of solvent to sample	Organic precipitants 3:1 or higher. Aqueous precipitants dependent on material (commonly 3% for SSA, 10% for TCA)
Internal standard (IS) equilibration time	IS can be included in precipitating solution or added before precipitant and allowed to equilibrate
Mixing time after addition of precipitating reagent	Seconds to many minutes
Temperature of precipitant or sample/precipitant mixture	<0°C to room temperature
Centrifugation time/force	Dependent on laboratory equipment and sample vessel
Evaporation/reconstitution	Volume and solvent choice; dependent on detection limit and proper chromatographic retention of analyte in solution
Transfer or direct injection of supernatant	Stability of extracts with or without retained protein content

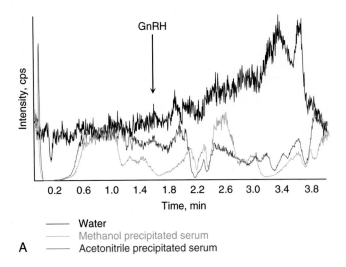
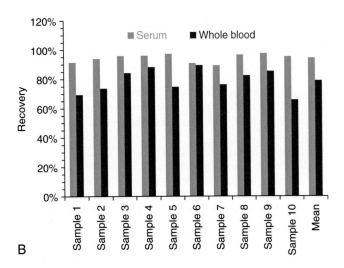
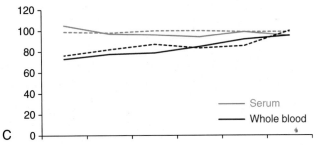

FIGURE 5.11 A, Postcolumn infusion data for gonadotropin-releasing hormone (GnRH) (*arrow* = retention time) after injection of water, methanol precipitated serum, and acetonitrile precipitated serum samples. **B,** Recovery of serotonin in serum versus whole blood. **C,** Recovery (*y*-axis) in serum and whole blood after mixing between 10 to 60 minutes, sampled in 10-minute intervals (*x*-axis).

the suspected cause of variable recovery. A subsequent timing study was performed, adding IS and analyte preextraction into 12 serum and 12 blood samples (Fig. 5.11C). Individual supernatant samples were tested from 10 to 60 minutes in 10-minute increments, and total efficiency was assessed as a ratio of analyte to IS peak areas versus the SST (sample C/sample A approach). Relative recovery of the ratio of analyte to IS was corrected after 60 minutes of mixing in whole blood and thus optimized. During the time-course study, IS peak area response decreased over time and the analyte peak area response increased; the culprit was determined to be platelets in whole blood.

Liquid-Liquid Extraction

LLE represents the most "tunable" preparative scheme that can be used in clinical sample analysis and one of the techniques most often used to affect sample concentration. LLE extracts are often the cleanest sample to inject,[27] with the ability to eradicate phospholipids and salts from the final extract, thus enabling flexibility in the HPLC separation and the ability to accelerate the HPLC gradient and maintain LC-MS/MS instrument uptime. A greater number of variables require optimization versus PPT; IS equilibration for equivalent recovery between analyte and IS solution (miscibility), pH control of the analyte (neutralize to enable partitioning), additives to enable partitioning (salt forcing), the LLE extraction solvent (just sufficiently polar for 20:1 S/N at LLMI and easy to evaporate), ratio of solvent to sample (minimum 10:1 starting conditions, titrate down after optimal solvents are determined), mixing time during LLE (~30 minutes, shorten later), separation of supernatant (freeze/pour off ~5 minutes in a dry ice bath ~−70°C for 5 minutes), backwashing to clean extracts (saponification with 50 μL of 1% sodium hydroxide, freeze, and pour off again), evaporation (time, temperature, gas flow, and gas composition—ideally nitrogen [N_2] to prevent oxidation) and reconstitution (injectable solvent chemistry, transfer volume, and recovery losses). At first glance, that is a lot of variables (Table 5.12). However, many may require optimization only if sample volumes are considerable and throughput is a concern (freeze/pour off, evaporation time/temperature) or recovery is suboptimal (salt forcing, lossless freeze/pour off, chaser solvent added to organic phase after pour off to prevent total evaporation [glycerol ~5 to 10 μL] and reconstitution).

An example of the power of LLE was published for urinary catecholamines[40] and uses the following ingenious process. On review of the analytes (dopamine, epinephrine, and norepinephrine), the first step in all decisions for preparative directions and planning, the common vicinal diol (catechol) moiety was observed (a benzene ring with two hydroxyls [OH] adjacent to each other). This functionality is rather rare in biochemistry and provides for the opportunity to extract selectively. Urine specimens were titrated to basic conditions (IS in aqueous base ~pH 9.0) and a cocktail of phenylboronic acid (PBA) and tetraoctylammonium bromide (TOAB) was added in octanol. The phenylboronic moiety forms a bond across the two vicinal diols in the catecholamines, with a phenyl ring at the other end at basic pH. The TOAB forms a secondary complex (van der Waals interaction) with the phenyl functionality and enables the three-component complex (TOAB-PBA-catecholamine) to partition into the octanol phase. After mixing for approximately minutes and centrifugation (~10 minutes at 1000 × g), the tube is placed in a methanol dry ice bath, the aqueous portion freezes, and the octanol is poured into a second tube containing aqueous acid (1% formic acid). Under acidic conditions, the PBA-catecholamine complex bond is disrupted;

TABLE 5.12 Liquid-Liquid Extraction Variables

Variable	Notes
Extraction solvent	Ethyl acetate, hexane, heptane, methyl tertiary-butyl ether (MTBE), diethyl ether, chloroform, toluene, dichloromethane, N-butyl acetate
Ratio of solvent to sample	Provisionally 10:1 with titration to optimum
Internal standard equilibration time	Seconds to minutes to hours
Mixing time post addition of solvent	Seconds to minutes to hours
pH of sample	1-14 to drive analyte to neutral, buffer capacity of sample with/without anticoagulants is important to control
Centrifugation time/force	Dependent on laboratory equipment and sample vessel
Evaporation/reconstitution	Time, temperature, gas flow rate
Separation of supernatant	Freeze and pour off, aliquot or possible direct injection (in HILIC mode)
Salt forcing	100 mmol/L to mol/L concentrations, dependent on desired pH
Reconstitution solution volume	Volume and solvent choice; dependent on detection limit and proper chromatographic retention of analyte in solution
Full evaporation prevention	Glycerol or other high boiling point solution

HILIC, Hydrophilic interaction liquid chromatography.

TABLE 5.13 Liquid-Liquid Extraction Experimental Recommendations

Variable	Notes
pH of sample	Adjust sample to +2.0 pH units from pK_a (bases) or −2 pH units from pK_a (acids).
Mixing time after addition of solvent	30 minutes (experimentally reduce time for efficiency gains).
Centrifugation time/force	Test plates or vials prior in centrifuge to ensure safety, 10-30 min (experimentally reduce for efficiency gains).
Pour off of supernatant	5-10 min in dry ice/methanol bath or 15-30 min in a −70°C freezer; ensure aqueous layer freezing. Transfer to clean tube.
Evaporation	35-50°C max initially. Low gas flow. Check often to establish expected time. Temperature can be modified experimentally for efficiency gains.
Reconstitution	Solvent determined from on-column detection limits and adsorptive loss studies. Volume dependent on on-column detection limits.

thus TOAB/PBA is retained in the octanol phase and the catecholamines partition into the aqueous phase. After a second mixing (2 minutes) and freeze/pour off step, the catecholamine-rich aqueous extract is ready for transfer into a vial or plate for injection. This example is illustrative of knowledge of chemistry to facilitate the selective extraction of highly polar molecules by a mechanism typically employed with nonpolar analytes, providing leverage for more selective analyses.

This detail is provided for the following consideration to set generic conditions: neutralization of ionizable groups—set pH of IS addition appropriately (analytes pKa + 2 pH units for bases, − 2 pH units for acids), LLE partitioning/mixing for 30 minutes, centrifuge to enable phase separation (the LLE container and centrifuge capacity are the key drivers—test dry containers first), snap-freeze (~5 minutes) and pour off to facilitate lossless LLE and relatively easy manual processing of tube-based LLE (Table 5.13).

Fig. 5.12 describes a solvent chemistry screen incorporating neat and mixed solvent chemistries covering a wide polarity index.[33] Notably, all mixtures are less dense than water; thus the removal of LLE solvent is performed by transfer of the upper layer rather than attempting to capture a bottom portion (the difficulties of this should be readily apparent when performed in high-volume assays). Polarity indices of mixtures are approximated as the sum of polarity indices for individual mixture components as a fraction of the total volume (ie, well B8—75:25 hexane-to—ethyl acetate approximate polarity = $0.75 \times 0 + 0.25 \times 4.4 = 1.1$) and provide coverage from a polarity index of 0 (hexane) to 5.8 (acetonitrile). The plate layout is designed around the spike-and-recovery approach[38] and incorporates four individual positions for the neat SST injection (sample A, wells 1, 8, 49, and 96); these samples are assessed for instrument drift across the batch and may be used to normalize response drift when rationalizing recovery differences. The top half of the plate (wells 2 to 48) comprises blank matrix extracts overspiked at the reconstitution step (sample B). The lower half of the plate (well positions 50 to 95) contains prespiked paired samples (sample C). For further consideration, this may be performed in tubes with transfer to a plate for injection or directly in a 96-well plate, as described here. The latter requires consideration of the total well volume and well-to-well wicking of organic solvents. Because of volume considerations, these studies are performed with 1.2 mL per well round well plates (well edges do not contact), sample volume = 100 μL, extraction solvent volume = 500 μL (900 μL in tubes). Addition of solvent is performed carefully, and the plate is chilled before sealing (aluminum foil heat seal), mixed (end over end, 30 minutes), centrifuged (10 minutes, 3000 ×g), frozen (methanol dry ice bath ~5 minutes), then unsealed for solvent transfer to a fresh 96-well plate for evaporation (2 hours at 50°C, 30 psi N_2—these conditions are optimized later). Samples are

A

	1	2	3	4	5	6	7	8	9	10	11	12
A	—	H	O	T	N	I	M	E	AC	H:E 90:10	H:E 70:30	H:3 65:35
B	H:I 25:75	H:I 75:25	H:M 25:75	H:M 75:25	H:N 25:75	H:N 75:25	H:E 25:75	H:E 75:25	H:T 25:75	H:T 75:25	H:O 25:75	H:O 75:25
C	T:I 25:75	T:I 75:25	T:M 25:75	T:M 75:25	T:N 25:75	T:N 75:25	T:E 25:75	T:E 75:25	T:AC 25:75	T:AC 75:25	T:O 25:75	T:O 75:25
D	AC:I 25:75	AC:I 75:25	AC:M 25:75	AC:M 75:25	AC:N 25:75	AC:N 75:25	AC:E 25:75	AC:E 75:25	AC:E 25:75	H:E 15:85	H:E 5:95	—
E	—	H	O	T	N	I	M	E	AC	H:E 90:10	H:E 70:30	H:3 65:35
F	H:I 25:75	H:I 75:25	H:M 25:75	H:M 75:25	H:N 25:75	H:N 75:25	H:E 25:75	H:E 75:25	H:T 25:75	H:T 75:25	H:O 25:75	H:O 75:25
G	T:I 25:75	T:I 75:25	T:M 25:75	T:M 75:25	T:N 25:75	T:N 75:25	T:E 25:75	T:E 75:25	T:AC 25:75	T:AC 75:25	T:O 25:75	T:O 75:25
H	AC:I 25:75	AC:I 75:25	AC:M 25:75	AC:M 75:25	AC:N 25:75	AC:N 75:25	AC:E 25:75	AC:E 75:25	AC:E 25:75	H:E 15:85	H:E 5:95	—

B

Polarity Index	Sample ID	Extraction Efficiency (%)	Matrix Effect (%)
4.4	Ethyl Acetate	85.3	8.8
1.875	75% MTBE: 25% Hexane	80.9	7.6
0.625	75% Hexane: 25% MTBE	105.3	−6.8
2.925	75% N-Butyl Acetate: 25% MTBE	118.5	4.3
0.975	75% Hexane: 25% N-Butyl Acetate	117.0	8.4
3.3	75% Ethyl Acetate: 25% Hexane	102.7	1.0
0.6	75% Hexane: 25% Toluene	106.6	−7.2
2.25	75% Isopropyl Ether: 25% Toluene	84.6	−6.3
2.35	75% Toluene: 25% Isopropyl Ether	81.7	5.5
2.425	75% Toluene: 25% MTBE	105.6	3.5
3.525	75% N-Butyl Acetate: 25% Toluene	121.8	−0.2
2.775	75% Toluene: 25% N-Butyl Acetate	102.3	8.4

FIGURE 5.12 A, A 96-well plate map for mixed organic solvent cocktails (hexane [H], octanol [O], toluene [T], N-butyl acetate [N], isopropyl ether [I], methyl-tert-butyl ether [M, MTBE], ethyl acetate [E], and acetonitrile [Ac]) incorporating pre/post spiking paired extracts. **B**, Exemplar data for plate based liquid-liquid extraction of progesterone from serum.

reconstituted in 100 μL injection solvent (no concentration factor to consider in calculations), mixed (∼10 minutes), centrifuged (10 minutes, 3000 × g), and subsequently injected.

The goals of the screen are to ensure matrix effects are controlled (<10%), that recovery is acceptable (>80%), and, where appropriate, removal of interferences has been enabled (phospholipids/isobars, etc.). Exemplar data are shown for the extraction of progesterone from serum in Fig. 5.12B, after analysis of samples and simple data filtering in Excel. Many LLE chemistries are provisionally acceptable based on the proposed selection criterion; the next considerations are selectivity, polarity index, and time to evaporate. In LLE, lower polarity index leads to better selectivity (generally) and faster time to evaporate. Further, solvent chemistries with polarity indices below 3 yield less than 1% phospholipid extraction (unless isopropanol is used, this is a rule not a law).[41,42] The data would be reviewed as follows: chemistries with polarity index above 3 are excluded, notably ethyl acetate is considered a universal LLE solvent, extracts will contain a significant degree of water from the sample, and evaporation times suffer. Further, isopropyl ether and n-butyl acetate can be difficult to handle, often requiring additional safety controls. Toluene has a relatively high boiling point (111°C), as does n-butyl acetate (126°C), leading to extensive evaporation time and/or temperature (evaporation temperatures >60°C should be carefully considered in the context of molecule or material degradation). In the absence of isobar selectivity information, three solvent chemistries are left for further optimization of solvent-to-sample volume ratio and minor changes to composition: 75:25 methyl tert-butyl ether (MTBE) to hexane (polarity index = 1.875), 75:25 hexane-to-MTBE (polarity index 0.625), and 75:25 hexane to ethyl acetate (polarity index = 1.1).

Supported Liquid Extraction

One significant drawback to LLE is the opportunity for scalability in terms of sample number when using tube-based extraction. A recent addition to the suite of preparation technologies, SLE,[42] provides for the ability to perform automated tube/96-well plate–based liquid extraction. The variables that require optimization are similar to those with LLE (see Table 5.13); IS equilibration for equivalent recovery between analyte and IS solution (miscible), pH control of the analyte (neutralize to enable partitioning), the LLE extraction solvent (just polar enough for 20:1 S/N at LLMI, easy to evaporate, and immiscible with water), ratio of solvent to sample (minimum 1.5:1 to 2:1 starting conditions, titrate down after optimal solvents are determined), evaporation (time, temperature, gas flow, and gas composition—ideally N_2 to prevent oxidation), and reconstitution (injectable solvent chemistry and recovery losses). Importantly, the ratio of solvent to sample is typically much less at optimal recovery than LLE, reducing hazardous materials as well as drying time of the extract. However, the technology obviates the need for determination of extraction time (use values that follow as generic), the use of high salt concentrations to force phase separation, backwashing (still an option but complicated), freeze/pour off, and losses from reconstituted sample transfer to the final vial or plate. The technology involves the incorporation of diatomaceous earth (celite) into a packing with membranes above and below. As with all solvent flow technologies with media-based extraction, gravity not excessive pressure is used when performing liquid addition steps (optimize later).

SLE comprises five steps and is performed with a collection device underneath the SLE cartridge or plate. After preparation of SST and fortified samples (as described in the LLE development step) and appropriate equilibration of the overspiked analyte, the following procedure is recommended:

Step 1: Addition of sample to the bed with a small (<5 psi for 5 seconds) positive or negative pressure pulse to initiate sample flow into the SLE bed.

Step 2: Wait 5 to 10 minutes until all liquid has passed into the media. During this step, water within the sample is absorbed into the diatomaceous earth (swelling the bed) and analytes are retained on the outside of the media.

Step 3: Addition of immiscible extraction solvent at a minimum of 1.5 to 2 times the bed volume ($\geq$300 µL and $\geq$600 µL for the 200- and 400-µL sample beds, respectively) with a small pressure pulse (<5 psi for 5 seconds) to initiate solvent flow into the bed. The solvent must be immiscible with water such that the entrained water is not reextracted from the diatomaceous earth. Options of extraction solvents and mixtures are shown in Fig. 5.12.

Step 4: Wait 10 to 15 minutes until all solvent has passed into the media.

Step 5: Complete extraction solvent recovery with a pulse of high pressure (>30 psi for >15 seconds). The extract solvent collected may then be evaporated and reconstituted as per LLE.

The significant gains in preparative efficiency (particularly with automated plate-based extraction technologies)[43] are also matched by the ability to incorporate extraction solvent mixtures that have a higher density than that of water but can provide cleaner extracts (chlorinated solvents). These solvents would normally be the lower layer of LLE-prepared samples, requiring the use of special laboratory equipment to facilitate aspiration and collection. Additionally, the extraction solvent volume requirement is significantly reduced relative to LLE. One can consider SLE as a clean elution solvent passing over a separation media versus mixing generating a solvent extract that becomes increasingly contaminated. Often, the solvent chemistry required for SLE is more polar for an equivalent tube-based LLE for the same analyte.

Fig. 5.13A describes a hybrid preparative solvent extraction scheme (scouting and stepwise) for analysis of estradiol in serum using a Biotage SLE+ 400 µL plate. Addition of 400 µL sample comprising 300 µL serum + 100 µL of water or 20:80 methanol to water was also performed (to potentially disrupt estradiol binding to sex hormone—binding globulin in serum samples). The historical LLE workflow for estradiol was validated with 80:20 hexane to ethyl acetate as a primary solvent chemistry with 90:10 hexane to dichloromethane as a backup option. These protocols were titrated in 10% steps to evaluate recovery differences with SLE versus LLE. Prespiking and postspiking reduced data (Fig. 5.13B) indicate that all extraction chemistries are ionization effect free; note that ionization suppression/enhancement is only one part of matrix effects. Further, optimal recoveries (>80%) require a greater component of ethyl acetate or dichloromethane than LLE (to date, all analytes follow this observation). As noted for LLE, just enough strong solvent is desired, thus approximately 50% to 60% ethyl acetate or dichloromethane is optimal.[43,44]

Solid-Phase Extraction

Off-line SPE represents another technique commonly used to concentrate samples before injection. The routine use of SPE involves desalting and deproteinizing samples during washing steps with subsequent elution of analytes to (ideally) a directly injectable solution, although evaporation and reconstitution are often incorporated. As with all solvent flow technologies with media-based extraction, use of gravity and not excessive pressure when performing liquid addition steps is advised (optimize later, slower is better). As per SLE, the analytes should be interrogated to ensure retention of analyte to the appropriate stationary phase.

A significant number of variables require optimization for SPE (Table 5.14); IS equilibration for equivalent recovery between analyte and IS solution (miscible), pH control of the analyte (and the SPE bed), SPE tube volume and bed size (related to expected sample size), SPE retention mechanism (functionality), wash solvent composition and number of washes, elution solvent composition and volume, and if concentrating extracts, the evaporation (time, temperature, gas flow, and gas composition) and reconstitution (injectable solvent chemistry and recovery losses). As noted in Table 5.15, ion exchange is a preferred mode of SPE because of the ability to perform extraction and HPLC that are orthogonal in nature together. Additionally, ion-exchange SPE offers the ability to use aggressive washing steps (enhancing selectivity) incorporating organic solvent and pH changes before elution. The use of generic hydrophobic SPE phases incurs significant limitations for selectivity (limited washing) into the analytical method, particularly when coupled to hydrophobic-based HPLC (reverse phase). Consideration should be made to the limitations of the manufacturers recommended conditions for operation. Almost exclusively these conditions are provided to maximize recovery (washing with 5% to 10% organic solvent, elute with 100% organic solvent) and not selectivity. Typically, the manufacturers' recommendations do not call for resolving phospholipids in the extraction

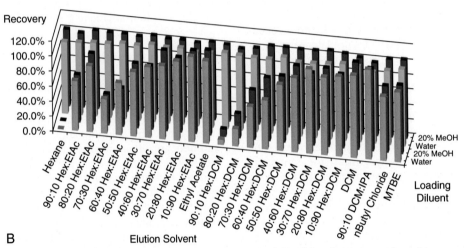

FIGURE 5.13 A, A 96-well plate map stepwise optimization together with solvent screening of hexane *(Hex)*, ethyl acetate *(EtAc)*, dichloromethane *(DCM)*, isopropanol *(IPA)*, N-butyl chloride *(N)*, and methyl-tert-butyl ether *(MTBE)*. **B,** Exemplar estradiol data indicating matrix effects and recovery as a function of disruption and extraction solvents.

process; empirical experimentation is needed to balance recovery and selectivity for the complete method.[41] Preference for polymeric SPE phases should be given. The ability to dry the SPE bed between steps is significantly more robust (practically) than silica-based phases in which the bed must be maintained wet.

The volume of sample (tube capacity) and bed volume also require consideration. The goal is to wash SPE media with at least 10 bed volumes for each step (1 mL for a 30-mg bed) and elute with 2 bed volumes (200 μL for a 30-mg bed), particularly with direct injection of the extract.

For strong ion-exchange media, the pH maximum required to neutralize (99% at $pK_a \pm 2$ pH units) the SPE bed and elute analytes is generally quite excessive because the stationary phase is usually permanently charged in aqueous solutions (<pH 1.0 for strong cationic phases usually incorporating aliphatic sulfonic acid moieties, pH greater than 14.0 for strong anionic phases usually incorporating quaternary ammonium moieties). The pK_a of the analyte is the deciding factor. An analyte that is a strong base requires a weak cation exchange SPE bed; elution is performed by neutralizing the SPE medium (usually aliphatic carboxylic acid) at approximately pH 3.0 or lower. Weakly basic analytes are extracted using a strong cation exchange SPE bed; elution is performed by neutralizing the analyte at approximately pH 11.0 or greater. Weak anionic exchange media usually incorporate amines and are also neutralized at pH 11.0 or greater.

SPE incorporates more than 5 (hydrophobic) or more than 6 (ion exchange) steps. Here, the latter will be described. Step 1 is the addition of a cleaning solvent (typically a strong elution solvent) to remove impurities from the SPE bed. This, as well as all wash steps, are transferred to waste. Step 2 involves conditioning the bed with aqueous solution to ensure that the charge state of the bed is controlled. This is generally neutral pH with weak buffering for strong ion exchange or high-purity water for weak ion exchange. Step 3 involves loading the sample to the SPE bed. As before, analyte charge state should be controlled through appropriate dilution with a buffer and pH control to enable opposite charge states of the SPE bed and the analyte and thus retention. This step must be combined with disruption of analyte from binding partners and equilibration of analyte/IS. Step 4 involves addition of solvent to elute unwanted interferences to waste (methanol, acetonitrile, etc.). Step 5 involves the

TABLE 5.14 Solid-Phase Extraction Variables

Variable	Notes
Stationary phase functionality	Ionic (anion, cation; weak or strong), lipophilic, hydrophilic, mixed-mode
Stationary phase bed volume	Dependent on sample size
Stationary phase tube volume	Dependent on sample size
Internal standard equilibration time	Seconds to minutes to hours
pH of sample	Dependent on desired interaction (ie, charged for ionic interaction); disruption of interaction by anticoagulants in sample is possible
pH of stationary phase	Critical in ionic interactions
Washing of stationary phase, presample addition	Removal of contaminants, preparation of stationary phase
Washing of stationary phase, postsample addition	Number of discrete washes for selective removal of interferences and matrix components while retaining analyte
Flow rate of solvents through stationary phase	Pressure driven or gravity driven; high flow rates can create channels through the bed
Elution of analytes	Volume and solvent selection for optimal recovery of analyte, retention of interferences and matrix components
Number of simultaneous extractions	Individual tubes vs. plate format. Both require manifolds for waste collection
Evaporation	Time, temperature, gas flow rate
Reconstitution solution volume	Volume and solvent choice; dependent on detection limit and proper chromatographic retention of analyte in solution
Full evaporation prevention	Glycerol or other high boiling point solution

TABLE 5.15 Example Ion-Exchange Solid-Phase Extraction Procedure Steps

Step	Process
1	Prewash SPE media with strong eluting solvent.
2	Condition SPE media with buffered aqueous solvent. pH dependent on exchange mode.
3	Load buffered sample (controlled for analyte-protein binding and internal standard equilibration). pH dependent on analyte and exchange mode.
4	Wash with organic to remove interferences and matrix components.
5	Wash with pH-modified organic to remove interferences and matrix components.
6	Elution of analytes with pH-modified solvent. Neutralization of SPE media, ion replacement with buffer or neutralization of analyte is an option, dependent on analyte and exchange mode.

SPE, Solid-phase extraction.

incorporation of pH and organic solvent to elute potential interferences and is generally performed at approximately pH 3.0 or 11.0 for strong cationic and anionic ion-exchange media, respectively, or approximately pH 6.0 or 8.0 for weak cationic and anionic ion-exchange media. As the determination of pH in organic solvents is impossible, the pH range can be compared to that in water or determined empirically via titration of the pH modifier in the solvent. Additional organic solvent and increasing stringency washes may be added after step 5 for ion exchange and are encouraged to enhance selectivity. Step 6 incorporates elution of analytes through organic and pH modification of solvent into the final plate or vial to minimize transfer losses (ideally, particularly if extract is subsequently injected). Relative pK_a and mode of ion exchange dictate the pH for elution to neutralize the appropriate component in the assay (bed or analyte); weakly basic analytes (SPE media) neutralize at approximately pH 11.0, weakly acidic analytes (SPE media) at approximately pH 3.0. The nature of the organic solvent is determined by recovery efficiency, adsorptive loss considerations, compatibility with the HPLC method (peak fronting at a sufficiently large injection volume), or volatility when evaporation and reconstitution is required.

Fig. 5.14A describes a multisolvent SPE protocol for a decapeptide (gonadorelin-related peptide, GnRH) on a weak anionic exchange 96-well plate (*left* for columns 1 through 6 is shown, postextraction spike; the right side [not shown] is a facsimile with preextraction spiking of GnRH). Preconditioning was performed with methanol, then water (1 mL each) before addition of 400-μL sample (300 μL serum + 100 μL water). The first number denotes the composition of methanol in wash 1 (50 to 5% from *row A to H*, remainder is water). The second number denotes the composition of acetonitrile in wash 2 (*columns 1 to 3 and 7 to 9* are washed with acetonitrile, *4 to 6 and 10 to 12* are not). The third number denotes the composition of methanol in aqueous bicarbonate buffer at approximately pH 9.0 in wash 3, *columns 1, 4, 7, and 10* are washed with aqueous bicarbonate buffer, *2, 5, 8, and 11* are washed with 30% methanol in aqueous bicarbonate buffer, the remainder (*3, 6, 9, and 12*) are washed with 50% methanol in aqueous bicarbonate buffer. The final number denotes the composition of methanol in elution at approximately pH 3.0 (10% to 80% from *rows A to H*, remainder is water). The SPE plate design has many internal comparators, and each extraction step was collected, evaporated (consideration for scaling down sample and bed volumes to enable faster evaporation), reconstituted, and injected to ascertain the presence of GnRH (four injected plates from this design of experiments).

Fig. 5.14B indicates the recoveries observed after aqueous methanol washing (*green*), step 3 with 30% methanol at pH 9.0 (*light blue*), 50% methanol at pH 9.0 (*red*), and finally increasing methanol elution composition from at pH 3.0 (*rows A to H, dark blue*). Note that neat acetonitrile did not

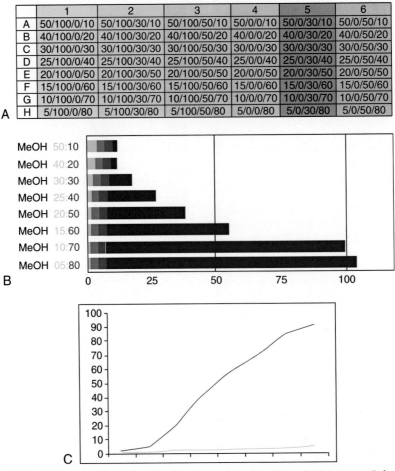

FIGURE 5.14 A, Example weak anion exchange solid-phase extraction 96-well plate map *(left side only)* stepwise optimization. The first number denotes the percent of methanol in water for the first wash. The second number denotes the percent of acetonitrile in water for the second wash (columns 4, 5, and 6 received no acetonitrile washing). The third number is the percent of methanol in an ammonium bicarbonate buffer at pH approximately 9.0. The fourth number is the percent of methanol in water at pH approximately 3.0 for elution. **B,** Exemplar data for gonadotropin-releasing hormone recovery of each fraction are shown with the fractions denoted by color. **C,** Breakthrough data for elution with increasing methanol (*x-axis,* 10% to 80%) with neat methanol *(green)* and acidified methanol (pH approximately 3, *blue*) indicating and increasing recovery (*y*-axis, as measured against postextraction spiking derived from columns 7 to 12).

recover GnRH and represents a good washing solvent, particularly between final basic washing and acidic elution to maximize the effectiveness of the elution solvent (lower final volume). Fig. 5.14C indicates a breakthrough plot as a graphical representation of the recoveries for aqueous methanol *(green)* and acidified methanol *(dark blue)*. The data generated enable consideration of the appropriate wash and elution compositions. The protocol described here incorporated solvent washing (steps 1 and 2), pH modification with solvent washing (step 3), and subsequent elution. More is always better for selectivity and LC-MS/MS instrument uptime but should be balanced against overall selectivity needs. To maximize the effectiveness of SPE, recovery should be sacrificed during washing; a composite of 20% to 30% recovery losses is acceptable unless sample volume and LLMI confounds. For the example shown, 50% methanol, 50% methanol pH 9.0, and acetonitrile were used as final wash solvents, with 70% methanol at pH 3.0 for elution at 200 μL (directly injectable to 75 μL).

Online Extraction (Online Solid-Phase Extraction and Turbulent Flow Chromatography)

Online extraction technologies require multiple HPLC pumping systems to enable recovery, washing and elution from an initial extraction column or cartridge, followed by transfer of the analytes to an analytical column (often requiring chromatographic refocusing, or chromatofocusing), with subsequent elution to the mass spectrometer for detection. It is often too complicated to perform prespiking or postspiking experiments (fraction collection after elution from a cartridge). The comparative evaluation in method development is generally made by total efficiency (sample C/sample A method, as noted earlier) where both analyte and IS are added before extraction to neat solutions and matrix samples, together with PCI to ascertain regions of ionization effects. Most of the off-line SPE variables still apply to online SPE; the technique is primarily used to inject relatively large sample volumes (50 to 200 μL) onto disposable cartridges for the analysis of endogenous analytes. The benefits of online SPE are speed of method development (often real time and screening multiple extraction media systematically), the potential for improved recovery (no evaporation or reconstitution), and reduced preparative time (dilution of sample with IS and buffer). Perhaps the only downsides to online SPE are the costs of hardware, particularly when performing ion-exchange SPE; these require a four- to six-solvent pumping system for the SPE cartridge and a secondary two- to four-solvent pumping system for the HPLC separation. Because of the similarity between SPE modes of development, it is recommended that a review of the details in Chapter 1 is undertaken. Additionally, an exemplar plasma metanephrines assay (metanephrine, normetanephrine, and 3-methoxytyramine) performed using a disposable cartridge-based platform with weak cationic exchange extraction and subsequent HILIC HPLC separation before MS/MS detection[45] is an excellent framework. The remainder of this section will detail the development and application of turbulent flow chromatography (TFC) with chromatofocusing and LC-MS/MS.[46] Many of the subsequent concepts are applicable to online SPE and should be considered during development.

The predominant difference between online SPE and TFC when deployed clinically is the use of disposable cartridges (SPE) versus a reusable extraction column (TFC), albeit SPE cartridges can be both reused (cleaned) and used repeatedly. The general rule is that online SPE cartridges are not reused to ameliorate carryover concerns. Online extraction is highly automatable for both sample preparation and analysis; thus it is advantageous to assay many molecules with this platform. However, the development challenge is to enable different analytes to be assayed using a generic solvent and column setup. The variables for consideration are sample diluent for disruption of binding and equilibration of IS (SPE also), injection volume used, sample loading flow rate, elution volume, composition and flow rate, TFC column functionality, chromatofocusing solvent composition and flow rates, analytical column functionality, HPLC solvent composition, and gradient (Table 5.16). For targeted development after the preliminary HPLC screening approaches, one part of the development process has been defined, namely, knowledge of the optimal column, solvents, and gradient. Thus the next challenge is to determine the extraction and focusing aspects of the TFC-LC method. It is common to exclusively deploy TFC-LC as a generic approach to combinations of drug analytes by class (ie, benzodiazepines, tricyclic antidepressants, etc.) using state-of-the-art triple quadrupole instruments (very sensitive) and not for endogenous molecules. The latter tend to require a more targeted approach to development, together with high-end MS platforms.[47]

The system configuration shown in Fig. 5.15A and B incorporates two valves and two pumps. The left-hand valve incorporates a fixed loop (100 μL), a three-groove rotor seal, the TFC extraction column (50 × 0.5 mm, 50-μm particles, Cyclone P), and the inlet from the loading pump/autosampler. Of note, the Cyclone P material provides broad extraction coverage and thus is not selective. Typical TFC column dimensions allow for extraction of more than 1 mg of material and have a dead-volume of 5 μL. The right-hand valve incorporates a waste outlet, a two-groove rotor, the inlet from the eluting pump, and the outlet through the analytical column to the MS system. Fig. 5.15C describes a TFC-LC method steps for sample loading, (1) transfer (chromatofocusing), (2) decoupling columns for HPLC gradient separation and TFC column/valve cleaning (3 and 4), loop filling and HPLC column clean-off (5), and reconditioning (6). Solvent chemistries are 0.1% formic acid for both loading and eluting pump weak solvents (A), 1:1 acetonitrile to methanol with 0.1% formic acid for loading and eluting pump strong solvents (B).

System operation is as follows: injected crude samples are passed through the left-hand valve and TFC column (*right to left*) at 2 mL/min for 15 seconds, this equates to more than four system volumes (including 100 μL injection loop) and 100 TFC column volumes (step 1). Unretained materials pass through the right-hand valve to waste. Both valves are switched (20.15B, step 2 in the method). The loading pump drives the prefilled solvent composition in the loop (also 100 μL, 20 TFC column volumes) through the TFC column in reverse elution mode (*left to right*) and into the right-hand valve groove for mixing. The elution step is usually performed at a lower flow rate to dilute organic solvent to enable analyte refocusing from the TFC column onto the HPLC column (50 × 2.1 mm, 5-μm Agilent XDB C18, a generic HPLC column) with the eluting pump. In the example method shown (Fig. 5.15; tricyclic antidepressants), eluting solvent composition for focusing (37.8% B) incorporates the HPLC starting conditions from HPLC screening minus the additive organic composition from the TFC elution step. It is recommended to start at 100% A for the elute pump to enable focusing and titrate up to starting conditions of the truncated HPLC gradient empirically (described later). After transfer and focusing, both valves are switched again. The loading pump solvent chemistry is

TABLE 5.16	Online Extraction Variables
Variable	Notes
Internal standard equilibration time	Seconds to minutes to hours.
Disruption of analyte binding partners	Acidic/basic solutions or organic solvents.
Sample diluent	Dependent on on-column detection limit/expected LLMI. Protein content may be deleterious to autosampler performance.
Loading flow rate	Chromatographic determination.
Elution volume	Chromatographic determination.
Elution composition	Chromatographic determination.
Chromatofocusing for transfer of analytes from extraction media to analytical column	Chromatographic determination.
Modification to predetermined LC separation	Interaction between two LC streams may result in changes to chromatographic performance from a one-dimensional method.

LC, Liquid chromatography; *LLMI*, lower limit of measurement interval.

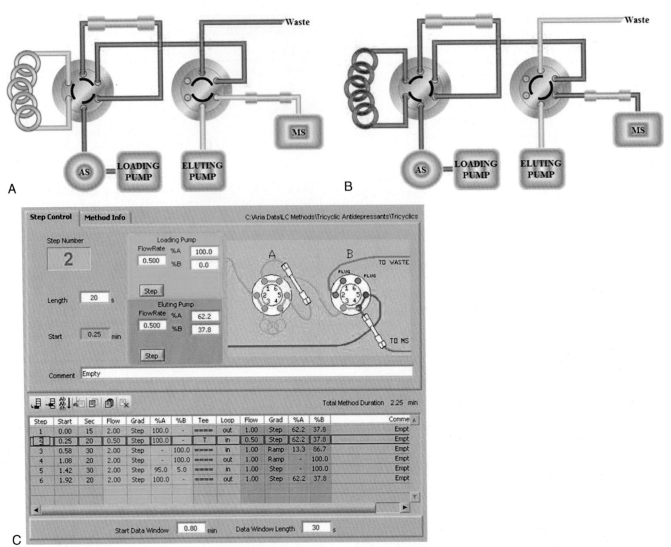

FIGURE 5.15 Turbulent flow liquid chromatography–liquid tomography (TFC-LC) plumbing configuration for sample loading **(A)** and chromatofocusing **(B)**. Exemplar TFC-LC method indicating loading pump *(blue)* and eluting pump *(magenta)* compositions, flow rates, and timing **(C)**. The transfer of analytes from the extraction column to the analytical column (step 2, *highlighted*) shows the reduction in flow rate of both pumps as solvent from each enters the analytical column. *AS*, Autosampler; *MS*, mass spectrometer.

changed to the strong solvent, and the flow rate is increased to clean the TFC column and left-hand valve (switched again in step 4) while the eluting pump provides gradient separation through the HPLC column to the mass spectrometer (including a 100% column clean at step 4). The loop is prefilled with the appropriate organic composition for TFC extraction in step 5 while the HPLC column is cleaned at 100% B. The left-hand valve is switched again, and both columns are equilibrated for the subsequent injection.

Initial experiments require injection of SST (100× LLMI) using the HPLC component only (sample A). Analytes are overspiked into samples at 1000× LLMI and diluted 10-fold with aqueous solution containing IS at 100× LLMI, (final concentration = 100× LLMI for analytes, 10× LLMI for IS) to create crude samples for injection (sample C). If complete or partial resolution of analytes and phospholipids was observed during HPLC screening, the overspiked matrix sample should be precipitated (10:1 acetonitrile or methanol) and injected as a secondary assessment of matrix effects or comparative recovery of phospholipids and analytes/IS. Aqueous diluent (with IS) should be prepared with diluent screening in mind by preparation of samples in 1% and 0.1% formic acid, neutral buffer (or water) and 1% and 0.1% ammonium hydroxide. The polymeric nature of Cyclone P is resistant to a wide range of pH conditions (1.0 to 14.0); be considerate of the injector and solvent lines when injecting samples of extreme pH. As always, prepare blank diluent samples to assess selectivity and clean the system between injections if necessary. For the majority of drug analytes, the addition of 1% formic acid at 10:1 diluent to sample yields a final injection solution at approximately pH 3.0 and enables release of analytes from albumin (isoelectric point ∼pH 4.7 in water at 25°C).

For many analytes, 0.1% to 1% formic acid dilution is universally used unless IS materials are chemically unstable when stored refrigerated, and a 1% formic acid diluted sample is usually used for initial development. Other diluents will be evaluated shortly after determination of extraction loop composition and eluting pump flow rate. Fig. 5.16A (*left*) shows injection of a 10:1 precipitated serum sample measuring benzodiazepines (B, nordiazepam, diazepam, clonazepam, norchlordiazepoxide, and chlordiazepoxide) and phospholipids (P). Partial resolution was observed from HPLC screening; however, only chlordiazepoxide and norchlordiazepoxide are observed because of the partial or complete coelution of phospholipids in positive ion ESI mode.

The first step in the TFC-LC scouting gradient involves extraction loop fill at 100% B after three injections of the neat diluent (ignore the first, column, and loop priming). Subsequent injection of the aqueous 1% formic acid 10:1 serum solution with 100% A eluting pump focusing eradicated phospholipids from the transfer and allowed focusing and detection of all five benzodiazepines (see Fig. 5.16A), clearly demonstrating the value of TFC under the method conditions described previously.[41,47] The first set of screening experiments involves ensuring that chromatofocusing occurs. The two variables to consider are the elute pump flow rate at 100% A and prefilled loop composition being transferred. Starting at 100% B in the extraction loop and 0.5 mL/min flow rate for both pumps, neither of the earliest eluting analytes (chlordiazepoxide or norchlordiazepoxide) are observed in the chromatogram (Fig. 5.16B, *left*). Notably, the acquisition window started at 0.8 minutes (after transfer) to maintain cleanliness of the source/ion optics; starting the acquisition earlier provides the opportunity to observe breakthrough (unresolved peaks during the transfer step). This observation was not unexpected; the HPLC column experiences a 100-μL injection of 50% organic from the blending of loop composition with weak solvent A; peak fronting was observed when injecting less stringent conditions during HPLC screening for these analytes. The next step is to reduce the effective organic composition; where HPLC system pressure overhead allows (hence the 50 × 2 mm, 5-μm HPLC column) increased elute pump flow can be performed. Fig. 5.16B demonstrates the influence of increasing dilution flow rate with effective retention and chromatographic control observed at 1.5 mL/min flow rate for the eluting pump during focusing.

The influence of loop composition on TFC-LC chromatofocusing and recovery is shown for a tricyclic antidepressant panel in Fig. 5.16C; the panel comprises nortriptyline, desmethyldoxepin, desipramine, amitriptyline, doxepin, imipramine, desmethylclomipramine, and clomipramine. The experiment employed the method detailed earlier with 0.5 mL/min 100% A for focusing with the elute pump. Reduction of the composition in loop from 30% to 20% marginally improved chromatofocusing for the early eluting analytes (doxepin and desmethyldoxepin). Further reduction in loop compositions leads to a reduction in overall recovery, but not much. Just enough extraction loop organic composition is needed for accurate and precise quantification of LLMI (and S/N >20:1, ideally); this consideration fundamentally enables selectivity, even using generic solvents and TFC-LC column chemistry. In practice, in excess of 100 analytes (in 5 to 12 analyte panels) are assayed using this technology approach from serum, plasma, blood, and urine, often validated in all four matrices.

The final step of preliminary TFC-LC development involves injection of the diluted matrix samples (with different diluents) using the same TFC-LC steps. Four key observations are made with these samples; first, qualitative determination of the presence of phospholipids as a function of diluent. Second, PCI determination of coeluting ion suppressing and enhancing composition transferred and eluted at the retention time of analytes (±20 seconds from earliest or latest eluting analytes). Third, in the absence of matrix effects, the relative recovery of ISs dissolved in various diluents versus neat injected SST (sample C/sample A method, previously, and ISs peak area ratio only). This step is included (and repeat injected after 24 hours) to ensure that the diluent and residual activity in the crude diluted specimen (enzymes) do not chemically alter the ISs (improperly labeled deuterated ISs materials; that is, those that use deuterium to replace acidic hydrogen atoms are prone to hydrogen-deuterium exchange in extreme pH conditions). Finally, with proven IS stability, total efficiency calculations are performed using the ratio of analyte and IS peak area (sample C/sample A method, previously) to enable final decision on sample diluent for direct injection.

Reduction of the stringency of the extraction loop solvent leads to improved selectivity but yields recovery losses. Fortunately, this is corrected by the appropriate choice of IS. Many of the panels described previously were validated with less than 40% recovery. Importantly, matrix effects were less than 5%. The addition of organic solvent to samples, in either loading mobile phase A or final autosampler wash solution should be avoided to prevent PPT in the injector port or syringe. Partial precipitation induces micro-seeding in the sample and will rapidly create blocked syringes, injector ports, inline frits, and TFC/LC columns (often all of these). Loading crude extract in nonaqueous solution induces proteins to precipitate on the TFC column, leading to degradation in TFC column lifetime. As an example from current TFC-LC assays, 10:1 aqueous diluted samples centrifuged at 3000 × g and injected (10 to 20 μL) with partial organic composition in the extraction loop (<50% B) allow for more than 20,000 injections before TFC and HPLC columns are replaced.

Assay Calibration and Carryover Assessment

Preparation of test solutions (SSTs) should be performed with the analytes at the LLMI and upper limit of measurement interval (ULMI, also known as upper limit of quantitation [ULOQ]) accounting for concentration and dilution factors and recovery losses determined in sample extraction experimentation (deriving expected on-column amounts). In most MS applications, the expected clinical utility will inform the concentration difference between the lower and upper limits of reference intervals. In the absence of informed values, a preferred starting concentration for the ULMI would be 500- to 1000-fold higher than the LLMI and adjusted as needed. Higher values for the ULMI often will contribute to substantial carryover in an autosampler. As an example of SST preparation, consider an assay with a 10-fold concentration factor and 80% recovery; SST would be prepared at 8× LLMI (and 8× ULMI) in neat injection solution. From development studies, short-term stability of SST in the

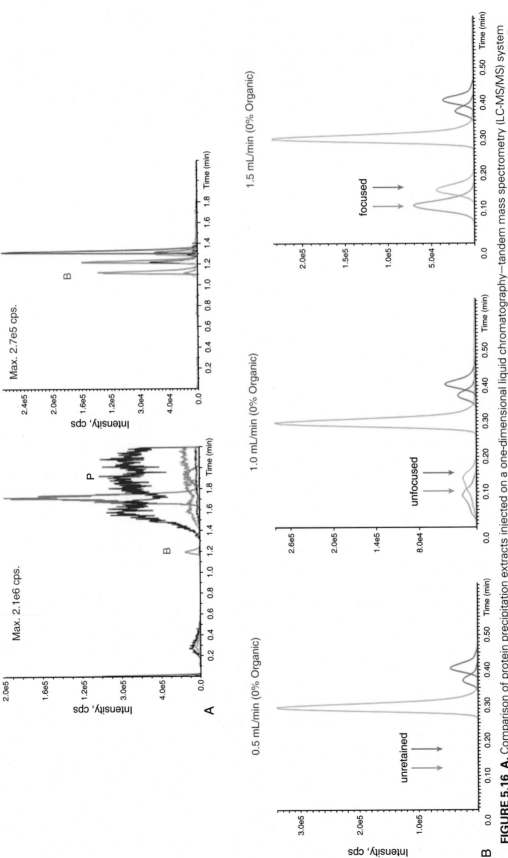

FIGURE 5.16 A, Comparison of protein precipitation extracts injected on a one-dimensional liquid chromatography–tandem mass spectrometry (LC-MS/MS) system versus turbulent flow liquid chromatography–LC-MS/MS (100% B in extraction loop) indicating reduction in phospholipids (*P*) when analyzing benzodiazepines (*B*). **B,** TFC-LC development indicating increased chromatofocusing flow rate (weak solvent) provided by eluting pump for a benzodiazepine profile. **C,** Impact on recovery seen through reduction in eluting solvent composition in the loop for tricyclic antidepressants.

CHAPTER 5 Development and Validation of Small Molecule Analytes

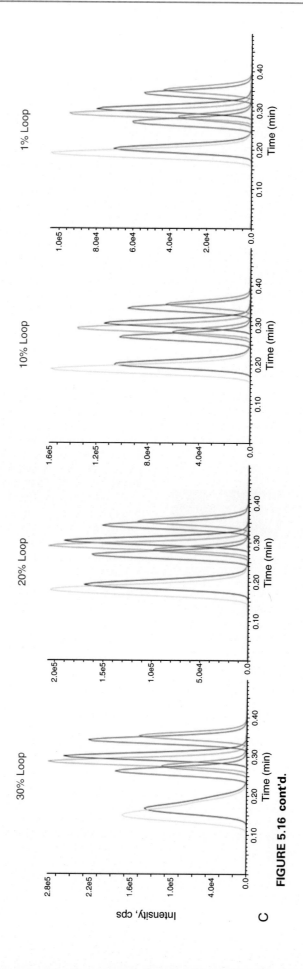

FIGURE 5.16 cont'd.

autosampler should be somewhat known. If in doubt, prepare SST fresh (from stable stored stock solutions or freshly prepared stocks) until autosampler or storage stability is proved. A second SST will include analytes at LLMI and IS at 10× to 25× LLMI for analyte (accounting for preparative factors described). The IS addition to the SST is used to confirm the retention characteristics of the analytes (particularly in complex multianalyte assays with isobars). Prepare a neat solution of analytes at the ULMI (ULOQ, accounting for preparative factors). Finally, prepare a neat solution of IS only (10× to 25× LLMI of analyte, again accounting for preparative factors). Injections of blank diluent, analyte LLMI/IS SST, analyte ULMI SST, blank diluent again, and IS SST (three times for each solution) are performed.

Careful critique of this injection series should be performed. Data analysis is as follows and shown in Table 5.17; the first set of blank samples should demonstrate no response for analytes or IS after the first injection. Analyte/IS SST should demonstrate coelution of analytes and IS (if labeled version is used) and identical peak shape for all transitions (if the qualitative transition is sensitive enough). All retention times should be consistent for analytes/IS in the triplicate injections to ensure appropriate gradient reconditioning is occurring. Analyte ULMI SST should demonstrate no contribution to IS transitions. The second set of blank samples should demonstrate no response for IS; analyte response is compared to analyte LLMI SST response to ascertain the degree of carryover in the first blank (ideally <20%). IS SST should demonstrate no response for the analyte (or minimal, assuming carryover is mitigated. The IS should be evaluated for conversion to the analyte (by hydrogen-deuterium exchange) by analyzing the area ratio of the analyte to IS over a number of days compared to the initial injections. Second pass review focuses on the LLMI and ULMI SST samples. As noted previously, the least sensitive analyte (LLMI SST) incorporating expected specimen volume and the need for repeat injections (ideally two, with residual volume left behind) should be the most critical analyte to focus on in assay panels. A minimum S/N ratio of 20:1 is desired for this analyte for at least one transition. Modification to sample injection volume (or concentration factor in preparation) should occur in optimization.

Review of the ULMI SST is performed, whereby transitions for each analyte should describe the same peak shape, particularly at the apex. Modification of CE is performed to reduce the instrument response for intense transitions, particularly when assaying analyte panels. CE or CE values can be informed from analyte infusion experiments and comparing peak shape of less-sensitive transitions for the same analyte to determine the degree of signal reduction required for identical peak shape across all transitions ($n \geq 4$). Some degree of knowledge of the MS detector linearity supports this refinement. However, linearity is empirically assessed in the following section in conjunction with sample extraction. After modifications to the CE, a reinjection of the LLMI SST should be performed to confirm preservation of appropriate response at the modified CE.

After transition normalization, the sample series is reinjected, assessing as detailed earlier. Finally, the precision of analyte LLMI/IS SST injections is calculated as a ratio of peak areas for all analyte transitions after 10 injections. Where two IS transitions are included, the matched transition is used for the respective analyte transition (usually top two) and the most intense transition for the other transitions. The precision of the area ratio goal is less than 10% (CV). This target is tighter than the acceptable variation at the LLMI (<20% CV) but is a good guide for what could be achievable with the transitions selected because it represents a best case scenario (no matrix effects, pipetting variance, or recovery losses). Modifications to the injection volume can be considered, reinjecting the sample series and reviewing again, to provide precision confidence for at least two transitions per analyte (ideally).

At this stage, manufacturing of assay calibrators from a fresh preparation of stock solution/lyophilized powder is performed. This is also time zero for stored stock and frozen calibrator stability studies. As described previously, gravimetric approaches must be used. Prepared calibrators (and stock solutions) should be sub-aliquoted and frozen (stability to be determined unless noted in literature in the calibrator matrix/stock solution). For exogenous analytes (drugs), analyte free matrix for the major expected sample type should be used. For endogenous analyte assays, either charcoal-stripped or synthetic matrices are routinely used; be considerate of isotonic and protein content for both matrices to minimize adsorptive losses between calibrators and samples. The number of nonzero calibrators is determined by the measurement range, minimum six levels (evenly spaced) or eight levels for $3*\log^{10}$ curves (1000-fold range).

TABLE 5.17 Data Analysis for Preliminary Calibration and Carryover Assessment

Sample	Data Component	Criteria
Blank solution	Analyte and IS contribution	No response at expected retention time for analyte(s) or internal standard(s)
System suitability solution (SST), all concentrations	Coelution of analyte and IS	Analyte and stable labeled IS coelute; similar peak shape across all transitions
ULMI SST at extracted concentration	Detector saturation	Refer to text
IS SST at extracted concentration	Analyte contribution from inefficient labeling/hydrogen-deuterium exchange	No response at analyte retention time
LLMI SST at extracted concentration	Signal to noise	Minimum 20:1

IS, Internal standard; *LLMI*, lower limit of measurement interval.

Selectivity

The next step in the process is to assess the compendium of potential extraction and analyte/IS transitions to reduce the experimental space moving forward. The number of specimens used for assessment expands further as the variables space gets smaller. A matrix pool (or separate disease and healthy pools) is generated for endogenous analytes (spiked for exogenous drugs analysis), sub-aliquoted, and stored for assessment of stability frozen/freeze—thaw conditions. For assays designed to measure more than one matrix (serum [red top tube] and serum separator tube, ethylenediaminetetraacetic acid [EDTA] plasma and lithium heparin plasma, EDTA whole blood, etc.), a pool of each matrix type is prepared, sub-aliquoted (as earlier), and assayed. Finally, one pool (seven replicates) is spiked with one of seven potential drug interference solutions (Drug Interference Mix, Ceriliant, Round Rock, Texas) to assess selectivity for common over-the-counter medications and imbibed substances (caffeine, acetaminophen, etc.). These materials are provided as "snap and spike" solutions at appropriate testing levels based on expected circulating concentrations observed in patient samples.[48]

The assessment of selectivity is performed with the following test samples: blank calibrator matrix (diluent added instead of IS), blank calibrator matrix + IS (10× to 25× analyte LLMI, accounting for volume differences between sample and IS), calibrators ($n \geq 6$ nonzero), carryover assessment blanks (determined from previous study), negative control (blank + IS), matrix pools (unspiked and spiked with interference test compounds), and samples to fill the remainder of the plate (Table 5.18). The nature and number of samples is an important consideration. For exogenous drug assays, a mixture of expected drug negatives and positives is desired. For endogenous analytes, putatively normal and abnormal or diseased population samples should be included ($n \geq 12$ per disease state; more is always better). Analysis of blank SST (clean system) and analyte LLMI/IS SST injections should be performed to ensure appropriate performance of the LC-MS/MS instrument before injection of the assay batch; consider performing SST injections between each set of extraction conditions to assess instrument drift. For further consideration, the optimal HPLC conditions will be used in the first iteration of the experiment; residual samples can be reinjected on second and third HPLC column/solvent chemistries in the event of method failure. Where on-column detection limits (and thus injection volume) are confounding, extract multiple plates using the same plate layout and combine extracts for repeat injections.

Following successful injection of the plate (review SSTs before, during, and after injection to assess instrument performance) a very detailed review of each individual subset of conditions is performed, with comparison between conditions as a secondary review (Table 5.19). Data review is as follows. Each extracted blank sample should contain no analyte or IS response at the retention time of the analytes/IS (maximum interference <20% analyte LLMI, <5% IS response).[49] This confirms that the solvents and calibration matrix are clean of interferences. Corrective action includes sourcing alternative matrices or walking back through the preparative solvents (evaporate, reconstitute, inject) to isolate contaminated materials. Where excessive contamination is observed, confirmation that the analyte/IS is present as a

TABLE 5.18 Selectivity Assessment Development Plate

Selectivity Test Samples	Replicates	Injection Order
Blank calibration matrix, no IS	1	1
Blank calibration matrix, with IS	1	2
Calibrators	1	3 through 8
Blank calibration matrix, no IS	1	9
Matrix pool, unspiked	1	10
Matrix pool, spiked with interference compounds	7	11 through 17
Blank calibration matrix, with IS	1	18
Individual samples	Remainder of 96-well plate positions	19 through 96
SST solutions	9	Pre-, mid-, and post-batch

IS, Internal standard; *SST*, system suitability test.

TABLE 5.19 Development Phase 4: Assay Qualification

Experiment	Outcome
Assessment of calibration scheme	≥6 Nonzero calibrators distributed from the LLMI to the ULMI.
Assessment of calibration linearity	Defined calibration curve fit and weighting. Truncate calibration range or deoptimization of MS/MS transitions that induce detector saturation.
Carryover assessment	No contribution from highest known concentration (ULMI) to a following blank injection.
Selectivity, interfering substances	Provisional screen of common compounds demonstrates no response change in the analyte or IS. The effect on quantitative measure is negligible.
Selectivity, final MS/MS transitions	Selective transitions with requisite sensitivity selected and used for transition ratio monitoring for both analyte and IS. Reassessment of scan times for the new method.

IS, Internal standard; *LLMI*, lower limit of measurement interval; *MS/MS*, tandem mass spectrometry; *ULMI*, upper limit of measurement interval.

contaminant is determined by transition ratio analysis (discussed later). The second sample, blank + IS, is compared to the first blank to assess whether the IS converts to analyte during the sample preparation (or is also contaminated). There should be no analyte response in this specimen (or <20% analyte LLMI, consistent with first blank). The calibration curves are analyzed to determine the degree of linearity of the analyte/IS peak area ratio response (y-axis) versus concentration (x-axis) for each transition and response relative to the double blank and blank + IS. Notably, only nonzero calibrators are used. Alter the weighting ($1/x$, $1/x^2$) or the fit (linear, quadratic) until the calibration curve demonstrates the lowest residual error (accuracy at each point is between 85% and 115% (80% to 120% at the LLMI), with a well-correlated data set ($r = 0.99$ or better).

Review of the correlation coefficient and shape of the calibration curve is performed to determine the requirement for further transition detuning (CE modification). Fig. 5.17C indicates the effect of optimized versus detuned transitions for oxycodone, where the optimized transition clearly demonstrates quadratic behavior and an unacceptable correlation coefficient ($r > 0.995$ is optimal). The detuned transition demonstrates a greater degree of linearity and calibrator agreement ($r = 0.9993$ vs. 0.9822). Quadratic-fit curves are acceptable and often necessary based on the wide dynamic range of expected specimens (normal and disease). Fitting a linear regression to inherently quadratic data is observed by negative bias (accuracy <100%) at the LLMI and ULMI and positive bias (accuracy >100%) in the middle portion of the calibration curve. It should be noted that a linear regression may exhibit a quadratic fit, which may represent the most appropriate curve shape to adequately address the needs of the assay. Evaluation of the back-fit accuracy of the lowest three calibrators (bias >20%) and measured peak areas for analytes versus blanks is the first clue toward transitions that are inherently not sensitive enough for further consideration. Additionally, the relative agreement of back-fit accuracy and curve shape should be conserved between transitions (with appropriate sensitivity). Quadratic behavior in all curves points to ion source saturation (reduced IS response with increasing concentration of analyte). When one curve is linear and another quadratic, this points to detector saturation, which may require modification of CE (reduction in product ion yield) to ameliorate. It is advisable to reduce CE rather than increase; less energy imparted often reduces the potential for interfering ions to fragment to provide the same product ion as the analyte.

The peak area response in carryover blanks ideally should be less than 20% analyte LLMI and less than 5% IS observed in the blank + IS sample.[49] Where unacceptable, carryover is observed, increase the number of autosampler wash cycles and solvent stringency until resolved, ensuring that the last wash solvent used does not affect the retention of analytes, which may lead to peak fronting or pressure increase at injection if inappropriate.

Review of the negative control (blank + IS) includes assessment of increased background noise (vs. samples 1 and 2) and consistency of IS response (vs. sample 2 and calibrators, ideally within 10%).

Review of matrix pools with and without fortification of concomitant medications, together with patient samples, is a key determinant between final transitions, extraction conditions, and HPLC conditions. Fig. 5.18A shows the peak area response for four temazepam transitions observed in the third calibrator. The peak area responses for each transition are generated exclusively from temazepam and serve as a fingerprint for the unequivocal presence in a sample. The interrelationship (ratio) of ions is used to confirm confidence in assignment when analyte is present.[29] In the example shown (see Fig. 5.18), the transition ratio between the most sensitive transition, m/z 301 to 255 and m/z 301 to 193 is 10:1, approximately 6:1 to m/z 301 to 177 and 20:1 to m/z 301 to 125. Further, the ratio of the second most sensitive transition m/z 301 to 177 to the other transitions is 6:1, approximately 2:1 and 3.5:1 (and so on). Fig. 5.18B indicates an interfering peak response for a drug-free sample; this is obvious because of the different peak shape versus calibrator (*left*) and lack of coelution with temazepam IS (not shown), but it is primarily due to the lack of transition ratio concordance between transitions. In this case, the m/z 301 to 177 (second most sensitive) transition should present a clearly observable peak with a peak height of approximately 2500 counts/s (scale is 1000 counts/s). Failure in transition selectivity, coextraction and complete or partial coelution of interferents may be resolved using different extraction chemistry or alternative transitions (ie, a return to a previous section of method development). The example shown indicates results from the TFC-LC-MS/MS analysis of temazepam; thus the extraction column and solvent chemistry was somewhat fixed. The obvious solution is to assess the second transition that was selective for all drug-negative samples and demonstrated the appropriate back-fit accuracy within the calibrators.

Finally, the IS response and transition ratio are evaluated within calibrators and among sample types. The precision of IS peak area response should be less than 15% within the calibrators and between calibrators and samples. Identify differences in IS response between different matrices and calibrators. These differences point to uncontrolled IS mixing, adsorptive loss differences, or coextracted interferences. Each may affect accuracy and precision (evaluated later). Second, evaluation of the IS peak area ratio is required. The two most sensitive transitions are compared by ratio of the lowest intensity (qualifying ion) to the highest intensity peak (quantifying ion). The ratio of IS transitions ions is generated using calibrators (or a neat solution). The ion ratio of the IS in the remaining samples should be within ±10% because of the relatively high concentration of IS added. Outliers should be reviewed carefully, particularly those from addition of concomitant drugs or alternative matrices, and IS transitions that are selective should be chosen. The selective transition will be a point of focus for the remainder of the assay refinement and, if acceptable, will be used to release patient samples after validation. This is the quantifying IS transition. The second IS transition is retained and the ratio of ions evaluated (<10% bias from calibrators) for all samples such that the second transition qualifies the selectivity of the quantifying ion (second transition is termed the *qualifying ion*). Notably, test samples containing drug and all endogenous samples should be reviewed for analyte transition ratio agreement when compared to the ratio generated from known samples (SSTs, fortified blank matrix, etc.). Tolerances for an assay

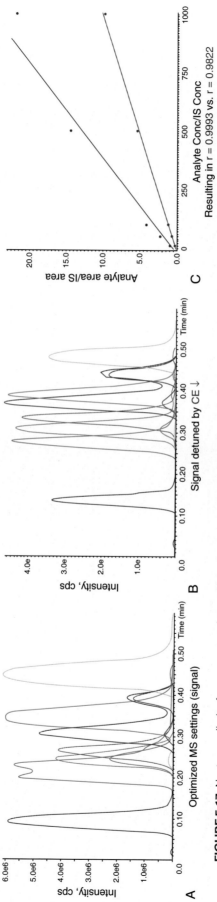

FIGURE 5.17 Neat upper limit of measurement interval injection for opiates before **(A)** and after **(B)** detuning collision energy *(CE)*. Note detector saturation observed in *light blue trace*, indicated by apparent peak splitting. Resulting improvement in calibration curve linearity for oxycodone following CE detuning **(C)**. The *blue calibration curve* demonstrates a quadratic fit (negative bias at highest concentration standard) caused by nonlinear detection. *MS,* Mass spectrometer.

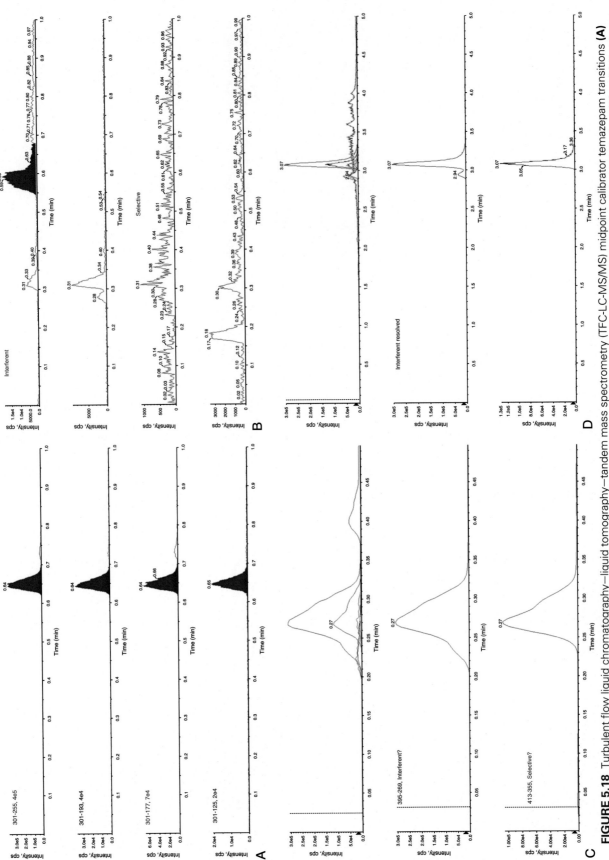

FIGURE 5.18 Turbulent flow liquid chromatography–liquid tomography–tandem mass spectrometry (TFC-LC-MS/MS) midpoint calibrator temazepam transitions (**A**) and extracted drug-free sample (**B**). 25-hydroxyvitamin D_2 quantifying/qualifying transitions in a dosed patient sample from the scouting LC gradient (**C**) and following fivefold gradient pitch reduction (**D**). Observation of peak fronting differences between different transitions of the same analyte initiates the gradient change. Resolution of unknown interfering molecular species demonstrates a nonselective transition.

in production can be wider than ±10% for analyte transition ratios because of different concentrations of analyte and thus S/N ratio. However, transition ratios will be monitored henceforth. After detailed review using the optimal HPLC column/solvent combination, if transition ratio failure persists, consider reducing the gradient pitch or reinjecting this sample set on the second and third best HPLC column options identified previously.

Fig. 5.18C indicates analysis of 25-hydroxyvitamin D_2 from a dosed patient sample measuring potential quantifying (m/z 395 to 269) and qualifying (m/z 413 to 355) ions. The different precursor ion masses arise from partial dehydration of 25-hydroxyvitamn D_2 in the ion source (500°C), which generates more than one precursor ion population. The extracted ion chromatograms demonstrate significant differences in peak shape, with peak fronting for the sensitive transition and peak tailing for the second transition (smaller for the first transition)—the question is which one is correct (transition ratio failed). Notably, the D_6-labeled IS in the sample eluted slightly before the analyte[23] and could not be used to differentiate which peak shape was correct. The gradient part of the HPLC screening method was extended fivefold and the sample reinjected (see Fig. 5.18D). Resolution of an interferent was observed for the most sensitive transition and chromatographic peak shape agreement, and the transition ratio was acceptable after gradient modification for the second peak in the chromatogram.

Having reduced the number of transitions to those demonstrating selectivity, retaining at minimum a single qualifying transition in addition to the quantitative transitions, the scan times should be readjusted (see HPLC development section on peak width scan time relationships). Reinjection of a calibration curve to ensure appropriate performance after this reduction in the number of transitions is recommended.

Final Method Refinement

Extracted blanks, calibrators, and patient samples are used to finalize the interface conditions, focusing on reduction in chemical noise in the chromatogram without sacrificing measurement quality at the LLMI and ULMI. This is generally performed with a full 96-well plate (allowing for reinjection). As previously, the quality of the assay is assessed for contamination in precurve blanks, linearity, and back-fit accuracy of the calibration curve (both transitions), carryover and negative control contamination, precision (and accuracy for exogenous analytes) of replicates of the QCs ($n = 3$) and patient samples. This batch tends to be the first time a full plate is injected as the "final" method, and reinjection of the calibration curve at the end of the assay should be performed to assess instrument drift. Here, both curves are overlaid in data review. Curve divergence may be indicative of stability issues in either samples or the LC-MS/MS method. A systematic evaluation (empirical and developmental review) of each step with consideration to time should be performed, from the repeat injection of samples within a plate all the way back to sample thawing. A lack of drift provides confidence that a full 96-well plate can be assayed in subsequent experiments. Review of analyte and IS chromatograms should focus on S/N (baseline noise particularly) to highlight specimens that may be used to refine interface conditions and HPLC gradient pitch. Repeat injections of specimens should be performed, with a repeat of the baseline conditions in the same time frame to ensure that observed improvements are not a function of time (or interface cleanliness). Review of transition ratio outliers is performed to assess the potential for untested interferences to affect measurement accuracy. At this stage, the assay has ideally minimized the influence of known interferences such as drugs, phospholipids, salts, and deuterium losses. Review of transition ratios thus becomes an exercise in generating knowledge on observed but unidentified interferences,[50] as well as rules for when to apply transition ratio tolerances and how wide they should be. Often the qualifier transition lacks sensitivity and cannot be used near the LLMI or has wider acceptance criteria at lower concentrations.[29] As noted earlier, sample preparation modification, HPLC gradient (or column) modification, and alternative transition selection are tools to use in solving assay failures. Finally, the precision of the QC pools is assessed within this first batch (CV <15% for three replicates of each QC and, if analyte-free matrix was spiked, accuracy of QCs also should be determined also [bias <±15%]).

After provisional determination of selectivity, precision, and method drift, efforts should turn to accuracy assessment and generation of QCs spanning the calibration curve range. QC pools should be prepared in the matrix (or matrices) considered for clinical testing. It is best practice to prepare QCs from a separate stock preparation, ideally with accuracy targets (for exogenous analytes) and use of gravimetry/class A volumetric glassware. When available, reference materials with assigned concentrations should be included to assess the trueness of the assay. Commercial controls are rarely an authentic representation of the test sample, so they should be used with caution. The specific concentrations and number of QCs is related to the clinical use of the assay. As an example, therapeutic drug monitoring assays usually require three concentrations of QC, a negative control (blank), a therapeutic normal, and a toxic/medical alert level when reporting results. The key component is ensuring QCs meet the criteria for medical decisions. In endogenous analyte measurement, this can become much more complicated. For instance, testosterone concentrations vary widely based on age and gender, with 21 discrete normal reference intervals spanning a 300-fold concentration range.[51,52] With abnormal results significantly higher, it is clearly unfeasible to create 21 separate QCs for validation. The position of QCs relative to the measurement range has been considered extensively for drug development assays[53] with QCs in the appropriate test matrix at 3× LLMI and 80% ULMI, plus one in the midrange. Where multiple reference intervals are expected within the same matrix, it is prudent to prepare QCs, including a pool for the predominant test population, one at 3× LLMI (by dilution with negative calibrator matrix if needed), one at 80% ULMI, and at least one other at the midpoint of the calibration curve range (or representing expected values for disease), thus a total of four QC specimens for the predominant testing matrix. If alternative matrices are expected to be tested (ie, plasma, whole blood, and urine samples measured by a serum-based calibration system), at least one level of QC should be prepared for the alternative matrices by pooling (for endogenous) or two levels by spiking for exogenous drug analytes (therapeutic/toxic). If in doubt, consultation with the medical director for guidance is recommended. The goal of QCs in the final development process is to ascertain matrix influence on measurement; the assay should measure analytes selectively, precisely, and accurately in all matrices.

Confidence in measurement accuracy (vs. calibrators) is determined by spike and recovery and dilutional mixing studies for healthy and diseased specimens together with QCs. These studies will be further detailed in prevalidation and extensively in validation studies described later. As a rule of thumb, the more studies the better and early assessment of accuracy and precision is favored. Additionally, qualified reference materials (with an associated higher order reference method) should be tested in the assay at this point. There is a general dearth of these materials; external proficiency samples are a poor substitute because they are rarely provided with accuracy targets (and sometimes are not in human matrix). That being said, they should be considered for inclusion in method refinement and assessment at this juncture because they will play a significant role for independently ensuring method performance when analyzing clinical specimens. This may be an appropriate time for method comparison with another validated assay, if available, particularly when looking to transfer a reference interval from a predicate assay.

Assay Prevalidation Studies

Although it may not be evident in either this chapter or in most publications, method development is largely an exercise in failure. Indeed, laboratory notebooks should be filled with examples of how not to analyze a particular analyte by LC-MS/MS, with the end result being, at minimum, a single pathway to a successful assay. The proof of this success is shown through validation; however, validation can elucidate errors not specifically addressed in method development. Validation is an exhaustive undertaking and is often associated with deadlines to other individuals; the initiation of validation should be delayed until certain prevalidation studies are performed.

Prevalidation studies are a small subset of experiments performed at the completion of method development and before the drafting of the validation plan. In large part, these experiments are constructed from the framework provided in the US Food and Drug Administration's (FDA's) Guidance for Bioanalytical Method Validation.[53] In three small batches, the probability of successful validation will be largely increased. The individual assay characteristics to be evaluated are shown in Table 5.20. Additionally, the experiments performed in the three small batches will be replicated on a larger scale during the course of validation; acceptable results during prevalidation can indicate that a failure during validation was caused by a random event as opposed to a fundamental flaw in the analytical procedure or platform.

The three small batches share many of the same characteristics. Each batch should have a calibration curve at the beginning and end of each sample set, with a double blank (extracted sample without analyte or IS) and a blank with IS to precede the first calibration series in a batch. Additionally, separate blank with IS samples, called the *carryover blanks* hereafter, should immediately follow the highest calibrator in each series, beginning and end. Also included in each batch are the LLMI and ULMI in six replicates. These additional replicates are measured as unknowns and are not included in the calibration curve regression. These samples are used to provisionally assess the accuracy and imprecision, both intraassay and interassay. Each QC should also be included in each batch at 6 replicates. Accordingly, a total of 18 replicates (6 replicates in three batches) for the LLMI, the ULMI, and each of the QCs are assayed.

Also important for these three batches are additional lots of the blank matrix (four to six, if available) that will be used in the generation of calibrators. They should be purchased from the vendor at a small volume (enough to extract at least 10 times; the actual volume depends on the sample aliquot needed for extraction). These additional lots shall be used to demonstrate that materials are available to prepare calibrators during stability evaluations to be performed in validation. These additional lots are primarily for assays for endogenous analyte that use charcoal-stripped matrix. The charcoal-stripping process may not remove the residual analyte of interest because the endogenous concentrations vary by donor. Using testosterone as an example, serum donated largely by females would have much less endogenous testosterone than a separate lot donated exclusively by 20- to 25-year-old males. Residual testosterone may be present in the latter example even after numerous charcoal-stripping steps (in which case the matrix is now largely water anyway). If the latter lot were to be used in calibration preparation, a fixed concentration bias would be observed.

Twenty to forty patient samples, preferably analyzed by a comparative assay (if used in interassay comparisons) should be accrued, together with the comparator assay calibrators. In the absence of method comparison results, these samples may be used to initially assess the agreement with literature-based reference intervals for endogenous analytes. The generation of new reference intervals is discussed more completely elsewhere (discussed later in "Assay Validation"[54]). Most importantly, these samples provide details regarding the potential range of transition ratio results expected in validation, together with additional confidence in the transitions selected for analyte and IS measurement. Finally, these samples should be analyzed again using a brand new column (from a different manufacturing lot) so that a single batch can be analyzed on the column used during method development as well as a new stationary phase.

Having sequestered these materials (ideally purchased earlier during method development to prevent delays), the three batches are constructed. In batch 1, only the samples described earlier are included: a double blank, a blank with IS, the calibration curve, a carryover blank, six replicates each of the LLMI, QCs, and ULMI, followed by another blank with IS, a final calibration curve, and finally a second carryover blank (Table 5.21). Calibration curves should be processed in order of increasing concentration; the six replicates of QCs also should be ordered in a similar fashion (eg, LLMI, low QC, mid-QC 1, mid-QC 2, high QC, ULMI). The blank immediately after the six ULMI samples is to prevent carryover in the subsequent LLMI of the calibration curve.

The entire batch 1 should be injected in singlicate. The first double blank should demonstrate no response for either analyte or IS (demonstrating that the extraction process is

TABLE 5.20 Prevalidation Experiments

Test	Samples/Frequency	Expectation
System suitability test solution	3 neat solutions at LLMI before each run	Provisional LC-MS/MS system prerun determination
Double blank	2 replicates in 3 runs, 6 individual lots including calibrator matrix	Response <20% analyte LLMI, <5% IS
Blank with IS	2 replicates in 3 runs, 4 replicates calibrator matrix	Response <20% analyte LLMI
Calibration curve	Minimum 6 levels bracketing curves in 3 runs	Back-fit bias <±15%,* CV <15%,* R-value >0.99
Carryover	2 in each of 3 runs	Response <20% analyte LLMI
Intra-assay accuracy and precision	6 replicates in 3 runs of LLMI, ULMI, spiked QCs	Bias <±15%,* CV <15%*
Interassay accuracy and precision	18 replicates across 3 runs of LLMI, ULMI, spiked QCs	Bias <±15%,* CV <15%*
Matrix effects	Singlicate 1:1 admixtures of QCs and ULMI	Bias <±15% from expected mean results*
Sample stability	Singlicate calibrators and QCs stored on bench top for 12-24h before assay	Bias <±15%*
Assay correlation	Comparative assay calibrators, 20-40 split specimens	Deming regression slope 0.9-1.1, r-value >0.9 accounting for interassay calibration bias
Selectivity	Split specimens (patients)	Transition ratio consistent between calibrators, QCs and specimens, concentration dependency
New column/autosampler stability part 1	Double, blank, blank, calibrators, carryover, QCs	Consistent with performance for column 1
New column/autosampler stability part 1	Patient samples	Deming regression slope 0.9-1.1, r-value >0.95, mean bias <10%, intercept <LLMI

*LLMI acceptance criterion bias <± 20%, CV <20% unless total allowable error (accuracy and precision requirement) is known.

interference free); the first blank with IS should have no response for the analyte (demonstrating that the IS solution has not degraded). Calibration curves should demonstrate back-fit bias less than ±20% at the LLMI and less than ±15% throughout the remainder of the range for quantifying transitions with a correlation coefficient (Pearson R-value) greater than 0.99. Carryover blanks should demonstrate a response less than 20% of the LLMI. Calculation of the mean, standard deviation, and CVs of the six replicates of QC should indicate acceptable performance. For calibrators and accurately spiked QCs, bias ±15 or less (±20% at the LLMI) and precision ±15% or less (±20% at the LLMI) are acceptable unless method performance needs are more stringent (discussed in section "Assay Validation").

In batch 2, replicates of blank matrix with and without IS added are analyzed. It is recommended that six individual lots of blank matrix (including the calibrator lot) are assayed without IS and four replicates of the blank calibrator (same lot) spiked with IS are assayed in the second run. Analyte response in five of six blank matrix lots should exhibit 20% or less analyte LLMI response and IS response in all lots less than 5% mean IS response in the run are acceptable (ie, there is no unlabeled analyte in the blank matrix or in the IS, and there is nothing in the blank matrix that would interfere with the signal from the IS). Bench-top stability of QC materials also should be performed in the second run.

To do this, single replicates (minimally) of calibrators and QCs should be stored overnight at room temperature and be assayed as unknowns in the second batch. When calibrator matrix differs from actual patient samples (eg, when using charcoal-stripped matrix calibrators; multiple patient sample matrices vs. a single matrix calibrator system), a 1:1 mixing study is performed. More specifically, a single level of QC material for each matrix is mixed 1:1 with the ULMI and assayed in triplicate. The target concentration is calculated from the mean of the six replicates of the ULMI and QC from the same run, with bias ±15% or less considered acceptable. In summary, this batch contains a double blank; a blank with IS; the first calibration curve; carryover blanks; individual lots of blank matrix without IS; replicates (four) of calibrator matrix with IS; six replicates of LLMI, ULMI, and QCs; room temperature–stored calibrators and QCs (singlicate); and 1:1 admixtures of QCs and ULMI (triplicate). Finally, a carryover blank, a blank with IS, the second calibration curve, and the last carryover blank complete the batch (Table 5.22). Data reduction for batch 2 is similar to that of batch 1. The six replicates of LLMI, ULMI, and QC are compared to data from the first batch to generate initial interassay accuracy and precision characteristics. Review of measurement bias in 1:1 admixed samples provides evidence that the assay is not influenced by differences in matrix composition. Comparison of stressed calibrators and QCs

TABLE 5.21 Prevalidation Batch 1

Sample Order	Replicates	Injection Order	Use
Blank calibration matrix, no IS (double blank)	1	1	Determine cleanliness of analytical system
Blank calibration matrix, with IS	1	2	Determine lack of contribution from IS to analyte
Calibrators	1	3 through 8	Generation of Standard Curve
Blank calibration matrix, no IS (Double Blank)	1	9	Determination of carryover
LLMI	6	10 through 15	Intraassay inaccuracy and imprecision, LLMI
QC 1	6	16 through 21	Intraassay imprecision, QC 1
QC 2	6	22 through 27	Intraassay imprecision, QC 2
QC 3	6	28 through 33	Intraassay imprecision, QC 3
ULMI	6	34 through 49	Intraassay inaccuracy and imprecision, ULMI
Blank calibration matrix, no IS (double blank)	1	50	Determination of carryover
Calibrators	1	51 through 56	Generation of standard curve

TABLE 5.22 Prevalidation Batch 2

Sample Order	Replicates	Injection Order	Use
Blank calibration matrix, no IS (double blank)	1	1	Determine cleanliness of analytical system
Blank calibration matrix, with IS	1	2	Determine lack of contribution from IS to analyte
Calibrators	1	3 through 8	Generation of standard curve
Blank calibration matrix, no IS (double blank)	1	9	Determination of carryover
LLMI	6	10 through 15	Intraassay inaccuracy and imprecision, LLMI
QC 1	6	16 through 21	Intraassay imprecision, QC 1
QC 2	6	22 through 27	Intraassay imprecision, QC 2
QC 3	6	28 through 33	Intraassay imprecision, QC 3
ULMI	6	34 through 49	Intraassay inaccuracy and imprecision, ULMI
Blank calibration matrix, no IS (double blank)	6	50 through 55	Determination of blank matrix availability
Room temperature stored (overnight) calibrators	1	51 through 56	Determination of short-term stability
Room temperature stored (overnight) QCs	1	57 through 59	Determination of short-term stability
1 to 1 mixed matrix samples, QC 1: ULMI	3	60 through 62	Determination of matrix effects*
Blank calibration matrix, no IS (double blank)	1	63	Determination of carryover
Calibrators	1	64 through 69	Generation of standard curve

*Not required if patient matrix, QC matrix, and calibrator matrix are the same matrix type.

against expected concentrations is performed with bias ±15% (±20% at the LLMI) and considered acceptable.

Batch 3 is one of the most important experiments in method development. Individual samples ($n = 20$ to 40) and comparator assay calibrators (if available) are used to evaluate whether one should return to method development and to determine how much work will be needed for interassay comparisons and the development of the reference interval. Batch 3 is prepared in a fashion similar to the first two, with a double blank and a blank preceding the first calibration curve, which is followed by carryover blanks. Individual specimens are analyzed, followed by the six replicates of the appropriate calibrators and QCs (again, measured as unknowns). Finally, a blank, the second calibration curve, and the carryover blank are assayed. The last six replicates of the LLMI, ULMI, and QCs are included in accuracy and precision calculations (Table 5.23). Interassay precision (CV) and accuracy (bias) should be 15% or less ($\leq$20% at the LLMI) for all sample types. Intraassay precision (CVs) and accuracy (bias) should be 15% or less ($\leq$20% at the LLMI) and consistent across the three batches for all sample types. Comparison of the sample results against a predicate assay should be performed on a sample-by-sample basis. If one were to calculate the mean bias for these samples, imprecisely inaccurate results may generate a mean bias of near 0%. As the samples are expected to have imprecision associated with both assays, a Deming regression could be used. Each sample's bias should be reviewed and rationalized against the comparative assay. Acceptable agreement is Deming slope 0.9 to 1.1, correlation coefficient greater than 0.9, and mean bias less than 10% after accounting for calibration biases between assays. In certain cases, known limitations to the established assay can explain unacceptable agreement between observed results. Fig. 5.19A indicates back-calculated concentrations of calibrators prepared for LC-MS/MS analysis for progesterone assayed by a predicate radioimmunoassay (RIA). Initially one would consider that the calibrators were incorrectly prepared, demonstrating agreement only between 50 and 1000 ng/dL (see Fig. 5.19B). Prior knowledge of the strengths (and weaknesses) of comparative assays are often required. In this case, the predicate assay is linear between 50 and 1500 ng/dL (sigmoidal calibration plot); thus agreement is not expected outside of these ranges. Fig. 5.19C demonstrates a systematic negative bias for the analysis of 25-hydroxyvitamin D_2 by a competitive binding assay with approximately 25% underrecovery or preparative error (see Fig. 5.19C). Analysis of calibrators from the comparator assay by LC-MS/MS back-calculated accurately (<10% bias for all levels tested). Repeat preparation of calibrators was performed, retested (same result), and sent to an alternative laboratory for testing (LC-MS/MS assay), indicating accurate preparation; therefore the comparator assay demonstrated underrecovery (see Fig. 5.19D).

Also important in batch 3, a thorough review of the transition ratios observed in individual samples is used to underpin assay selectivity. If four total transitions were maintained through development to this stage, this is also an opportunity to exclude transitions that are of little value. Poor analytical sensitivity results in substantial deviations in the transition ratio at low analyte concentrations because of S/N. Transitions that demonstrate no observable response until analyte concentrations reach into the upper half of the calibration curve may be excluded, as long as there are still other qualitative transitions to use. Acceptable ranges of transition ratios can be generated from the calibrators, providing that the tolerances are developed across a range of analyte concentration. Confirmation of selectivity in all QCs and individual samples is then based on the acceptable ranges. This review is essentially an extension of the previous refinement of methodology, although with a larger data set.

Finally, reinjection of batch 3 using a second HPLC column (ie, from the same manufacturer with the same dimensions and solid phase, but from a different manufacturing lot) after storage in the autosampler for approximately 24 hours is warranted because this represents an opportunity to demonstrate ruggedness and help plan validation studies. Review of

TABLE 5.23 Prevalidation Batch 3

Sample Order	Replicates	Injection Order	Use
Blank calibration matrix, no IS (double blank)	1	1	Determine cleanliness of analytical system
Blank calibration matrix, with IS	1	2	Determine lack of contribution from IS to analyte
Calibrators	1	3 through 8	Generation of standard curve
Blank calibration matrix, no IS (double blank)	1	9	Determination of carryover
LLMI	6	10 through 15	Intraassay inaccuracy and imprecision, LLMI
QC 1	6	16 through 21	Intraassay imprecision, QC 1
QC 2	6	22 through 27	Intraassay imprecision, QC 2
QC 3	6	28 through 33	Intraassay imprecision, QC 3
ULMI	6	34 through 49	Intraassay inaccuracy and imprecision, ULMI
Blank calibration matrix, no IS (double blank)	1	50	Determination of carryover
Patient samples, 20-35, comparative assay standards	1	51 through 89	Intermethod correlation, specificity, provisional reference interval information
Blank calibration matrix, no IS (double blank)	1	90	Prevention of carryover
Calibrators	1	91 through 96	Generation of standard curve

LC-MS/MS calibrators in progesterone RIA Assay

Expected concentration (ng/dL)	Measured concentration (ng/dL)	Bias (%)
10	14	40.0
50	57	14.0
100	105	5.0
500	461	−7.8
1000	935	−6.5
2500	2070.9	−17.2
5000	3075.3	−38.5

A

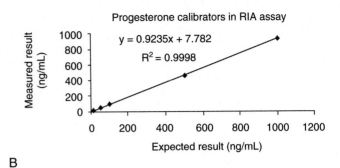

B

LC-MS/MS calibrators in 25-hydroxyvitamin D CBP assay

Expected concentration (ng/mL)	Measured concentration (ng/mL)	Bias (%)
5	2.5	−50.0
20	16.0	−20.0
50	29.3	−41.4
200	135.4	−32.3
300	229.1	−23.6
400	271.6	−32.1
500	387.6	−22.5

C

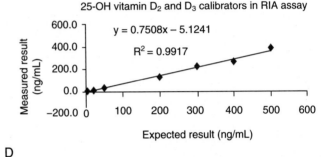

D

FIGURE 5.19 Analysis of progesterone liquid chromatography–tandem mass spectrometry *(LC-MS/MS)* calibrators by a predicate radioimmunoassay *(RIA)* **(A)** indicating agreement within the ED20-ED80 (∼50 to 1500 ng/dL) region of the RIA **(B)**. Analysis of 25-hydroxyvitamin D_2 LC-MS/MS calibrators via a predicate competitive binding immunoassay *(CBP)* **(C)** indicating manufacturing error or underrecovery **(D)**. Acceptable *r*-squared value indicates precise recovery in both assays. Slope deviation from unity demonstrates inaccuracy.

SST sample sequence is required to ensure the HPLC column performance is consistent with previous runs. Data reduction is identical to that described earlier, ensuring appropriate contaminant-free performance (double blank, blank), calibrator back-fit accuracy (using the same weighting and fit), system carryover, and chromatographic resolution from known critical pairs. The original run becomes the predicate assay for comparison and, as stated earlier, should be assessed on a sample-by-sample basis, including QCs. QCs should demonstrate equivalent intraassay precision and accuracy (≤15% for all samples, except ≤20% at the LLMI) or within 2 standard deviations of previously determined results (whichever is more stringent). A Deming regression should be used to compare repeatability of the assay with a naïve column. The acceptance criteria incorporating QCs and sample unknowns are Deming slope between 0.9 and 1.1, correlation coefficient greater than 0.95, mean bias less than 10%, and intercept lower than LLMI. Each sample's bias should be reviewed and rationalized against the initial injection, with a thorough review of appropriate transition ratio and S/N differences when individual samples demonstrate bias greater than 15% (>20% at the LLMI). Failure of this experiment requires consideration on clinical impact and may require modifications to the method before repeating the prevalidation studies.

As a final word of caution, MS assays have the potential to be accurate, precise, and selective. As demonstrated earlier, considerable effort is needed to ensure that the fundamentals of the assay are properly considered and designed. Fig. 5.20 demonstrates comparative measurement of specimens between a TFC-LC-MS/MS assay for serum cortisol and a predicate assay (extraction followed by RIA). Comparison of calibrators via both assays was acceptable; however, a number of sample results did not agree. For example, the more

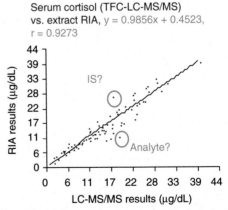

FIGURE 5.20 Analysis of serum cortisol turbulent flow liquid chromatography–liquid tomography–tandem mass spectrometry (TFC-LC-MS/MS) results versus predicate radioimmunoassay *(RIA)* with extraction indicating samples that disagree. *Top circled sample* indicates either overrecovery of the analyte in the RIA or underrecovery of the internal standard in the MS assay. *Bottom circled sample* indicates underrecovery of the analyte in the RIA or overrecovery of the internal standard in the MS assay.

concentrated sample has a concentration determined by LC-MS/MS that is lower than the RIA. This could result from underpipetting (repeat analysis could confirm), matrix effects (ruled out by demonstrating that the coeluting, labeled IS peak area was consistent with the IS peak areas observed for the calibrators), or selectivity of IS transition. The latter was evaluated, and the IS transition ratio was marginally acceptable (lower than expected). This points to contribution to the quantifying IS transition selected, the HPLC method was extended (as per Fig. 5.18C and D), and an interference was observed for the proposed quantifying IS transition. Recalculation of prevalidation studies was performed after inverting the IS transitions (ie, the qualifying transition became quantifying IS transition). As another example of results not agreeing between the two assays, a few samples had concentrations determined by RIA that are lower by TFC-LC-MS/MS. After the same principles discussed previously (ie, retesting, evaluating IS peak area, extending the HPLC gradient), together with evaluating dilutional linearity and spike/recovery studies, it was determined that the MS assay was accurate. The key thing to remember is that burden of proof is required to have confidence in MS assays; good analytical detective work is needed in every step of the process.

Assay Validation

Validation of a newly developed assay is, simply, the production of objective evidence that the test platform fulfills the claims of the test.[55] Although a simple definition, proper test validation is hardly a trivial experience. Appropriate experimentation and documentation is key to ensuring that the developed assay can be rationally justified as a diagnostic test. It should be noted that validation is not an exercise in successful analysis, but a process in which the limits of the assay are scientifically defensible.

Having performed prevalidation checks for common sources of error (and possibly revisiting method development as needed to address discovered errors) the laboratory should proceed with a validation plan. This document serves as a roadmap for experimentation; each component of the validation should be explicitly defined (replicates of samples, periodicity of testing, etc.), as well as the acceptance criteria. Note that this document should be drafted and finalized before the initiation of validation. Each subchapter included in the validation experimentation section can be used as a framework for drafting this plan and includes experimental design, criteria for determination of acceptance, and recommendations in the event of failure to meet the acceptance criteria.

Although this section is written in a linear manner, many of these validation experiments are performed concurrently. The breadth of a well-executed validation seems overwhelming, yet it can be achieved, with the exception of long-term stability, in approximately 20 working days.[56,57] A complete listing of experiments and putative acceptance criteria are provided in Tables 5.24 and 5.25. Certain additional experiments are recommended to demonstrate the validity of facets seen only in MS analysis, such as transition ratio analysis. Other standard experimental paradigms frequently used in clinical chemistry have been deviated from, to either address limitations in MS analysis or to accentuate the value of the information provided by MS technologies. The changes in the experiments are rationalized in each experimental section.

At the end of validation experimentation, typically concluding with the establishment of reference intervals for endogenous analytes or interassay comparisons, particularly for exogenous analytes (to use previously published therapeutic and toxic ranges or confidence in measuring the presence above a cutoff), the totality of the reduced validation data should be included in a validation report.[55] This report should describe the experiments performed and the reduced data and should explicitly define any limitations the assay presented as a function of the experiments performed. Such a report is often used to draft the constraints of the analytical methodology as defined in the laboratory's SOPs and specimen collection and handling considerations. Specific points of note from validation experiments to be included in the assay SOP are detailed within the sections.

The core validation experiments are Inaccuracy (Relative Recovery), Imprecision, Hydrolysis/Deconjugation/Derivatization, Carryover, Interferences, Ion Suppression/Enhancement, Matrix Effects, Selectivity, Spike and Recovery, Calibration Curve Fit and Linearity, Dilutional Linearity, Preprocessing Stability, Postprocessing Stability, Ruggedness, Long-Term Stability, Interassay Comparison/Correlation, External Quality Assessment/Certified Reference Material Evaluation, and Reference Interval Generation.

General Validation Rules

All validation runs are preceded by SST injections to assess instrument performance. The process of validation generates confidence in measurement, together with acceptance criterion for the SST when deployed in clinical use. When validation batches are acceptable, the SST characteristics (retention time, peak shape, area response, selectivity, etc.) are also acceptable and should be retained for the final SOP as prerun instrument performance verification. Construction of validation batches should have consistent features, comprising a double blank (blank matrix extracted without IS), a blank with IS (blank matrix extracted with IS), the calibration curve, a carryover blank (blank with IS), and the QCs. It is recommended that these samples be included in this sequence at both the beginning and the end of each validation batch performed, thus bracketing the experimental samples for validation experiments with calibrators and QCs. The maximum batch size should be determined in a prevalidation experiment as previously described and proved in validation.

Standard curves should be calculated with each validation batch. Linear least-squares regression is preferred to generate the analyte-to-IS peak area ratio versus analyte concentration best fit line, with parameters for fit and weighting provisionally established in method development. These parameters should not be modified during validation or subsequent clinical use of the assay. As acceptable calibration variance would not have been adequately determined at the onset of validation, acceptable back-calculated concentrations of the calibrators shall be within ±15% at all concentrations with the exception of the LLMI, which is acceptable within 20% of the expected concentration,[53] or more stringent if the assay

TABLE 5.24 Validation Experiments Part 1

Test	Samples and Frequency	Expectation
System suitability test solution	3 Neat solutions at LLMI before each run	Final LC-MS/MS system prerun determination
Intraassay accuracy and precision	20 Replicates in 1 run of LLMI, ULMI, spiked QCs	Bias <±15%,* CV <15%*
Interassay accuracy and precision	1 Replicate across 20 runs of LLMI, ULMI, spiked QCs	Bias <±15%,* CV <15%*
Carryover	5 Runs (all runs)	Response <20% analyte LLMI
Double blank	2 Replicates in all runs, 6 individual lots including calibrator matrix	Response <20% Analyte LLMI, <5% IS in 5 of 6 lots and all runs
Blank with IS	2 Replicates in all runs, 4 replicates, calibrator matrix	Response <20% analyte LLMI in all runs
IS selectivity	3 Replicates ULM and 20 samples without IS added	Response <5% IS
Drug interferences	10× Upper limit of normal	Response <20% analyte LLMI, <5% IS
Ionization effects	10 Specimens (normal/abnormal) per matrix	IS recovery 80%-120%
Matrix effects	Matched draw specimens (3 donors) and admixing (3 replicates per admixture) for each matrix type against calibrators/high QC	Bias <±15%,* CV <15%*
Sample composition	3 Replicates of supraphysiologic hemolysis, icterus, lipemia spiked into high QC	Bias <±15%,* CV <15%*
Spike and recovery	10 Specimens (normal/abnormal) per matrix and 3 levels per matrix type	Bias <±15%,* CV <15%*
Transition ratio (selectivity)	Minimum 40 patient samples clinically distributed versus calibrators/QCs; all validation samples	Analyte and concentration specific, samples consistent with calibrators and QCs
Calibration curve	Minimum 6 levels bracketing curves in 5 runs	Back-fit bias <±15%,* CV <15%,* r-Value >0.99
Dilutional linearity	3 Replicates per dilution level, high QC and/or ULOR	Bias within ±15%, define CRR

CRR, Clinically reportable range; *CV*, coefficient of variation; *IS*, internal standard; *LC-MS/MS*, liquid chromatography–tandem mass spectrometry; *LLMI*, lower limit of measurement interval; *QC*, quality control.
*LLMI acceptance criterion bias <±20%, CV <20% unless total allowable error (accuracy and precision requirement) is known.

has predetermined criteria for total allowable error (TAE) by biological variation or general consensus[58]. Exclusion of unacceptable data points in validation should be undertaken judiciously (maximum 25%[18,53]) because removal of calibrator values from standard curves or unacceptable QC recovery is an early indicator of problematic analytical performance characteristics.

Historically, LC-MS/MS assays have consistently demonstrated performance accuracy and precision meeting bias less than ±15% and CV 15% (except LLMI).[18,48,53] Thus these performance standards are promulgated in the absence of more stringent expectations and should be the goal of all well-developed and validated LC-MS/MS assays. The conditional concentrations of the QC materials are provisionally accepted if within 15% of the expected concentration (exogenous analytes fortified into naïve matrix) or defined by TAE. All previous considerations for overall batch quality (calibration curve fit, IS recovery, carryover, etc.) should be critically assessed in each validation batch as QC accuracy targets and precision may not have been generated (endogenous analytes). If TAE is known, the requisite accuracy and precision should be considered as performance targets in validation[58]. When using external QCs (ie, vendor-provided lyophilized powders reconstituted in the laboratory), the concentration stipulated by the outside vendor must be confirmed in the prevalidation phase for use as accuracy targets during validation.

The first two experiments—accuracy and precision—are the cornerstones for some of the experiments used to further validate the analytical characteristics of the assay, as well as many facets of the assay's lifespan. Where multiple measures of a sample are performed to generate quantifiable results between the LLMI and ULMI (spike and recovery, dilutional linearity, sample mixing, etc.), a minimum of three replicates is recommended, with precision considered a requirement (CV <20% at LLMI, <15% throughout the remainder of the assay range unless more stringent requirements are required). Additionally, the standard deviation values generated during interassay precision determination are used to generate QC measurement ranges for batch acceptance using QC rules[59] for clinical use of the assay. Accuracy and precision details should be summarized in the test SOP as a

TABLE 5.25 Validation Experiments: Part 2

Test	Samples/Frequency	Expectation
Preprocessing stability	3 Donors, whole blood and bench-top after RBC separation	Bias <±15%,* CV of recovery <15%*
Sample stability	All conditions expected for sample draw and transit (preprocessing)	Bias <±15%*
Freeze–thaw	3 Replicates fresh donors, LLMI, ULMI, and QCs 1-3 cycles	Bias <±15%,* CV <15%*
Hydrolysis	3 Individual specimens across 10-20 time points	Consistent time for yield plateau
Derivatization	All matrix materials and minimum 2 disease states with increasing derivatization reagent concentrations	Consistent time for yield plateau, CV of yield <15%
Bench-top stability	First run including blanks, calibrators, QCs, and >20 samples, second run left on bench top assayed after 4 h	Deming slope 0.9-1.1, r-value >0.95 after batch acceptance confirmed
Autosampler stability	First injection including blanks, calibrators, QCs, and >20 samples, reinjection 24-72 h later	Linear least squares slope 0.9-1.1, r-value >0.95 after batch acceptance confirmed
Assay ruggedness	Singlicate replicates of LLMI, ULMI, and QCs for each induced error	Bias <±15%
Maximum batch size	Bracketing curves and QCs, samples; 96-192 injections	Back-fit bias <±15%,* QCs bias <±15%/2 SD, carryover free, IS CV <20%
Column ruggedness	Double, blank, blank, calibrators, carry-over, QCs, and >20 patient samples	As per prevalidation studies
Long-term stability	3 Replicates/time-point for LLMI, ULMI, QCs	Bias <±15%* vs. fresh/expected, CV <15%*
Interassay comparison	40 Samples plus comparative assay calibrators	Deming regression slope 0.9-1.1, r-value >0.9 accounting for interassay calibration bias
Reference interval generation	120 Per interval for generation 20 per interval for verification	
External trueness assessment	As many as possible/feasible	Bias <±15% or as defined by TAE
Materials stability	All critical in laboratory materials (calibrators, QCs, IS, stock) solutions), continually updated (−20 and/or −80°X)	Bias <±15%,* CV <15%,* bias <±5% for stock solutions

CV, Coefficient of variation; *IS*, internal standard; *LLMI*, lower limit of measurement interval; *QC*, quality control; *RBC*, red blood cell; *SD*, standard deviation; *TAE*, total allowable error.
*LLMI acceptance criterion bias <± 20%, CV <20% unless total allowable error (accuracy and precision requirement) is known.

reference for laboratory staff to assess continuing assay performance.

Accuracy

The assessment of accuracy and recovery are fundamentally equivalent in regard to the experimentation performed. The wording is related to the availability of a certified reference material or method. The accuracy of an assay is defined by its measure of agreement between the assay's determined quantity and the true value of the analyte measured.[57] When materials to define accuracy are unavailable, the term *relative recovery* may be more appropriate. For the remainder of this chapter, however, accuracy shall be used exclusively for the purpose of describing the closeness of measure to an expected value, because recovery shall be used in other contexts. The mathematical expression of accuracy is bias, expressed as a percentage.

All samples related to accuracy should contain a known amount of the measurand(s). Such samples may be obtained from certified reference materials or may be created by fortification of blank materials with a known amount of compound. Refer to the discussion of material preparation earlier in this chapter for recommendations regarding matrix and spiking considerations. Samples should be stored in conditions determined to be provisionally stable; if degradation of the analytes is observed during method development, validation must not be performed until at least 20 working days of stability is obtained to cover the duration of validation. Alternatively, freshly prepared calibration curves must be prepared for each batch.

At a minimum, determinations of accuracy are performed at the extremes of the analytical range when calibrators and samples are in the same matrix (note that charcoal-stripped serum and serum are not the same matrix). This makes the assumption that if trueness is acceptable at the LLMI and ULMI, then all values in between those limits will approximate to the same measure of accuracy. Because the dose-response function of MS is generally linear and MS exhibits a dynamic range of up to 10^6, this assumption is acceptable. If, for some reason, a curve generates a nonlinear or nonproportional function, accuracy should be determined at appropriate points, typically at or near inflection points of the standard curve. Best practice includes determination of accuracy for accurately spiked QCs for exogenous drug analytes into all test matrices.[53] Because QCs and calibrators are prepared from different starting materials (ideally), the inclusion of accuracy assessment for QCs may be considered as the relative accuracy of metrologic preparation, agreement ($\pm$15% bias unless TAE is known) provides confidence in calibrator manufacturing.

The first assessment of accuracy is performed as an intraassay experiment, using 20 replicates[56,57] of LLMI, ULMI, and QCs extracted and bracketed by duplicate calibration curves. Both curves should be used in linear least-squares regression fitting with the 20 replicates at each level treated as unknowns. In the absence of bias from TAE criteria, bias $\pm$20% or less at the LLMI and $\pm$15% or less throughout the remainder of the measureable range is acceptable.

Interassay accuracy is determined using singlicate replicates of LLMI, ULMI, and QCs assayed over 20 days to emulate a month of operation, which is considered a provisional robustness assessment.[56,57] Identical acceptance criteria are used (see Table 5.24).

Although generally discouraged in MS quantitation, if an assay requires an upper limit of reporting (ULOR) greater than the ULOQ and dilution into range will not be performed (such as STAT testing with immediate results required), samples prepared at that ULOR should be evaluated under the same criteria (bias $\pm$15% or less). It should be noted that the clinical decision (relevancy of accuracy) may override the bias observed from measuring above the calibration curve range without dilution.

It should be noted here that neither a lower limit of detection (LOD) nor a limit of blank (LOB) studies are included as a validation experiment for quantitative assays.[18] In MS quantitation, the blank solution must be truly blank. The response of the blank solution, whether it is blank matrix, charcoal-stripped matrix, solvent, or some equivalent, should be noise, with no peak for each transition selected or retention time of the analyte of interest.[49,53] However, noise can be integrated as a peak, because the scanning function and subsequent averaging of signal(s) to generate counts per second can create artificial and irreproducible noise. Protocols to determine the LOD or LOB have been defined that rely on the quantitative result of the LOD and LOB samples to set such values.[60] However, these concentrations would be below the lowest reliable value for quantitation (the LLMI); the occurrence of type I error (or a blank sample indicating a response) as used to rationalize an LOB can therefore generate a quantitative value with no verifiable concentration. The LOD calculations are such that the only claim is reliable detection of signal. In the case of quantitative results, LOD serves no clinical purpose because no value can be appropriately released. Instead, reports would list the concentration as "less than the LLMI or LLOQ" or an equivalent statement.

Failures in assay accuracy require an investigation of the cause, with the possibility of redevelopment of the method. Inaccurate measures at the LLMI are often related to poor S/N ratio or excessive background causing poor integration of peaks. Optimization of MS parameters and injection volume may provide corrective avenues. Inaccuracy at the ULMI is frequently associated with a quadratic curve generated by linear least squares, such that a small change in the area ratio (y-axis) results in a substantial change in the concentration (x-axis). Truncation of the analytical range or reduction in signal (through MS parameter detuning or lower injection volume) is recommended. Further considerations for accuracy failures are pipetting accuracy and appropriate mixing of analyte and IS together with uncontrolled matrix effects.

The accuracy of the calibration materials can be used as targets during the preparation of new calibration standard lots. These accuracy experiments are designed for the measurement of the calibrators and QCs (if appropriate) and do not include the comparison of the assay's results with external quality assessments or certified reference materials. Those experiments should be performed only after the described accuracy evaluation such that confidence in the result is appropriate.

Precision

Precision (or imprecision) is the degree to which independent measures agree with each other.[61] Evaluation of precision will generally be expressed as CV (standard deviation/mean), expressed as a percentage for ease of data review. Precision evaluations are performed using the samples and replicates from accuracy evaluation, LLMI, ULMI, and QCs.[56,57]

Intraassay precision (20 replicates in one run) and interassay precision (one replicate repeatedly assayed over 20 days) acceptance criteria is 20% or less at the LLMI and 15% or less throughout the remainder of the analytical range (see Table 5.24) unless the allowable imprecision from TAE determination is more stringent. Calibration curves from MS analysis are largely heteroscedastic, and thus it is not unusual that QCs at the lower end of the analytical range have higher CVs than QCs with higher concentrations. Acceptance criteria failures in the assessment of precision are generally consistent with those described for accuracy, with the addition that pipetting volume may be confounding (too low).

Carryover

Carryover is the contribution of the analyte from one sample to another in the analytical system. The impact of carryover should be determined after the ULMI. Carryover is observed in blank samples immediately after the ULMI in five runs for assignment of carryover; however, it is monitored in all validation runs for acceptability and the potential to influence subsequent samples in a run. The contribution of analyte in the carryover blank is then compared to the lower limit of quantitation to determine effect on the assay's performance. This measure is the response of the analyte (counts per second or equivalent), not concentration because calculated concentrations less than the lower limit of quantitation are not accurate. General acceptance criteria are carryover less than 20% LLMI analyte peak area response.[53] This criterion should be considered within the context of clinical utility. For most clinical assays, carryover less than LLMI is acceptable and is highly dependent on the relationship between an LLMI and the lowest medical decision point (potential for bias). When observation of response is considered abnormal, carryover must be unobserved. For example, the detection of 6-monoacetylmorphine, the primary metabolite of heroin, can present actionable clinical information just by virtue of detection, thus the requirement for no observable response in a blank after a high concentration sample.[62]

When samples are expected to exceed the analytical measureable range (AMR, LLMI to ULMI), the influence of carryover and remedial actions must also be determined and included in the assay SOP. Additional study samples are prepared by fortification to 10× to 100× ULMI (or highest observed value). After injection of increased analyte sample, replicates of the lowest QC (or LLMI, $n \geq 6$ usually) are made. Assessment of the number of subsequent samples demonstrating unacceptable carryover is made through bias comparisons to QC target concentrations (acceptable bias <15%, 20% at LLMI). The aim of this validation study is to determine the number of samples after an increased (greater than ULMI) result that are affected when reporting clinical results (see Table 5.24). Unacceptable carryover may require the selection of different solvents in the autosampler to wash the syringe and injection port. Adding additional volume or time to wash cycles also may remediate carryover. Absent these solutions, the SOP must note the number of samples immediately after an increased concentration specimen (sufficiently high to produce carryover) that are potentially clinically inaccurate and the remedial actions of reinjection or reextraction (particularly if subsequent sample contamination is suspected).

Analytical Interferences

Assessment of interferences includes all aspects of material selectivity and anticipated concomitant medications or analytes. A clinical assay should demonstrate a lack of interferences in the following examples: (1) contribution to the analyte and IS from extraction reagents and mobile phases, (2) contribution of calibrator matrix to analyte and IS materials, (3) contribution to the analyte from the IS and vice versa, and (4) contribution to the analyte and IS from analytes anticipated in patient samples.

The determination of analyte and IS response from assay reagents is performed using double-blank samples and calibrator matrix with diluent added (no IS). Analysis of extracted double-blank samples should be performed in one batch with six sources of blank calibrator matrix for matrix-based calibrators.[49,53] The replicates should be placed after the carryover samples (after the first curve), with the assurance that carryover is negligible (ideally unmeasureable). Acceptance criteria are analyte peak area response 20% or less of the LLMI in five of six lots of blank calibrator matrix and IS peak area response less than 5% of the mean IS response in that specific run for five of six lots.[49,53] Acceptable results demonstrate that the assay process is free from interferences that could lead to measurement bias and further that the calibrator matrix is free of interferences (or residual analyte from charcoal stripping). This provides confidence that blank matrices may be readily acquired when preparing new calibrators (but should be checked first), plus the assigned concentrations of calibrators are appropriate.

Further consideration should be given to the double blank in each validation and clinical run for continuing assurance of assay selectivity. Acceptance criteria failure requires assessment of source of impurity and possibly redevelopment.

Contribution to the analyte signal from the IS is determined after evaluation of contributions from analytical reagents and calibration matrix. A minimum of three replicates of clean blank calibrator matrix are fortified with IS and assayed in a run immediately after the blank matrix evaluation above (blank + IS). The singular added variable is the purity of IS materials either intrinsically or through conversion in the analytical process. Acceptable results are analyte responses less than 20% LLMI in all replicates.[49,53] Causes of such a contribution are related to impure labeling of the IS, resulting in measurement of native analyte. This may be corrected by simply lowering the amount of IS added or by selection of a discrete lot/vendor of the IS material. Deuterated ISs may demonstrate hydrogen-deuterium exchange during storage in solution, by ionization, or by gas-phase interactions. In the absence of a distinctly labeled IS by virtue of label location or availability of a ^{13}C, ^{15}N, or oxygen-18 (^{18}O) IS, redevelopment under different ionization conditions, alternative solutions, or storage conditions is required. Because this sample is also included in all assay runs, continual monitoring of IS purity should be performed.

Contribution of analyte to IS response is performed by extracting three ULMI samples without IS and assessing peak area response for IS. Acceptance criteria are peak area response less than 5% IS mean response in blank plus IS samples above. The net effect of analyte contribution to IS results in calibration curve nonlinearity. In the instance that IS degrades, the contribution (and region of nonlinearity) changes over time; this should be designed out of the method, and analyte should not contribute to IS within the AMR.

A host of patient samples should be used to assess possible contribution to IS signals from common endogenous components. At minimum, 20 diverse samples from the expected testing population should be extracted without IS. After analysis, these samples should be assessed for response at the expected retention time of the IS (interference response <5% of calibrator IS mean in the run). Reduced data should be critically reviewed to ensure that high concentrations of the analyte within these samples do not contribute by virtue of naturally occurring isotopes from the analyte of interest (as noted earlier). Failure requires sourcing the contribution. Possible solutions include selection of a different MS/MS transition, a differently labeled IS to provide a discrete precursor ion or precursor-product ion pair, or higher chromatographic fidelity.

Contribution to the analyte with the potential to generate false positives and contribution to IS resulting in false negatives should be considered in detail. Interferences from drug compounds, both prescribed and over the counter, should be measured. A full list is provided in Clinical and Laboratory Standards Institute EP7-A2.[48] Concentrations of potentially interfering substances are provided at 10× maximum concentration (drugs and metabolites) and should be tested at 10× the upper limit of the reference interval for endogenous compounds. Analysis of snap and spike drug interference solutions (Drug Interference Mix, Cerilliant, Round Rock, Texas) should be performed through spiking into clean calibrator matrix (no IS) and/or QCs (assessing for inappropriate bias >15%). Spiking drug mixtures and potential endogenous interferences into blank matrix should yield responses less than 20% analyte LLMI and less than 5% IS mean in the run. Where interferences are observed, reduction of the interferents test concentration is warranted to provide a cutoff for observed concentration or assay redevelopment.

The previous experiments are used when the materials to stress contributions to the analyte measure are available. Interferences are an ongoing challenge for the clinical laboratory; therefore the ability to discriminate interferents through deviation in observed ion ratios (analytes and IS) should be included in the assay SOP and test reporting literature to annotate clinical results that may be erroneous. The use of transition ratio monitoring is described in the section "Selectivity."

Ion Suppression and Enhancement

An important phenomenon in mass spectrometry is the modification of ionization efficiency as a function of closely or coeluting compounds.[27,28] Protocols for the rough determination of ion suppression via PCI have been established in the method development section of this chapter. For validation, however, PCI is insufficient to accurately determine ionization effects and is used to provide confidence in the overall selectivity from the combination of extraction and chromatography. PCI alters both the postcolumn dwell volume and the solvent composition during the ionization of the compounds of interest. Instead, a different approach to determine ionization effects and recovery efficiency has been described.[38] This protocol is particularly useful with complex extractions or in extractions in which 100% recovery is not feasible, such as SPE or LLE.

A minimum of 10 samples should be assessed for matrix effects, 5 normal and 5 abnormal (lipemic, hemolyzed, icteric, and disease symptomatic). The current College of American Pathologists (CAP) checklist[63] provides for the following consideration: matrix effects measured (PCI or spiking) should not exceed ±25%, and the CV of matrix effects should be 15% or less. As a matter of best practice, this should be performed using spike and recovery studies (see later) and review of IS peak area response in these specimens relative to the calibrators/QCs in the run (or comparison to neat solvents). Acceptable IS variance in samples versus calibrators/QCs is bias ±25% or less.

Failure requires consideration of the clinical use of the assay; all efforts should be made to reduce matrix effects in measurement. If the analyte LLMI response is robust relative to the noise, a 50% loss of signal in real samples may not hinder the quantitative result produced. This form of acceptance criteria should be made in conjunction with the response of the IS in real samples, providing acceptable tolerance between IS response in samples verses matrix effect–free measurement as a trigger for detailed review of analyte concentration and clinical impact when assaying clinical specimens (75% to 125% IS response versus calibrators, for instance). Review of IS peak area response is an ongoing component of test validation and clinical utility of the assay, and clear acceptance criteria must be included in the assay SOP.[63] Excessive ion suppression or enhancement can produce IS responses with poor precision (yielding inaccurate reporting of analyte-to-IS area ratio) or analyte measurement outside of MS detector linearity resulting in the same phenomenon. Transition ratio differences (discussed later) often serve to highlight the latter example if improper aliquoting is ruled out as the cause.

Matrix Effects

Although often combined as a single moniker, matrix effects are differentiated from ion suppression in validation. Both the origin and the experimental paradigm to demonstrate assay validity in the presence of matrix are discrete. In matrix effects evaluations, certain preanalytical errors are appraised for their effect on the analytical result. The experiment addresses the quality and integrity of the specimen for analysis, including hemolysis, lipemia, icterus, type of anticoagulant or preservative, and other variables as appropriate.[48] Three distinct experiments are used for determination of the impact of matrix effects with subtlety driven by the nature of analytes (endogenous vs. exogenous). For endogenous analytes, many studies are performed by comparison of circulating concentrations versus the preferred sample type. Samples are drawn from a minimum of three individuals into the preferred sample tube type (baseline, ie, red top serum) and other potential sample tube and anticoagulant types that may be clinically measured (ie, serum separator tube, Li heparin plasma). Importantly, the "matched draw" comparison incorporates both the composition of final sample before extraction (coextracted matrix composition), together with additional bias that may occur through adsorptive losses (to the serum separator) and residual enzyme activity (plasma vs. serum, stabilizer vs. none, etc.). Acceptance criteria is bias ±15% or less for each matched pair (or more stringent if TAE is known) together with consistent transition ratio tolerance (when compared to calibrators and QCs at the equivalent concentration).

This particular experimental design can be performed via spiking the same concentration of analytes into each sample type immediately after sample collection into the collection tube ($n = 3$ donors per tube type, <5% dilution into blood with dimethyl sulfoxide [DMSO] or water [H_2O], if possible). After sample mixing and processing as per the matrix type, comparison relative to the preferred sample type is performed as previously detailed. The goal is to emulate the potential adsorptive losses and residual enzyme activity as earlier and is often confounded by ensuring that the sample collection volume was identical for each tube collected.

An important process for validation is determination of matrix equivalency or the ability of the assay to be matrix insensitive. This must be performed when the assay is intended to measure disparate sample types, such as the analysis of urine, plasma, and CSF for amino acids. Independent calibration schemes for these three matrices can become unwieldy in a laboratory. Measuring all three matrices against a single calibration matrix requires proof that the matrix does not affect measurement. Notably, this process does not account for adsorptive losses with gel-based separator tubes; the previously mentioned approach or matched draws should be performed in that regard. Further, often the QCs (in alternative matrix types) are stored sub-aliquoted and are usually assayed as aged samples.

The principle is to determine the deviation from expected results by increasing the amount of matrix effect—causing material in a sample of known concentration. This process is used for both endogenous assays (true matrix-based high QC and low-level charcoal-stripped matrix calibrator) and exogenous drug assays and is *not* a substitute for spike and recovery studies (see later). Incorporation of a 5-point mixing scheme[48] is routinely used to determine matrix equivalency, whereby the highest QC from the preferred sample type is admixed with the lowest QC (if stability is not a concern) or freshly drawn blank samples from potential alternative sample types (matrix effects). In a single batch, the alternative matrix samples (low level) and the samples of known concentrations (high QC) are measured unmixed and after mixing at 3:1, 1:1, and 1:3 by volume to generate samples at five levels. Each admixture and the unmixed samples are assayed in triplicate (minimally) if using a single donor and singlicate if using three donors, with the exception of the high QC (triplicate). Evaluation of the data can be performed in two separate modes. The first is using the means of neat known samples and neat alternative matrix samples to calculate the expected concentrations of admixes of those specimens. Acceptance criteria are mean bias ±15% or less from expected for each admixture and precision less than 15% for analyte measurement within the AMR.

The second is to plot the measured concentrations versus the expected concentrations and evaluating the slope and correlation coefficient of a least-squares regression with a linear fit, unweighted. Acceptance criteria are a slope of 0.9 to 1.1 with a correlation coefficient greater than 0.9. The choice of data reduction would depend on the initial concentration of the available alternative matrix samples. For example, if a low-level alternative matrix sample is admixed with a low-level calibrator, the difference in concentration across neat and mixed samples may be quite small, resulting in a line with a slope of near zero (if the concentrations are equivalent, the slope would be exactly zero). The linear least-squares approach is sensitive only when sufficient concentration differences exist between the known sample and the alternative matrix sample. In those cases, bias from expected is a more reliable measure of the possible matrix effects of the assay.

If the matrix used in calibration material preparation is not equivalent to the test matrix, this same evaluation should be performed by mixing of the calibrators with the intended test material. Even in the case of charcoal-stripped matrix, as is common with many endogenous assays, the charcoal-stripping process removes fundamental sample components. Fig. 5.21 demonstrates the assessment of matrix equivalency via the 5-point mixing scheme for synthetic urine calibrator admixed with spiked pooled high QC *(red)* together with the admixture of unspiked human urine pool and spiked urine pool *(blue)*. As expected, the slope of agreement is approximately 1, with minimal deviation from expected concentration for each admixture. Admixture biases were less than 6%, and CV at all measured concentrations was less than 5%. For further consideration, matrix agnostic performance provides insight into potential diluents to be evaluated in dilutional linearity studies (see later). Externally prepared QCs, often lyophilized, are popular with laboratories. Unfortunately, the matrix of these QCs is rarely identical to that of the test samples, and 5-point mixing studies with both calibrators and matrix QCs should be performed to determine measurement equivalency.

Hemolysis, lipemia, and icterus are the presence of lysed blood cells, fatty acids and triglycerides, and bile, respectively.[48,63] Two experimental workflows are appropriate for determination of the effect of poor-quality specimens, predominantly for serum and plasma assays. The first employs the 5-point mixing scheme listed earlier with samples comprising grossly hemolyzed, icteric, and lipemic pools (HIL pools) sub-aliquoted from the laboratory's current specimens. The concentration of interferents in each pool should be measured by calibrated clinical assays to determine the amount of each interferent in each pool, together with the QC samples (it is virtually impossible to get completely lipemia-free matrix). Admixtures of the pools and the influence on measurement of the analytes by extrapolation is performed as previously described and a concentration of interferents that results in interference (bias <±15% from expected for each admixture and precision <15% for analyte measurement within the AMR) is noted as the level of each interferent that can be tolerated. These details should be listed in the assay SOP; further, it is appropriate to generate high-quality color images for inclusion in the SOP to aid in differentiation of samples with inappropriate matrix composition.

A more streamlined approach incorporates spiking of concentrated solutions of these interferences into the high-QC pool (usually the predominant test sample type). Recently, a commercially available test kit has enabled this experiment (Interference Assurance Test kit, Sun diagnostics, New Gloucester, Maine). Table 5.26 describes the testing process by which high-pool QC is diluted with concentrated solutions of HIL interferents to determine acceptable maximum concentrations of these interferents.[48] Each spiked sample is assayed in triplicate, and the spiking solution is assayed after dilution into blank calibrator matrix. Measurement of HIL spiking solutions added to blank matrix enables determination of the presence of analytes in the spiking solutions to correct the expected concentration for bias calculations when spiked into high QC. Acceptable interference-free measurement is thus mean bias ±15% or less (CV percent of triplicates <15%) relative to the expected concentration of the high QC accounting for dilution. Interference-free measurement using this approach generates specimens with HIL interferents far in excess of clinically expected values, removing the need for visual assessment of sample integrity at the bench. Where interference occurs, reducing the absolute concentration of HIL interferents added and generating a cutoff with an image in the SOP is required. Failure to ensure matrix types or potential interferences do not yield biases usually requires one of two considerations: exclusion of sample type and interferents (clearly documented in the test ordering and SOP) or generation of matrix-specific reference intervals (see relevant section later). Table 5.27 shows some of the common causes of interferences and matrix effects observed in clinical laboratories.

Spike and Recovery

The ability of an assay to add a known amount of material to a sample and subsequently recover that amount is a core principle in analytical techniques. This experiment is particularly important in the assessment of human specimens in clinical

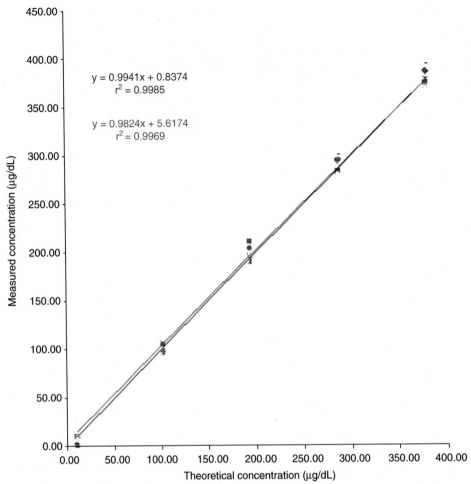

FIGURE 5.21 Urine cortisol five-point mixing scheme for charcoal-stripped low calibrator *(red)* and pooled urine *(blue)* admixed with pooled urine high quality control. Use of linear regression for this data analysis is predicated on the disparity of concentrations of the two sample types. Good agreement is observed between the two matrix types, indicating the assay is tolerant to both charcoal-stripped urine and true human matrix.

diagnostics. Simple recovery of the analyte in calibration solutions or from QCs is unsatisfactory evidence of the analytical recovery because those solutions are often self-fulfilling in their recovered values, particularly in the absence of certified reference materials with assigned trueness.

To perform spike and recovery, a minimum of 10 different true matrix samples for each proposed test sample type are sequestered from the expected test population (as noted earlier, five pristine samples and five samples with abnormal features such as hemolysis, lipemia, or icterus). Pooled material is not recommended. Each sample should have adequate volume to aliquot using class A volumetric glassware, although standard manual pipetting can be used in the absence of large volumes of test matrix. Each sample is split into two separate aliquots with volume to perform triplicate analyses for each sub-aliquot, allowing for excess because of loss in pipetting.[63] One aliquot is stored unmodified. The second aliquot is fortified with a known amount of material using accurate spiking techniques. It is also recommended that a neat solution (ie, high-purity water or other solvent amenable to the assay's extraction motif) be prepared at the same time to measure the absolute amount added (unless class A volumetric glassware is used). The fortification process should minimize the amount of additional volume used in fortification to less than 5% preferably. The dilution factor from fortification should be accounted for in the determination of recovery accuracy.

Certain additional factors may require extra steps to this protocol. If the analyte of interest exhibits protein binding (either specific or nonspecific), the fortified samples should be allowed to equilibrate before extraction and analysis. The amount of equilibration time should be roughly equivalent to the equilibration time afforded an isotopically labeled IS in the extraction process as determined in method development. Each of the samples, both unspiked and spiked, as well as the spiked neat solution should be bracketed by standard curves. Recovery is calculated as (concentration measured after spike − concentration measured before spike) divided by spiked amount, expressed as a percentage. The spiked samples are adjusted for any dilution accounted for by fortification ($\leq$5%). Mean recovery is thus calculated and should be less than ±15% or the allowable inaccuracy from biological variation (TAE), whichever is least.

Notably, the CAP checklist[63] does not explicitly state recovery study design, merely the provision for using 10 samples in matrix effects studies. The interpretation of recovery

TABLE 5.26 Interference Spiking Scheme for Hemolysis, Icterus, and Lipemia

Assurance Test Kit*

Interferent Provided	Test Concentration	Typical Concentrate Values
Triglycerides	3000 mg/dL (34 mmol/L)	>15000 mg/dL (170 mmol/L)
Hemolysate	0.5g/dL	>10.0 g/dL
Total protein (from albumin and γ-globulins)	12g/dL	25 g/dL
Bilirubin (conjugated)	20 mg/dL (342 μmol/L)	>400 mg/dL (6.8 nmol/L)
Bilirubin (unconjugated)	20 mg/dL (342 μmol/L)	>400 mg/dL (6.8 nmol/L)

Interassay Mean Value as Bias Target, Accounting for Dilution

Pool	Interferent	Analyte Concentration (Bias %)
4 Parts	1 Part triglyceride	80 (15)
19 Parts	1 Part hemolysate	95 (15)
1 Part	1 Part protein	50 (15)
19 Parts	1 Part conjugated bilirubin	95 (15)
19 Parts	1 Part unconjugated bilirubin	95 (15)

*Concentrations of the materials used for the interference test kit and example spiking scheme with expected results as a measure of bias (lower portion).
From Sun Diagnostics, New Gloucester, Maine.

TABLE 5.27 Examples of Matrix Effects

Source	Result	Resolution
Coeluting phospholipids	Ionization suppression	Redevelop LC or sample preparation conditions
Hemolysis	Differential binding of analyte vs. normal sample, change in clinically relevant concentration (if analyte sequestered in red blood cells), poor extracts reduce ruggedness	Reject all hemolyzed specimens or redevelop sample preparation conditions to control
Lipemia	Differential binding of analyte vs. normal sample, poor extracts reduce ruggedness	Reject all lipemic specimens or redevelop sample preparation conditions to control
Icterus	Differential binding of analyte vs. normal sample, poor extracts reduce ruggedness	Reject all icteric specimens or redevelop sample preparation conditions to control
Tube type difference	Change in circulating concentration of analyte, adsorptive loss to gel-based separation material, change in pH because of anticoagulant, introduction of interfering species from tube type	Reject inappropriately drawn specimen, redevelop sample preparation conditions to control. If source tube is not available (because of splitting of samples), matrix differences can be assessed by a separate assay (ie, calcium for serum/plasma differentiation).
Interfering species	False positive (analyte interference), false negative (IS interference), difficult integration (partially coeluted)	Monitoring of transition ratios, redevelopment of LC or sample preparation conditions to resolve

LC, Liquid chromatography.

in 10 samples to determine matrix effects and accuracy is a streamlined approach to validation. To fulfill the need to demonstrate concentration-specific recovery requirements for the European Medicines Agency's and FDA's validation guidance,[49,53] it is recommended to also perform recovery studies at three distinct concentrations, low range (<10× LLMI), midrange (ULMI/2), and high range (80% of ULMI) in all matrices, assayed in triplicate.

Failure of spike and recovery is a serious flaw in the analytical procedure. The veracity of calibration materials, QCs, and all patient results are questionable when spike and recovery is unsuccessful. Remediation is to address the poor recovery in a return to method development. Validation should be reinitiated after addressing the cause.

Selectivity

Validation of selectivity is a separate experiment from the determination of interferences in MS/MS analysis, although not feasible in most single-stage MS assays. For MS/MS platforms, the generation of more than one product ion from a single precursor is common and is used to generate the transition ratio tolerances for both analytes and ISs. The ratio of transitions is consistent for authentic peaks of a single compound. During the course of validation, calibrators, and QCs (in which only the analyte of interest should be represented in the integrated peak) are analyzed repeatedly and used to generate provisional transition ratios ranges. Comparison of transition ratios for real samples used during inter-assay correlation and reference interval assessments are compared against the expected range of transition ratios to assess the selectivity of the assay. Acceptance criteria can be constructed from the standard deviations of the transition ratio of the calibrators and/or QCs, with 2 standard deviations being the norm. At low concentrations of analyte, however, the acceptance criteria are usually less strict because the sensitivity of the qualifying ion transition can be such that the integrated peak area is imprecise or nonexistent. Determination of the allowable departure from the expected transition ratio based on concentration is an assay-dependent feature.

After assessment of the agreement of transition ratio in individual samples, a composite of all acceptable validation data (that generates quantifiable results) should be performed and acceptable tolerance flags should be incorporated in the final SOP with continual monitoring of assay selectivity in measurement of clinical specimens. Fig. 5.22 shows the transition ratio for testosterone in an exemplar assay calibrator (see Fig. 5.22A) indicating a ratio of approximately 1.2 derived as peak area m/z 289 to 97 divided by 289 to 109. As noted earlier, the mean transition ratio from calibrators should be computed with only concentrations that provide acceptable responses in the qualifying transition (approximately <10% CV across calibrators). After review of all measureable results against the transition ratios in calibrators (including mixing, recovery, etc., as earlier), a composite plot comparing transition ratio bias for each sample is generated. Notably, the expected transition ratio in calibrators is calculated in each individual batch

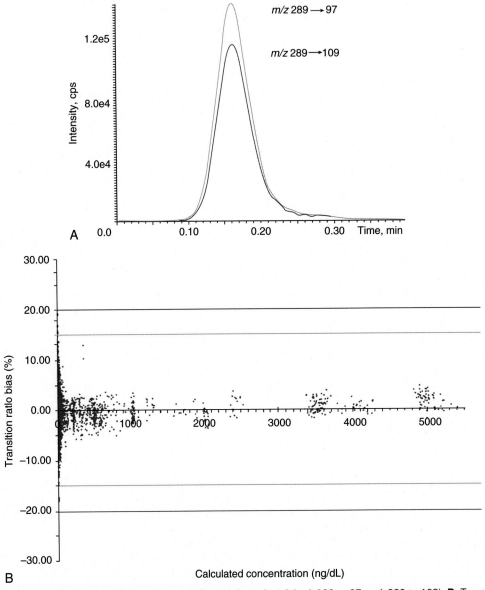

FIGURE 5.22 A, Transition ratio for testosterone of approximately 1.2 (*m/z* 289 to 97, *m/z* 289 to 109). **B,** Transition ratio bias observed for testosterone during validation studies. Samples at the low end of the concentration range indicate broader ratios resulting from the influence of noise on low-level signals. Acceptance criteria for transition ratios in this assay is concentration dependent.

to account for day-to-day instrument drift. Fig. 5.22B indicates the distribution of transition ratio bias in all validation study samples within the AMR ($n = 1798$ samples). As expected, increased bias is observed at lower concentrations, related to S/N ratio variation for both transitions. Two sets of tolerances are shown *(red and maroon lines)* indicating transition ratio biases less than $\pm15\%$ and less than $\pm20\%$, respectively. The results demonstrate that wide tolerances are of limited value when the entirety of the validation data is included and thus assay-specific tolerances with different ranges at different analyte concentrations are warranted. In the example shown, transition ratio tolerances less than $\pm20\%$ are used for concentrations less than 7.5 ng/dL ($3\times$ LLMI) and tolerances less than $\pm10\%$ are used for the remainder of the AMR.

Calibration Curve Fit and Linearity

Demonstration of appropriate fit of the calibration curve is a relatively brief experiment because each validation batch contains standard curves (used to generate interassay accuracy and precision data). This validation experiment requires only the reduction of data made available during the course of other experimentation. In quantitative MS assays, calibration curves are constructed by linear least squares regression of the ratio of the analyte to the IS peak area and expected concentrations of analytes. After the rules of calibrator exclusion, as discussed previously (minimum 75% calibrators used to generate the regression of the calibration curve), standard solutions should exhibit good agreement across the calibration curve, as indicated by an r-value and residual back-fit bias (or accuracy) within acceptable limits. The repeatability of calibration is assessed over 5 days.[49,53] Individual calibrator point and mean bias for each calibrator level should be less than $\pm15\%$ ($<\pm20\%$ at the LLMI) unless TAE requires more stringent accuracy. The r-value, also called a correlation coefficient, is a description of dependence of the quantities on the x-axis and y-axis. Simply, the increase in the response ratio (y-axis) is proportional to the increase in the x-axis. Note here that an r-squared value (coefficient of determination) is not proposed for data reduction. The correlation coefficient of each curve across 5 separate days is calculated from the calibration curves. Acceptance criteria are such that all correlation coefficients be greater than 0.99. This value may be considered atypically high, but in most cases of calibration curves, very few actual data points are used. Thus random variance and true error would be difficult to distinguish at a lessor criterion.

Dilutional Linearity

The ability of an assay to linearly dilute a specimen from outside the analytically measureable range to within the range established by the calibration curve is essential to most clinical assays. Often, biomarkers are chosen for upregulation as a function of a disease or disorder, generally such that the values far exceed the normal population's reference interval. Calibration curves for MS should be constructed such that the majority of the assay's values are represented within the measurement range, ideally with the lowest medical decision point greater than $10\times$ LLMI. Even with the dynamic range provided by the mass spectrometer (although constrained by carryover in the autosampler), samples greater than ULMI should be expected. To report an accurate result, the ability to dilute such samples should be validated.

This experiment consists of two separate evaluations of dilutional linearity: dilution of a high QC and a specimen fortified to a concentration greater than ULMI, or the ULOR. Additionally, the diluent must be chosen appropriately. Best practice notes that to dilute a specimen of a particular matrix type, the same matrix absent of the analyte of interest should be used to dilute the specimen.[49,53] Successful sample mixing studies may provide supporting data for a threefold dilution using blank calibrator diluent (see Fig. 5.21). Certain assays are not capable of calibration with the matrix of interest or a particular assay measures more than one matrix type, obviating a single matrix to serve as both calibrators and blank solution. For example, nicotine and metabolite analysis may be performed using a single extraction motif with equivalent calibration ranges for urine, saliva, whole blood, plasma, and serum. Blank matrices without nicotine may be difficult to sequester. Therefore the calibration scheme for this assay is standards diluted in methanol (note that validation of this matrix choice should be performed by recovery studies).[21] The blank solution used to assess possible contamination is methanol; dilution of a plasma or whole blood sample with methanol is inappropriate because the precipitation of proteins modifies the density of the sample as well as inhibiting pipetting by clogging with protein flocculants. In this case, distilled and deionized water can act as an appropriate diluent and measurement of residual contamination should be performed in the diluent used in validation and clinical sample analysis to calculate bias appropriately. Validation of dilution should be performed with triplicate replicates of sample at each dilution, not serially sampling from the single diluted sample. The samples are analyzed and the results compared to the mean concentration of the interassay precision, accounting for the dilution factor.

The dilution performed on the high QC demonstrates that a sample with a known quantity (as assessed in interassay and intraassay precision) can precisely recover that concentration on dilution. It is recommended that during clinical sample analysis, samples requiring dilution and the high QC be diluted using the same dilution factor, providing a quality assurance check for the additional dilution step. The amount of dilution can be rationalized by the highest expected concentration to be observed by real patients, as well as the absolute value of the QC. If the highest expected patient result is 25 times greater than the ULOQ, then validation of a 50- or 100-fold dilution of the QC is necessary. This may overdilute the high QC for measurement within the AMR (concentration less than LLMI). Consideration for these two elements is important in executing this experiment appropriately on an individual assay basis. Multiple steps of dilution should be considered (ie, 2-, 5-, 10-fold) for validation of the maximum allowable dilution using the high QC. The second experiment involves the accurate fortification of real human specimens with the analyte of interest to levels greater than the ULMI (ULOR). For both exogenous and endogenous assays, the baseline level of prespike matrix should be measured in triplicate to confirm no outliers (relative to intraassay imprecision); the addition of the spike concentration and the residual concentration is used to generate the expected concentration of the ULOR sample. Additionally, the spike volume for fortification relative to the absolute sample volume should be minimized to maintain the integrity of the matrix of choice before dilution (<5%). Class A volumetric glassware should be used where feasible.

This sample, similar to the diluted QC, is extracted in triplicate after dilution into the AMR. After accounting for dilution, the bias from expected is calculated for both QCs and ULOR samples. Mean dilution bias less than ±15% and precision less than 15% is acceptable unless TAE requires more stringent criterion.

Failure of sample dilutional linearity may be related to changes in protein binding (analyte and IS), adsorptive losses (diluent content), poor mixing (time-course study), or inaccuracy in the fortification procedure (repeat with class A volumetric glassware). Repeat of the experiment (including spiking) is recommended before revisiting validation. If there is no means to linearly dilute the matrix of interest, the inability to quantify values greater than the upper limit of quantification must be explicit in the validation report and assay SOP, together with the maximum validated dilution. The combination of AMR and validated dilution factor defines the clinically reportable range (CRR), calculated and expressed in the assay SOP as the LLMI to the ULMI multiplied by validated dilution.

Preprocessing Stability

The stability of samples before extraction and analysis provides information regarding the quality of the sample during draw, shipment, and storage. Each step the sample takes from the patient to the laboratory bench should be assessed to ensure that prejudiced results are not released. Thorough experimentation requires the ability to draw samples proximate to the laboratory to assess postdraw handling and storage. Other means of stressing samples may be performed on the bench. Because this experiment requires the evaluation of samples over time, preprocessing stability should be initiated at the beginning of a validation. For preprocessing stability, the following factors should be evaluated:

1. For serum and plasma, the length of time before spinning down or separating the blood cells
2. For acute urine collection, CSF, whole blood, serum or plasma, the duration of time spent at room temperature before refrigeration or freezing
3. For 24-hour urine collection, the general storage conditions for the vessel between depositions of the sample, including preservatives and additives to maintain stability
4. For all patient sample types and expected shipment and transit conditions
5. For all patient sample types, calibrators, QCs, reagents, expected laboratory storage conditions
6. For all patient sample types, calibrators, QCs, reagents, maximum number of freeze–thaw cycles

To address these conditions, a baseline must be established. For most exogenous compounds, the storage treatment of samples may be replicated on the bench by fortification of materials into the matrix and stressing the sample under each condition, calculating bias for each condition by comparing measured concentration versus that expected from fortification. However, compounds sequestered in partitioned matrices may require assumptions rather than experimental proof. For example, immunosuppressant drugs are largely retained in blood cells. Bench-top replication of immunosuppressant-sequestered red blood cells is impossible to perform. Because most immunosuppressant assays lyse the cells before analysis, the laboratory must make the assumption that the stability of the compounds in the red blood cells is equivalent to that in extracellular molecules in lysed samples.

The same approach can be taken for endogenous compounds, although it is recommended that postdraw stability analysis be performed for novel biomarkers. Thus the baseline for the first two conditions (postdraw separation of fluid from blood cells and room temperature storage before refrigeration or freezing for shipment or transfer) must be performed with volunteer blood draws (three donors minimum). Endogenous analyte measurements generally require no particular consideration for circulating levels because the assay has been designed to measure circulating concentrations within the validated AMR. Exogenous analytes require fortification with drug before separation of serum and plasma from blood, ensuring identical blood volume is used for matched samples before spiking. A series of multiple tubes is drawn from the same volunteer; the first tube drawn is immediately processed and extracted for analysis. Consecutive tubes should be drawn and stressed by allowing the tubes to sit before separation of blood cells from the extracellular fluid. After removal of the serum or plasma from the blood cells, sub-aliquots of those samples may be further stressed at room temperature conditions before analysis, noting the time that sample was stored as whole blood and bench-top after separation. The design of timing should emulate the expected handling conditions in sample collection, with some margin for error—usually up to 48 hours on whole blood (4, 8, 24, and 48 hours are usually tested) and up to 8 hours at room temperature conditions ($\sim 20°C$, at 2, 4, and 8 hours usually).

Each analysis should be performed in singlicate for each donor, for each condition (triplicate results per level). Mean bias less than ±15% between stressed and baseline samples (or target concentration from accurate fortification) is acceptable unless more stringent requirements are needed (TAE); however, the bias for each of the three samples should be reviewed to ensure that the precision of recovery (accuracy not bias) is less than 15%. It should be noted that final measurement accuracy and precision is a composite of all steps from collection through to MS/MS detection; ideally, mean bias should be minimal in all sample handling steps. This consideration holds true for all subsequent stability experiments; in most cases, biases within measurement error (<5% to 10%) are desired to reduce the influence of sample handling on compounding assay error.

Failures of acceptable bias of initial draw conditions are not typically addressed by further method development. Some molecules may benefit from the addition of inhibitors or distinct anticoagulants. In most cases, however, this experiment provides constraints in which to reduce preanalytical variation. The amount of time samples determined to be acceptable under the various stressed conditions must be included in the assay SOP and instructions for sample collection. The stability of samples in transit to the laboratory also should be evaluated in this experiment. If the setting is a reference laboratory to which samples are mailed or shipped for analysis, additional considerations for stressed stability must be considered, particularly if samples are to be shipped at ambient conditions. The temperature of airliner cargo holds in the heat of summer while the plane is waiting on the runway for takeoff must be taken into account. Actual replication of those conditions is difficult to perform in a laboratory because some assumptions must be made. In this case, it is recommended that the samples be stored in a water bath (or equivalent) for up to 8 hours at 55 to 65°C and compared to baseline materials. Lack of stability at these conditions can be corrected by rejection of samples shipped under ambient conditions.

In the laboratory environment, the stability of all materials for the assay should be undertaken across all expected conditions. This includes calibration standards, QCs, IS solutions, and critical assay reagents. Typical evaluations include retaining those materials at room temperature, refrigerated, and frozen conditions for up to 14 days. It is recommended that shorter time points be tested as well (ie, 3, 5, 7 days) in the event of 14-day failure. Having these additional samples available will reduce the amount of time required to repeat the stability experiment. After the allotted stress time and condition, samples are extracted in triplicate and compared to either freshly prepared calibrators or to calibrators of known stability that exceed the duration of the timing for stressed conditions. The storage of materials at long-term conditions (preferably 80°C ultrafreezer) will be appraised during the course of validation, and therefore the need to prepare fresh materials is not absolute. Such reasoning is proper only if the solutions are indeed stable at those conditions; research and observations made during method development (or information from the literature when identical materials are used) may help to engender confidence that the materials are suitable for long-term storage.

Freeze—thaw cycles are evaluated for analyte stability after consecutive storages at frozen conditions. All materials used for the assay that are to be frozen shall be addressed in this experiment, including patient samples, calibrators, and QCs. Samples are initially evaluated for baseline levels and immediately frozen. After a minimum of 12 hours of storage in the frozen state, samples are removed, allowed to thaw for at least 4 hours, and analyzed against fresh calibration curves. This process is repeated at least two more times, which would account for initial freezing—thawing (before a normal first analysis), and two more subsequent analyses as may be required to address batch failure, sample reextraction, or required dilution.

Stability data provide constraints for the handling of samples and materials in support of clinical sample analysis. If stability is inadequate for the laboratory's needs, method development evaluations for the addition of stabilizing reagents is recommended, with appropriate consideration of selectivity and recovery experiments after the addition of stabilizing reagents.

Hydrolysis and Derivatization

In certain analytical workflows, modification of the analyte through deconjugation (by acid-base hydrolysis or enzyme activity) or by derivatization should be proved to be optimal. This is particularly important in hydrolysis protocols in which neither the conjugate analyte nor conjugated ISs are available; quantitative recovery of the conjugated compound cannot be determined in this instance. To perform validation of the hydrolysis procedure, solution pH, temperature, and concentrations of deconjugating components (whether enzymatic or chemical) should not deviate from the conditions determined in method development. This is a standard experiment in chemistry, and thus details are not included in this chapter; however, the principles outlined in the following text apply to the developmental exercise. This experiment relies on standard Michaelis-Menten kinetics in which the plateau of recovery is established. Therefore this experiment will evaluate only the time variable for hydrolysis.

To perform this experiment, the amount of time determined to approximate the best recovery of the analyte (from method development) should be roughly doubled. Thus if deconjugation time of an analyte and its conjugate is 4 hours, the amount of time used is up to 8 hours for this experiment. The time-course is then divided into 10 to 20 time points for evaluation of yield. From the previous example, 30 minutes will be used, providing for 16 time points. At least three different unspiked samples should be evaluated for this experiment; three replicates are extracted for each unit (48 replicates in this case), and all are simultaneously extracted and assayed. At the appropriate time intervals, samples are removed from the deconjugation conditions, quenched (if necessary), and assayed against a single standard curve.

The data from the replicates of the individual samples are plotted as concentration of the analyte versus time. The data should indicate a leveling off of the recovery at a certain time point, which should agree with the prescribed time point in method development. Deviations between samples indicate poor reproducibility of the deconjugation. If recovery continues to increase over time without any observed leveling of response, method development should be reinitiated to establish appropriate conditions.

In addition to the concentration of the analyte in this determination, review of the IS recovery also must be performed. The hydrolysis conditions should affect both the analyte and the IS similarly; however, at the beginning of the hydrolysis condition, the IS is in a native state and the analyte is in a conjugated state. Stability of the IS under the hydrolysis conditions can be justified by measurement of the response of the IS across the time points and compared to the response of the IS in the unhydrolyzed calibrators. Should the IS degrade at a rate nonproportional to the analyte, the reported concentrations may be abnormally high, providing a false conclusion to analyte recovery. The clinical impact of overrecovery should be considered, and, if appropriate, alternative IS materials or addition of IS after hydrolysis may be considered. It should be recognized that additional variation in accuracy and precision is introduced using this approach and criteria described for the assay may need to be duly adjusted.

In assays that derivatize the analytes of interest, validation of appropriate stoichiometry should be performed with known amounts of analyte spiked into matrices. Particular consideration to the milieu of potentially derivatized species and the concentration differences in healthy/disease or drug-free/toxic concentrations should be considered.

To validate adequate derivatization, numerous samples representing the intended test population, including samples that would be considered abnormal for the test's indication, should be used to demonstrate valid yield/recovery. Because biological pathways in disorders can be shunted and additional unrelated metabolites generated in vivo, simple fortification is insufficient to represent a human sample. The derivatization reagent is prepared at increasing concentrations and the extraction is performed. The increasing concentration should be such that up to 100-fold the concentration determined in method development is tested. Samples are prepared in triplicate per concentration of the derivatizing reagent and assayed, bracketed, by duplicate calibration curves. The recovery of these specimens should be plotted for concentration of measured derivatized analyte versus concentration of derivatization reagent for each individual sample. Acceptable reagent concentration is indicated by a plateau in yield for all samples with precision of yield less than 15% (at most); particular attention should be paid to the abnormal samples in this protocol because insufficient derivatization of those patient specimens may result in false-negative results. Explicit details should be included in the assay SOP and at least one QC must be included as a process control for hydrolysis protocols.[63]

Postprocessing Stability

Postprocessing stability generally consists of two experimental protocols. The first is the assessment of bench-top stability of the extracts after sample preparation. Rarely are samples taken directly from the laboratory bench to the instrument without delay. Occasional oversights occur in which a batch of samples is left in front of the mass spectrometer, waiting for analysis. To reiterate, the goal of validation is to objectively prove that the assay's performance under normal operating conditions is acceptable. As humans interact with the extraction processes and instrumentation, error-free procedures must be considered. To perform this study, a single batch consisting of calibrators, QCs, and a minimum of 20 patient samples (with concentrations covering the clinical range) are extracted in duplicate. The first set of samples is immediately analyzed on the mass spectrometer. The second set of samples is left on the bench-top for a number of hours (recommended here to be between 3 and 7 hours) and then loaded into the autosampler and analyzed. The chosen amount of time should be indicative of a working day; extreme bench-top stability can replicate an overnight condition or even 72 hours (one weekend).

Data reduction is performed by distinct quantification of each batch and comparison of the patient samples by Deming regression. Deming regression is preferred over linear least-squares because both batches should exhibit some inaccuracy related to pipetting, extraction, and analysis. The assumption that the first batch (the x-axis in the comparison) is without error is by itself erroneous. QCs should be within 2 standard deviations (as assessed by interassay imprecision). Acceptance criteria for the statistics from the Deming regression should be a slope of between 0.9 and 1.1 and an r-value of greater than 0.95. Single outliers should be interrogated for cause; multiple outliers indicate a lack of stability. Shorter time points should then be explored until stable conditions are experimentally justified. Should brief storage at room temperature indicate a lack of stability of the extracts, stabilizing additives or engineering controls (temperature, plate conditions, etc.) will need to be assessed during a return to method development.

Autosampler storage stability is the second experiment in postprocessing stability. Experimental evidence is required to ensure that samples injected during the course of a large batch or reinjected after an instrument shutdown are not affected by the storage time. A batch is prepared in singlicate containing calibration curves, QCs, and at least 20 patient samples. The extracts for this batch must have sufficient volume for duplicate analyses. The batch is analyzed immediately after extraction. After analysis, the batch is then stored in the autosampler for an appropriate amount of time, generally not less than 72 hours, to mimic a weekend. The samples are then reinjected. This data set is reduced in a similar fashion to the bench-top stability experiment with each injection of the batch independently reduced, wherein QCs are determined to be acceptable by virtue of accepted QC rules. In this case, however, linear least-squares regression is the more appropriate comparison for the patient samples, with the same acceptance criteria of a slope of 0.9 to 1.1 and a correlation coefficient of 0.95 or greater. Failure of this experiment requires repeat of shorter time points until acceptable correlation between the freshly analyzed and the stored samples is reached. All details should be included within the assay SOP.

Ruggedness and Maximum Batch Size

An appraisal of the assay's ability to withstand change is a feature not required for validation, although it is recommended to support clinical utility of the assay. These experiments demonstrate that variations in mobile phase or solution preparation or the installation of new columns do not adversely affect the measurement of the analyte(s). Ruggedness against mobile phase or solution preparation is a mechanism to determine the influence of human error in the protocol; ruggedness against new columns is to determine the influence of a column manufacturer's ability to reproduce acceptable stationary phases.

One aspect of this experiment is driven by forced error.[64] The various solutions used in sample preparation and LC should be prepared with variations in their makeup. For example, an acidic solution used as an elution buffer in SPE is determined to be 2% formic acid in acetonitrile. To understand the influence of error in this preparation, a 1% solution of formic acid in acetonitrile and a 4% solution of formic acid in acetonitrile are used as the extraction solvent. To determine acceptance, QCs and real patient samples are extracted under these modified conditions. Unacceptable recovery of the analyte and IS is gauged by medical decision points and the assay's desired accuracy and precision budget. Because of the variety of possible extraction motifs being substantial, independent recommendations for each technique cannot be made. Scientists should, however, be prepared for the inevitable error of human interaction and understand the effect caused by improper solution preparation. The range in which to assess this error also depends on the assay and is left to the scientist's discretion. It is important to note the impact of these errors when a deleterious result is obtained, because troubleshooting an issue related to solution preparation may be shortened by having previously determined the consequence of aberrant mobile phase or solution preparation (and thus noted in the SOP).

Evaluation of column stationary phase material robustness is not an estimate of injections-to-failure. As noted previously, columns are consumables and differences in sample quality expected in the diagnostic laboratory may prevent such an evaluation from returning meaningful data. It should be shown that the columns are consistent across different manufacturing lots of the stationary phase, if possible. This experiment can reduce the significant costs and frustration of replacing a column on an assay only to discover that the characteristics of the column are different between the old and the new.

Column vendors can typically accommodate the request for a separate lot of stationary phase material for any particular column. Once the new column lot is in hand, a single batch consisting of 40 patient samples bracketed by calibration curves with interspersed QCs is assayed on each column. Deming regression is used to reduce the data between the two distinctly analyzed batches; a slope of 0.9 to 1.1 and a correlation coefficient of greater than 0.95 indicates commutability between the lots of columns (note that if the samples used do not exhibit concentrations throughout the AMR, r-squared is more appropriate to demonstrate commutability). Additional batch-specific acceptance criteria (as noted in prevalidation testing) should be met to provide further confidence in the longevity of an assay, with particular attention paid to agreement in transition ratio monitoring.

Failure can be remedied by one of two paths. The first is the return to the HPLC phase of method development in search of a column capable of the needed analytical performance. A second—and far less frequently used approach—is to purchase as many columns of the original lot the assay was validated on and hope the supply does not run out before the next generation of assay development. Risk-averse laboratories would do well to choose the former.

Finally, the maximum allowable batch size should be validated (before recalibration of the assay).[18,49,53] This particular study is preferably executed with reference interval and interassay correlation samples. The goal of the study is to stress the analytical platform with patient samples and assess the impact of drift on QC repeatability, carryover, and, more importantly, calibration. Generally, a single or double 96-well plate batch is validated depending on anticipated sample volume and frequency of clinical assay setup. The first consideration for acceptable batch size is the agreement of bracketing curves; back-fit bias at each calibration point should be less than $\pm15\%$ ($<\pm20\%$ at the LLMI), and curves should not be obviously divergent. Second, QCs placed after the first curve, throughout the run, and before the last curve should agree (bias $<\pm15\%$, CV $<15\%$). Double-blank and carryover blanks should demonstrate equivalent cleanliness throughout the assay run to demonstrate that the instrument and assay do not lead to a buildup of carryover. IS repeatability across the entire run should be critically reviewed for the appearance of drift (CV of IS response $<20\%$ is acceptable) because this may affect measurement accuracy and precision (back-fit differences in calibrators and bias in QCs should also highlight this phenomenon). Finally, transition ratios should be critically reviewed and consistent across the batch, not merely by concentration, as demonstrated in Table 5.21. Inappropriately labeled ISs often exhibit only reproducible hydrogen-deuterium exchange after the interface has reached a temperature plateau, thus affecting accuracy (analyte/IS peak area ratio) as batch sizes change.

Long-Term Stability

The agreement of assay materials over time should be initiated during the process of validation but may require subsequent updates to validation reports because the data are generated longitudinally. Immediately after the generation of calibrators and QCs, sub-aliquots of these materials should be stored at the most stable conditions available in the laboratory. Often, this is in an ultra-freezer ($-80°C$) or standard freezer ($-20°C$). These materials should be stored until the end of validation, at which point a freshly made set of calibrators and QCs using freshly weighed or aliquoted top stock materials (if stable) are prepared. The use of substock solutions manufactured for the preparation of the original lot of calibrators and QCs are not fit for long-term evaluation of the compound's stability. The new solutions should be prepared identically to the original lot of calibration materials, preferably in class A volumetric glassware. Long-term stability is an ongoing component to validation. Concurrent preparations of materials from freshly prepared top stocks and subsequent comparison to the current lot of calibrators should be performed until failure of the materials is observed. Storage of multiple sub-aliquots in long-term storage conditions is recommended to facilitate this determination.

Comparison of these new materials is performed by triplicate analysis of each of the new materials bracketed by duplicate calibration curves of the original lot, interspersing QCs as appropriate. The back-calculated means of the new materials should be within the interassay accuracy values established in validation. If the means are within the expected accuracy of the assay, the original lot of materials is considered stable at those long-term storage conditions for the duration in which they were stored.

Failure of these criteria is related to either compound degradation or the inaccurate preparation of the new lot of materials. The experiment should be repeated with another separate lot of calibrators prepared from freshly made solutions to confirm that solution preparation is not at fault. Unfortunately, the case of compound degradation means that data generated using the original calibrators are unsuitable for validation because biased calibration curves lead to biased experimental results. Remediation is a return to method development to establish appropriate conditions for long-term storage and a repeat of validation (unless fresh calibrators were prepared for each validation run). Method development should be of a length of time to determine appropriate stability such that revalidation of the long-term stability does not fail on repeat.

Interassay Comparison and Correlation

Having completed the experiments described previously, a general analytical validation has been completed. Validation of the clinical component follows, in which the assay is compared to predicate platforms and assays. Interassay correlation is performed by the coanalysis of specimens, preferably more than 40, distributed across the clinical range (25% in each quartile). This consideration prevents inappropriate comparisons through regression resulting from imprecision in both methods. This experiment also can be used to verify existing reference intervals if such values may need not be generated de novo.[54]

The comparison of assays also should include an evaluation of calibration standards to ensure commutability of the materials used, particularly in the absence of certified reference materials. Generally, MS is capable of accepting most matrices used for calibration. Other technologies, such as immunoassays, may have strict requirements for the sample matrix; charcoal-stripped sera or solvents may not provide accurate recoveries on certain test platforms. Calibration bias can elucidate a fixed percent bias across the compared range of concentrations.

The 40 samples should be bracketed by duplicate calibration curves with QCs within the batch as appropriate. The comparative assay and the MS assay should analyze samples on the same freeze–thaw cycle and/or after demonstrated laboratory stability (room temperature, postprocessing stability, etc.). The concentrations reported for the samples by both assays are assessed by Deming regression analysis. In general, an acceptable slope is 0.9 to 1.1; an r-value greater than 0.95 represents good correlation. Both criteria (slope and r-value) should be applied because sample sets that exhibit good mean precision but poor accuracy may demonstrate an acceptable slope, and sample sets that have a fixed bias across the concentration range may demonstrate excellent correlation with a deviated slope.

Ascertaining the cause of failure of interassay comparison can be troublesome. First, confirmation that all results are within the respective assay's measurable range should be provided (see Fig. 5.19). Samples outside the measurable range should be excluded from this analysis. Exclusion under any other condition is inappropriate because patient samples should be commutable. The validation of the MS assay previously performed can be used to justify acceptable results, whereas the alternative technology may not have been appropriately validated. The flaws inherent in certain technology, such as interferences in immunoassays or nonselectivity in LC–ultra-violet detection assays, can be used to rationalize particular outliers.

External Quality Assessments, Certified Reference Materials, and Reference Intervals

The assurance of trueness for diagnostic assays is evaluated through the use of materials created outside the laboratory. Trueness can be verified only when a certified reference material is available or when a reference method procedure (RMP) can be used as a comparative assay. In the case of an RMP used to establish trueness, the protocol for interassay comparison is appropriate, with 40 patient results used to inform error. If the only available material is that of external quality assessments (EQAs), it should be recognized that certain EQAs are based on interlaboratory imprecision (peer group agreement) and do not demonstrate a measure of trueness, only of comparative agreement across laboratories. That said, EQAs are a critical component of ongoing quality assurance and should be assayed during validation to preempt peer group disagreement.

Comparison to an RMP may be performed as a send-out to the RMP laboratory as part of an organized process sponsored by the reference laboratory. As harmonization has long been the goal of clinical laboratories, projects such as the Hormone Standardization Program (HoST) from the Centers for Disease Control and Prevention have been recently initiated for testosterone and estradiol.[65] In the case of the HoST project, an initial set of samples is received by the laboratory for calibration adjustment and consecutive sets (40 per quarter for 4 quarters) are sent for assessment of accuracy and precision.

Regardless of whether samples are sent to an RMP laboratory or obtained from an RMP laboratory, comparison data should be reviewed in two different contexts. The first context is precision. When plotted as a Deming regression, the slope is initially ignored because the mean bias of erroneous samples could indeed be zero, indicating unity. Focus instead on the correlation coefficient (R-value). The agreement between specimen concentrations should yield an R-value of 0.98 or greater (this will be sensitive only if analyte concentrations are normally distributed across the range). Substantial deviations from that value indicate a lack of individual sample accuracy, which may suggest poor selectivity or poor recovery of the assay. If the values relative to the RMP concentrations are higher, lack of selectivity is a probable cause. Lower than expected concentrations representative of poor recovery are typically related to the IS, either through overrecovery of the IS (relative to the analyte, equilibration again is key) or through nonselective IS transitions. If the correlation is indeed poor, irrespective of the slope of the analysis, additional method development and revalidation are required.

The second context should evaluate the data for a proportional bias between the expected (RMP value) and the measured concentrations. If multiple samples spanning the analytical range are available and there is a clear proportional fixed bias (ie, $>\pm10\%$ throughout the range), calibration values may not represent true concentrations and should be adjusted accordingly. If the bias reduces at lower concentrations (ie, more positive to less positive or positive to negative bias), adsorptive losses (appropriate carrier materials in blank calibrator matrix) or methodologic bias in manufacturing should be explored (serial dilution of calibrators yields this type of bias). Should the bias shift from lower to higher values (more negative to less negative or negative to positive), calibrator matrix contamination, selectivity, or methodologic bias should be explored.

Certified reference materials may be available with more than one concentration of the analyte of interest, although it is more frequent that a single solution is available. In instances in which the value is of sufficiently high concentration, accurate dilution of the CRM can serve to estimate trueness at various concentrations throughout the AMR. This dilution must be constrained to the amount shown to be valid through the previously described experimentation (dilutional linearity).

External qualification materials, such as proficiency test specimens, may be available to assess the assay's performance compared to that of other laboratories. Care should be taken in understanding the origin of these materials and the manner in which the target value(s) are generated. If possible, comparison to the same technology is preferred.

The evaluation of reference intervals, either by de novo generation or by reference interval verification has been exhaustively addressed elsewhere of this text. The experiments described therein are directly applicable to LC-MS/MS assays.

Full characterization of the analytical performance of the assay has been performed. All data should be included in a validation report for review and approval by appropriate medical, quality assurance, and technical management. All raw and processed data should be retained for at least 2 years after retirement of the assay to enable review by external auditors or longer for assays supporting clinical trials. SST data from analytically acceptable prevalidation and validation runs should be reduced into a results table, ignoring the first injection from the three injection series. SST acceptance criteria (tolerances for retention time, asymmetry, S/N, etc.) are generated. These data should guide decision making for acceptable LC-MS/MS instrument performance because the assay was demonstrated to be acceptable during these studies. SSTs are an invaluable asset for troubleshooting and verification of instrument performance before clinical analysis. Clinical assays are generally modified over time, and a partial validation may be considered. The framework of validation experiments described here represents a complete validation. The clinical and analytical impact of modifications to a clinical assay require consideration of the potential for error in reporting patient results and should be judiciously considered, including appropriate quality assurance and medical director and technical input; burden of proof should guide conscience.

POINTS TO REMEMBER

IS selection	Stable, isotopically labeled analyte highly preferred, no observable analyte, mass difference of +3 to +6, be aware of labeling position
MS development	Check for nonstandard precursors, provisional optimization in the presence of solvent
LC development, part 1	Establish optimal mobile phases for molecule
LC development, part 2	Screen many columns, reduce data according to chromatographic features (asymmetry, resolution, response)
Sample preparation development	Ensure appropriate equilibration of ISs and use neat solutions for calculation of recoveries in all experiments
Prevalidation	Test likely failure points and assess assay characteristics before initiation of validation
Validation	Address as many clinical and analytical components as possible; provide documentation, including validation plan and validation report

REFERENCES

1. Finkle BS. Drug-analysis technology: overview and state of the art. *Clin Chem* 1987;**33**(Suppl.):13B−7B.
2. Maurer HH. Role of gas chromatography-mass spectrometry with negative ion chemical ionization in clinical and forensic toxicology, doping control, and biomonitoring. *Ther Drug Monit* 2002;**24**:247−54.
3. Millington DS, Kodo N, Norwood DL, et al. Tandem mass spectrometry: a new method for acylcarnitine profiling with potential for neonatal screening for inborn errors of metabolism. *J Inherit Metab Dis* 1990;**13**:321−4.
4. Kushnir MM, Rockwood AL, Nelson GJ, et al. Liquid chromatography-tandem mass spectrometry analysis of urinary free cortisol. *Clin Chem* 2003;**49**:965−7.
5. Kushnir MM, Rockwood AL, Bergquist J. Liquid chromatography-tandem mass spectrometry applications in endocrinology. *Mass Spectrom Rev* 2010;**29**:480−502.
6. Lagerstedt SA, O'Kane DJ, Singh RJ. Measurement of plasma free metanephrine and normetanephrine by liquid chromatography-tandem mass spectrometry for diagnosis of pheochromocytoma. *Clin Chem* 2004;**50**:603−11.
7. Yang Z, Wang S. Recent development in application of high performance liquid chromatography-tandem mass spectrometry in therapeutic drug monitoring of immunosuppressants. *J Immunol Methods* 2008;**336**:98−103.
8. Decaestecker TN, Clauwaert KM, Van Bocxlaer JF, et al. Evaluation of automated single mass spectrometry to tandem mass spectrometry function switching for comprehensive drug profiling analysis using a quadrupole time-of-flight mass spectrometer. *Rapid Commun Mass Spectrom* 2000;**14**:1787−92.
9. Marin SJ, Hughes JM, Lawlor BG, et al. Rapid screening for 67 drugs and metabolites in serum or plasma by accurate-mass LC-TOF-MS. *J Anal Toxicol* 2012;**36**:477−86.
10. Struck-Lewicka W, Kordalewska M, Bujak R, et al. Urine metabolic fingerprinting using LC-MS and GC-MS reveals metabolite changes in prostate cancer: a pilot study. *J Pharm Biomed Anal* 2015;**111**:351−61.
11. Yin P, Xu GJ. Current state-of-the-art of nontargeted metabolomics based on liquid chromatography-mass spectrometry with special emphasis in clinical applications. *J Chromatogr A* 2014;**1374**:1−13.
12. Benton HP, Ivanisevic J, Mahieu NG, et al. Autonomous metabolomics for rapid metabolite identification in global profiling. *Anal Chem* 2015;**87**:884−91.
13. Botelho JC, Shacklady C, Cooper HC, et al. Isotope-dilution liquid chromatography-tandem mass spectrometry candidate reference method for total testosterone in human serum. *Clin Chem* 2013;**59**:372−80.
14. Vesper HW, Botelho JC, Shacklady C, et al. CDC project on standardizing steroid hormone measurements. *Steroids* 2008;**73**:1286−92.
15. Vesper HW, Botelho JC, Vidal ML, et al. High variability in serum estradiol measurements in men and women. *Steroids* 2014;**82**:7−13.
16. Vogeser M, Seger C. Pitfalls associated with the use of liquid chromatography-tandem mass spectrometry in the clinical laboratory. *Clin Chem* 2010;**56**:1234−44.
17. Clinical Laboratory Standards Institute. *Mass spectrometry in the clinical laboratory: general principles and practice—approved guideline*, vol C-50A. Wayne, PA: Clinical Laboratory Standards Institute; 2007.
18. Clinical Laboratory Standards Institute. *Liquid chromatography-mass spectrometry methods: approved guideline*, vol C-62A. Wayne, PA: Clinical Laboratory Standards Institute; 2014.
19. Clinical Laboratory Standards Institute. *Metrological traceability and its implementation: a report*. EP-32R. Wayne, PA: Clinical Laboratory Standards Institute; 2006.
20. Clinical Laboratory Standards Institute. *Characterization and qualification of commutable reference materials for laboratory medicine: approved guideline*, vol EP-30A. Wayne, PA: Clinical Laboratory Standards Institute; 2010.

21. Clinical Laboratory Standards Institute. *Evaluation of the linearity of quantitative measurement procedures: a statistical approach—approved guideline*, vol EP-6A. Wayne, PA: Clinical Laboratory Standards Institute; 2003.
22. Sargent M, Harte R, Harrington C. *Guidelines for achieving high accuracy in isotope dilution mass spectrometry (IDMS)*. London: Royal Society of Chemistry; 2002.
23. Wang S, Cyronak M, Yang E. Does a stable isotopically labeled internal standard always correct analyte response? A matrix effect study on a LC/MS/MS method for the determination of carvedilol enantiomers in human plasma. *J Pharm Biomed Anal* 2007;**43**:701–7.
24. Stokvis E, Rosing H, Beijnen JH. Stable isotopically labeled internal standards in quantitative bioanalysis using liquid chromatography/mass spectrometry: necessity or not? *Rapid Commun Mass Spectrom* 2005;**19**:401–7.
25. Yuan C, Kosewick J, He X, et al. Sensitive measurement of serum 1α,25-dihydroxyvitamin D by liquid chromatography/tandem mass spectrometry after removing interference with immunoaffinity extraction. *Rapid Commun Mass Spectrom* 2011;**25**:1241–9.
26. Jin Z, Daiya S, Kenttämaa HI. Characterization of nonpolar lipids and selected steroids by using laser-induced acoustic desorption/chemical ionization, atmospheric pressure chemical ionization, and electrospray ionization mass spectrometry. *Int J Mass Spectrom* 2011;**301**:234–9.
27. Bonfiglio R, King RC, Olah TV, et al. The effects of sample preparation methods on the variability of the electrospray ionization response for model drug compounds. *Rapid Commun Mass Spectrom* 1999;**13**:1175–85.
28. King R, Bonfiglio R, Fernandez-Metzler C, et al. Mechanistic investigation of ionization suppression in electrospray ionization. *J Am Soc Mass Spectrom* 2000;**11**:942–50.
29. Kushnir MM, Rockwood AL, Nelson GJ, et al. Assessing analytical specificity in quantitative analysis using tandem mass spectrometry. *Clin Biochem* 2005;**38**:319–27.
30. Williams TM, Kind AJ, Houghton E, et al. Electrospray collision-induced dissociation of testosterone and testosterone hydroxy analogs. *J Mass Spectrom* 1999;**34**:206–16.
31. Thevis M, Beuck S, Höppner S, et al. Structure elucidation of the diagnostic product ion at m/z 97 derived from androst-4-en-3-one-based steroids by ESI-CID and IRMPD spectroscopy. *J Am Soc Mass Spectrom* 2012;**23**:537–46.
32. Klee S, Derpmann V, Wißdorf W, et al. Are clusters important in understanding the mechanisms in atmospheric pressure ionization? I. Reagent ion generation and chemical control of ion populations. *J Am Soc Mass Spectrom* 2014;**25**:1310–21.
33. Kallal T, Sanchez A, Grant RP. *Accelerating bioanalytical LC-MS/MS method development using an automated modular strategy*. Nashville, TN: Presented at the American Society for Mass Spectrometry; 2004.
34. Gao S, Bhoopathy S, Zhang ZP, et al. Evaluation of volatile ion-pair reagents for the liquid chromatography-mass spectrometry analysis of polar compounds and its application to the determination of methadone in human plasma. *J Pharm Biomed Anal* 2006;**40**:679–88.
35. Rappold BA, Grant RP. HILIC-MS/MS method development for targeted quantitation of metabolites: practical considerations from a clinical diagnostic perspective. *J Sep Sci* 2011;**34**:3527–37.
36. Polson C, Sarkar P, Incledon B, et al. Optimization of protein precipitation based upon effectiveness of protein removal and ionization effect in liquid chromatography-tandem mass spectrometry. *J Chromatogr B* 2003;**785**:263–75.
37. Little JL, Wempe MF, Buchanan CM. Liquid chromatography-mass spectrometry/mass spectrometry method development for drug metabolism studies: examining lipid matrix ionization effects in plasma. *J Chromatogr B* 2006;**833**:219–30.
38. Matuszewski BK, Constanzer ML, Chavez-Eng CM. Matrix effect in quantitative LC/MS/MS analyses of biological fluids: a method for determination of finasteride in human plasma at picogram per milliliter concentrations. *Anal Chem* 1998;**70**:882–9.
39. Marney LC, Laha TJ, Baird GS, et al. Isopropanol protein precipitation for the analysis of plasma free metanephrines by liquid chromatography-tandem mass spectrometry. *Clin Chem* 2008;**54**:1729–32.
40. Kushnir MM, Urry FM, Frank EL, et al. Analysis of catecholamines in urine by positive-ion electrospray tandem mass spectrometry. *Clin Chem* 2002;**48**:323–31.
41. Rappold BA, Holland PL, Grant RP. *Balancing sample preparation and liquid chromatography to remove phospholipid-based matrix effects in positive ESI*. Denver, CO: Presented at The American Society for Mass Spectrometry; 2008.
42. Grant RP, Crawford ML, Bruton J, et al. *The phospholipid fix: quantitative measurement and analytical solutions for phospholipid depletion*. Salt Lake City, UT: Presented at The American Society for Mass Spectrometry; 2010.
43. Holland PL, Shuford CM, Green MK, et al. *Evaluation of the SPEware Cerex® ALD-III 192 for use in automating SPE and SLE methods in validated LC-MS/MS assays*. Baltimore, MD: Presented at The American Society for Mass Spectrometry; 2014.
44. Dee S, Holland PL, Shuford CM, et al. *Measurement of low level endogenous biomarkers for use in clinical diagnostics*. Baltimore, MD: Presented at The American Society for Mass Spectrometry; 2014.
45. de Jong WH, Graham KS, van der Molen JC, et al. Plasma free metanephrine measurement using automated online solid-phase extraction HPLC tandem mass spectrometry. *Clin Chem* 2007;**53**:1684–93.
46. Herman JL. The use of turbulent flow chromatography and the isocratic focusing effect to achieve on-line cleanup and concentration of neat biological samples for low-level metabolite analysis. *Rapid Commun Mass Spectrom* 2005;**19**:696–700.
47. Grant RP, Crawford M, Rappold BA, et al. *Design and utility of open-access liquid chromatography tandem mass spectrometry in quantitative clinical toxicology*. Canada: Presented at The American Society for Mass Spectrometry, Vancouver; 2012.
48. Clinical Laboratory Standards Institute. *Interference testing in clinical chemistry; approved guideline: second edition*. EP-7A2. Wayne, PA: Clinical Laboratory Standards Institute; 2003.
49. Guideline on bioanalytical method validation, European Medicines Agency, EMEA/CHMP/EWP/192217/2009 Rev.1 Corr, last updated 16/09/2014.
50. Little JL, Cleven CD, Brown SD. Identification of "known unknowns" utilizing accurate mass data and chemical abstracts service databases. *J Am Soc Mass Spectrom* 2011;**22**:348–59.
51. Bjerner J, Biernat D, Fosså SD, et al. Reference intervals for serum testosterone, SHBG, LH and FSH in males from the NORIP project. *Scand J Clin Lab Invest* 2009;**69**:873–9.

52. Clinical Laboratory Standards Institute. *Mass spectrometry for androgen and estrogen measurements in serum. C57-Ed1*. Wayne, PA: Clinical Laboratory Standards Institute; 2015.
53. US Department of Health and Human Services. US Food and Drug Administration, Center for Drug Evaluation and Research, Center for Veterinary Medicine. *Guidance for industry: bioanalytical method validation* 2001. <http://www.fda.gov/downloads/Drugs/GuidanceComplianceRegulatoryInformation/Guidances/UCM070107>.
54. Clinical Laboratory Standards Institute. *Defining, establishing and verifying reference intervals in the clinical laboratory*. 3rd ed. Wayne, PA: Clinical Laboratory Standards Institute; 2010. EP28-A3.
55. International Organisation for Standardization. *Quality management for the medical laboratory*. ISO/FDIS 15189. Geneva, Switzerland: International Organisation for Standardization.
56. Clinical Laboratory Standards Institute. *Evaluation of precision performance of quantitative measurement methods. EP5—A2*. Wayne, PA: Clinical and Laboratory Standards Institute; 2004.
57. Clinical and Laboratory Standards Institute. *User verification of performance for precision and trueness. EP15—A2*. Wayne, PA: Clinical and Laboratory Standards Institute; 2005.
58. Ricos C, Alvarez V, Cava F, et al. Current databases on biological variation: pros, cons and progress. *Scand J Clin Lab Invest* 1999;**59**:491—500.
59. Westgard JO. Assuring analytical quality through process planning and quality control. *Arch Pathol Lab Med* 1992;**116**:765—9.
60. Clinical and laboratory standards institute. *Protocol for determination of limits of detection and limits of quantitation. Ep17—A2*. Wayne, PA: Clinical and Laboratory Standards Institute; 2004.
61. International Organisation for Standardization. *Accuracy (trueness and precision) of measurement methods and results. II. Basic method for the determination of repeatability and reproducibility of a standard measurement method. 5725—2*. Geneva, Switzerland: International Organisation for Standardization; 2002.
62. Fugelstad A, Ahlner J, Brandt L, et al. Use of morphine and 6-monoacetylmorphine in blood for the evaluation of possible risk factors for sudden death in 192 heroin users. *Addiction* 2003;**98**:463—70.
63. College of American Pathologists. Chemistry and toxicology checklist. *Northfield, IL: College of American Pathologists* 2014.
64. Section 8. *Guideline, ICH harmonised tripartite: validation of analytical procedures—text and methodology. Q2(R1)*. 2005. Geneva, Switzerland.
65. Yun YM, Botelho JC, Chandler DW, et al. Performance criteria for testosterone measurements based on biological variation in adult males: recommendations from the Partnership for the Accurate Testing of Hormones. *Clin Chem* 2012;**58**:1703—10.

Proteomics

Andrew N. Hoofnagle and Cory Bystrom

ABSTRACT

Background
Clinical proteomics has traditionally referred to experiments that attempt to discover novel biomarkers for disease diagnosis, prognosis, or therapeutic management by using tools that measure the abundance of hundreds or thousands of proteins in a single sample. These discovery experiments began with protein electrophoresis, particularly two-dimensional (2D) gel electrophoresis, and have evolved into workflows that rely very heavily on mass spectrometry (MS). Using the workflows developed for discovery proteomics, clinical laboratories have developed quantitative assays for proteins in human samples that solve many of the issues associated with the measurement of proteins by immunoassay. The technology is changing clinical research and is poised to significantly transform protein measurements used in patient care.

Content
This chapter begins with the history of clinical proteomics, with a special emphasis on 2D gel electrophoresis of serum and plasma proteins. It then describes discovery techniques that use MS, including data-dependent acquisition and data-independent acquisition. It finishes with a discussion of targeted quantitative proteomic methods, both bottom-up and top-down, as replacement methodologies for immunoassays and Western blotting. Special attention is paid to peptide selection, denaturation and digestion, peptide and protein enrichment, internal standards, and calibration.

HISTORICAL PERSPECTIVE

The word *proteome* is a combination of the words *protein* and *genome*, first coined by Marc Wilkins in 1994.[1] Wilkins used the term to describe the entire complement of proteins expressed by a genome, cell, tissue, or organism, and *proteomics* refers to the comprehensive identification and quantitative measurement of these proteins. Today, the term encompasses separation science, protein microchemistry, bioinformatics, and mass spectrometry (MS) as the fundamental techniques used in the large-scale study of protein identity, abundance, structure, and function.

Analysis of proteomes, and the human proteome in particular, was a logical extension of the completion of the human genome, which revealed the genetic blueprint from which the proteome is constructed. The promise of proteomics for clinical research was an outgrowth of the understanding that many clinically relevant markers are present in blood and that disease-related aberrations in protein abundance might be quantified by comparative analysis of proteomes.

Early Proteomics
The earliest investigations of the human serum proteome preceded the completion of the human genome by 24 years and used two-dimensional gel electrophoresis (2D gel) to provide a high-resolution snapshot of serum proteins (Fig. 6.1).[2] In this technique a complex pool of proteins was first resolved by isoelectric point in the first dimension and by molecular mass in the second dimension followed by staining for visualization. This spread the proteins in the sample into an array in which each spot was associated with one protein. Applying this technique to serum illuminated its complexity, revealing over 300 spots, which were suggested to arise from 75 to 100 unique proteins. Some of these proteins were identified by comparison with the migration of purified protein standards, comparison with immunoprecipitated proteins, or immunoblotting.

Although powerful, 2D gels did not gain widespread use for comparative studies until the early 1990s. The main limitation was a lack of tools for the rapid identification of the proteins contained in a single spot, which could link the location, intensity, and identity of a gel spot to biology. Before the advent of MS-based tools, identification of a single protein took several weeks of dedicated labor starting with an effort to isolate sufficient protein for subsequent analysis. Once isolated, the extracted protein was proteolyzed with trypsin and two or more peptides were isolated using preparative HPLC separation. These peptides were then analyzed with Edman degradation to obtain short N-terminal amino acid sequences.[3] The sequence data, along with approximate molecular weight, isoelectric point, and any other available data, were used to search available databases and infer protein identity.

Proteome analysis accelerated dramatically through the 1990s largely because of the availability of an array of technologies that transformed the protein identification problem from an arduous task to a scalable, simple procedure that

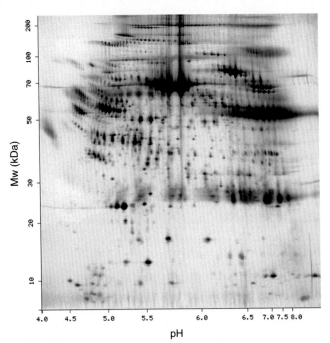

FIGURE 6.1 Two-dimensional gel electrophoresis. Before the advent of mass spectrometric methods to probe the proteome, proteins were resolved and quantified using two-dimensional polyacrylamide gel electrophoresis. Proteins were first separated based on isoelectric point (horizontally) and then based on size (vertically). Hundreds of spots were visible. (Reprinted by permission of the publisher from Hoogland C, Mostaguir K, Sanchez JC, Hochstrasser DF, Appel RD. SWISS-2DPAGE, ten years later. *Proteomics* 2004;4:2352–2356. Copyright 2004 WILEY-VCH Verlag GmbH. http://world-2dpage.expasy.org/swiss-2dpage/viewer&map=PLASMA_HUMAN&ac=all)

representing healthy and disease states, perform preanalytical fractionation, separate the individual protein fractions by 2D electrophoresis, and stain the gels (see Fig. 6.1). Using imaging techniques, comparative analysis of gels identified comigrating spots with different intensity or changes in spot position. Many studies reported statistically significant differences in protein abundance associated with disease, but validation of a primary discovery leading to a biomarker with sufficient promise for clinical studies has not been realized. Frequently, the biomarker proteins were acute-phase proteins that were known to generally associate with many diseases. Alternatively, high-abundance proteins with no clear biological significance to the disease were identified, suggesting experimental artifact. The constraints on proteome depth and breadth and the inability to perform sufficient technical and experimental replicates to overcome variability left many gel-based experiments significantly underpowered to successfully detect biomarkers.

As an analytical tool for proteome analysis, 2D gels suffer from variability, significant demands on labor, and limited dynamic range. Since biological samples had a tremendous range of protein abundances (7 to 12 orders of magnitude), the 1 to 2 orders visible on a 2D gel made a deep analysis of the proteome essentially impossible.[14,15] Unfractionated samples of tissue or blood often have 10 to 20 proteins that account for 80% or more of the total protein content of the sample.[16] As a result, the observable proteome on a 2D gel is severely limited. Loading greater amounts of sample onto 2D gels causes distortions during separation, dramatically reduces reproducibility, and can contaminate large portions of the gel, leading to ubiquitous identification of the abundant proteins, even in unexpected regions on the gel.

To improve the depth of analysis, creative fractionation and depletion strategies were developed to work around the protein abundance problem. These approaches focused on separating protein pools into subfractions based on chemical,[17] structural,[18] or biophysical features.[19,20] Depletion of highly abundant proteins using immunoaffinity[21] and semi-specific chemical affinity approaches[22] also became commonplace. These approaches were effective but added cost, complexity, and variability to an already challenging technique. Although this discussion is historical, challenges of proteome dynamic range are still a consideration in modern proteomics workflows.

It was soon recognized that complex mixtures of proteins could be directly ionized and analyzed with MALDI MS to give complex spectra comprising the masses and crude abundance of many proteins.[23] Although much more limited in resolution than 2D gels (as a result of reducing separation to mass alone), it was appealing to suggest that MALDI spectra could substitute for 2D gels and resolve at least some of the long-standing technical issues that compromised gel-based analysis: labor intensiveness and poor throughput. With the ability to apply the MALDI technique to biomarker discovery, efforts could be directed at a smaller set of technical challenges, namely sample preparation.

Biomarker Discovery in the Post–Two-Dimensional Gel Era

With the increased throughput and improved precision that appeared possible with MALDI analysis of complex mixtures, manufacturers developed integrated systems of reagents and

could be completed in a day or two. First, improvements in MS, which combined cost-effective, semiautomated, high-performance instruments equipped with gentle ionization methods (matrix-assisted laser desorption ionization [MALDI] and electrospray ionization [ESI][4,5]; see Chapter 2 for a discussion of ionization techniques) allowed laboratories to collect easily interpretable mass spectra. Second, protein microchemistry techniques allowed sufficient sample to be extracted from single gel spots to yield interpretable mass spectra. Finally, high-quality genomic databases and novel statistical algorithms could compare mass spectrometric data to database entries to provide protein identifications with unprecedented speed and accuracy.[6,7] Together, these technologies merged into a powerful analytical platform that enabled protein identification and characterization in a manner that matched the throughput and number of protein spots that could be resolved on a 2D gel. A number of public 2D gel databases were published on the World Wide Web, some of which are still available (eg, http://world-2dpage.expasy.org/list/).

Progress continued into the mid-2000s, with advances in 2D gels and associated technologies that allowed the reproducible resolution of hundreds to thousands of protein spots in complex samples. Efforts to apply the techniques in clinical research quickly expanded to more groups and more diverse sample types, including tissue,[8,9] blood,[10,11] tears,[12] and urine.[13] The general approach was to obtain samples instruments for performing discovery experiments. Although mass spectrometers are subject to constraints similar to those with 2D gels regarding the observable dynamic range of protein abundances in a single spectrum, the ability to automate upstream sample preparation or carry out microscale fractionation directly on MALDI targets and commercial availability of "chips" made the technical aspects of biomarker discovery much less daunting.[24,25] Although identification of proteins in this workflow was generally not possible without substantial effort, it was argued by some that the identity of proteins giving rise to individual discriminatory peaks in a spectrum was not important, and instead the pattern was the biomarker that linked phenotype to a snapshot of relative protein abundance. Critics pointed out that in a spectrum of a complex matrix such as serum or plasma, a peak was unlikely to be a single protein and instead would be composed of dozens, if not hundreds, of proteins, making it impossible to create the link between the biology and biochemistry of a disease process and the change in abundance of any specific peak.[26] The dawn of high-throughput proteome analysis led to a shift in thinking about biomarker discovery from hypothesis-driven to hypothesis-generating. In the former, biochemical pathways associated with disease are rigorously interrogated to link observations and biology. In the latter, an unbiased comparison of samples without consideration of biology is used to identify putative differences in protein abundance between sample types. At around the same time, the field also adopted the hopeful concept of multiprotein biomarkers or changes in abundance of multiple proteins, none of which might be individually valuable as a marker that could be combined to provide important clinical information. Unbiased discovery and reduced stringency promised to accelerate biomarker discovery and feed the pipeline of biomarker validation.

Given the apparent technical benefits, and ignoring the probable compromises, the approach was applied to many diseases that had defied biochemical diagnosis or those that would benefit from early detection.[27,28] Initial studies on breast and prostate cancer were promising, and results from one particular study electrified both the clinical and proteomics research communities. The results from that study suggested that ovarian cancer could be unambiguously identified, even at the earliest stages, in almost all patients.[29] Efforts to validate this study led to an acrimonious public debate, and eventually it was accepted that the apparent clinical utility of the test was merely an artifact arising from poor experimental design and fundamental misunderstandings of MS data and processing.[30-33] Biomarker discovery studies using these methods continued for several years after this upheaval, but the approach was largely abandoned by 2007.

A pivotal moment in the early days of proteomics was the development of integrated liquid chromatography–tandem mass spectrometry (LC-MS/MS) systems with automated data acquisition and automated data processing. Pioneering work by Washburn and Yates[34] provided a way to perform proteome analysis that truly overcame many of the shortcomings of 2D gels. The strategy started with a complex protein mixture, which was digested with trypsin. The peptides were then fractionated by multidimensional chromatography, with each fraction being directed to the mass spectrometer. Although it is counterintuitive to make a complex mixture even more complex via proteolysis, the technique was highly automated, scaled easily, and used to generate protein catalogs of thousands of proteins at a fraction of the labor required for a similarly complex 2D gel analysis. Although not originally developed to perform quantitative analysis, subsequent enhancements of this technique provided the innovations that are at the core of most proteomics discovery experiments carried out today.

BIOMARKER PIPELINE

The goal of proteomic biomarker discovery is to identify and then demonstrate clinical utility of a protein or combination of proteins that provide useful diagnostic or prognostic information regarding disease.[35-37] At each step of this process, experimental tools and performance expectations change. At the earliest stages of discovery, dozens of samples are interrogated at the level of thousands of proteins (Fig. 6.2). Discovery workflows from preanalytical sample preparation to complete data analysis are very time consuming, requiring hours to days of data collection for each individual specimen that is being processed. It is not uncommon for discovery experiments to require nonstop data acquisition for a month or longer, which constrains the size of the discovery experiment. After the raw data are collected, they are reduced to give a list of each peak by mass, retention time, intensity, and identity (if available). These lists are then mathematically manipulated to align common features between multiple specimens and to normalize the overall intensity between each acquired data set. Finally, the feature lists are compared to generate a list of peaks that are reproducibly and significantly different between the two clinical states. Frequently, these discovery experiments yield 10 to 100 candidate proteins suitable for evaluation in preclinical studies. The investigator may choose to use other data to help refine this list by focusing on known or postulated biological relevance.

In the next stage of the pipeline, specific targeted assays for selected candidate proteins are developed with methods that achieve higher throughput and precision, with the goal of verifying the potential utility of the candidate marker proteins in a series of pilot studies that will incorporate samples from hundreds of patients. This will often reduce the number of proteins to 5 to 20 from the original group.

In the final phase of biomarker development, the most neglected phase to date, an assay for one or a panel of markers is developed and rigorously validated to establish analytical performance. The validated assay is then used to analyze hundreds to thousands of samples from prospective and/or retrospective clinical studies to establish the utility (or lack thereof) of the biomarker or panel of biomarkers. On completion of this work, a successful biomarker or biomarker panel assay will be described in a standard operating procedure (SOP) that includes preanalytical requirements, a description of the intended use, and the expected clinical performance of the assay as described by metrics for sensitivity, specificity, positive predictive value, and negative predictive value.

Discovery Experiment

Similar to the first comparative studies with 2D gels, the aim of present-day discovery proteomics experiments is to

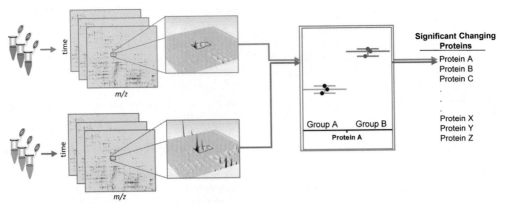

FIGURE 6.2 Proteomics biomarker discovery pipeline. To discover potential novel protein biomarkers, samples can be prepared for liquid chromatography–mass spectrometry (LC-MS) and analyzed as intact proteins. Replicates of pooled samples from two different disease states or unpooled samples from individual patients from two disease states may be compared (eg, healthy vs. control). Signals from the mass spectrometer are integrated and compared *in silico* to identify protein features that are significantly different among the disease states, which can be subsequently identified in later analyses. This type of analysis of intact proteins is limited to small proteins. Alternatively, samples can be prepared by using proteolysis before being analyzed as peptides by LC-tandem MS (LC-*MS/MS*). Using this approach, proteins are identified and quantified to find proteins that are different between pathophysiologic states. (Figure courtesy Dr. Tim Collier, PhD. Used with permission, copyright Tim Collier, 2015.)

identify and quantify all of the proteins in complex mixtures to find biomarker proteins that differ in relative abundance between two biological states (eg, diseased and normal). The discovery pipeline begins with the selection of useful samples. Ideally, clinically relevant samples are selected that have been collected, prepared, and stored in an identical fashion along with a dossier of information regarding demographic and clinical history. Preanalytical steps are generally applied to fractionate or otherwise modify the sample to improve depth of analysis to lend the greatest opportunity to measure proteins that may differ by disease state. Subsequently, the samples are then digested from whole proteins into a complex pool of peptides destined for LC-MS or LC-MS/MS experiments to provide data sets that can be compared to identify significant differences between samples.

Requirement for Separations

As mentioned earlier, given the enormous complexity of biological specimens and the additional complexity that is derived from proteolysis of proteins into peptides, separation technologies must be used to allow a mass spectrometer to specifically detect and identify peptides from among a complex mixture of peptides. This is due to the limited duty cycle of mass spectrometers and the need to acquire one tandem mass spectrum at a time (also called a fingerprint spectrum, further described later).

High-performance liquid chromatography (HPLC) using reverse-phase chemistry has been the chief tool for performing peptide and protein separations directly interfaced to mass spectrometers.[38] The compatibility of reverse-phase solvent systems with LC-MS, commercial availability of a wide range of column dimensions and packing materials, and the ease with which high-resolution separations are achieved with modest method development are key advantages compared to alternative separation technologies. Chromatography can be performed under a wide range of flow rates from 0.5 µL/min or less for maximum sensitivity but low throughput, to greater than 1 mL/min for maximal throughput at reduced sensitivity.[39,40] Alternative separation techniques such as capillary electrophoresis have successfully been interfaced with mass spectrometers[41] and offer very-high-resolution separations, but challenges in maintaining robust performance and complexities in developing routine separation methods have prevented wider adoption.

Data Acquisition Strategies

MS experiments in biomarker discovery studies are typically conducted using one of three data collection strategies: (1) full scan only LC-MS, (2) data-dependent LC-MS/MS, or (3) data-independent LC-MS/MS. Each workflow has different strengths and weakness, and laboratories select workflows based on experience and specific assay requirements.

In a full-scan acquisition experiment, a high-resolution mass spectrometer continually collects spectra over a wide mass-to-charge *(m/z)* range, capturing signals from all ions that are presented to the mass spectrometer as the chromatogram is developed. No tandem mass spectra are acquired, which are needed for the identification of the peptides eluting from the chromatographic column. Instead, very-high-resolution contour maps of mass spectrometric data are collected across three dimensions: time, intensity, and *m/z*. If a peak appears to be different between two clinically relevant groups in this type of discovery experiment, a second round of analysis is required to identify the peptide of interest.[42]

Data-dependent acquisition (DDA) is an algorithm-guided strategy in which a wide *m/z* range spectrum or survey scan is collected and used to identify candidate peaks for subsequent MS/MS in real time (Figs. 6.3 and 6.4A). The survey scan is rapidly processed to generate a list of observed *m/z* for each detected peak ranked in order of intensity. This list is then used to assemble a queue for directed tandem mass spectrometric experiments. Iteratively, the instrument then performs selection and fragmentation of each precursor ion stored in the queue. To prevent the instrument from repeatedly analyzing the same set of high-abundance precursors,

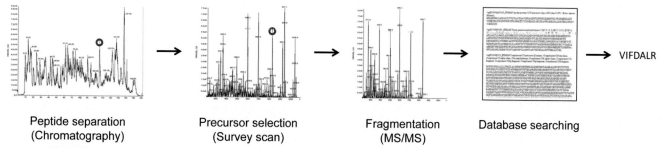

FIGURE 6.3 Data-dependent acquisition. Peptide identification in discovery proteomics experiments often uses software to drive the mass spectrometer in the selection of the precursor peptides to be fragmented. As peptides elute from the chromatographic column, the mass spectrometer first performs a survey scan to assess peptide precursor *m/z* and their abundance. In this theoretical example a survey scan is performed at 105 minutes and the most abundant peaks are selected for subsequent tandem mass spectrometry *(MS/MS)* analysis, including the specific peptide that is fragmented in the third panel. The resulting spectrum is compared against a theoretical database of proteins to identify the peptide from the spectrum using various statistical approaches. *VIFDALR* is the example peptide fragment identified (using single letter amino acid codes).

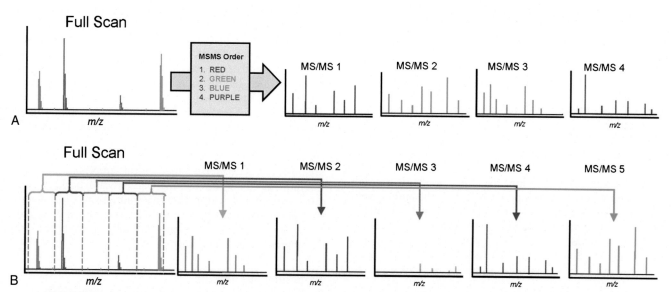

FIGURE 6.4 Comparison of data-dependent acquisition (DDA) and data-independent acquisition (DIA). During DDA and DIA discovery experiments, the mass spectrometer begins with a high-resolution, high mass accuracy survey scan of the peptides eluting off the chromatographic column. **A,** In DDA, the survey scan is used to build a list of the precursors (using a 0.7 to 2.0 Da window) that will be fragmented/analyzed by tandem mass spectrometry *(MS/MS)* in subsequent steps in the mass spectrometer (typically 2 to 8 fragments are targeted from each precursor survey scan). This cycle of precursor/survey scan and MS/MS steps (typically ≥ 7 Hz) is repeated throughout the chromatographic run. **B,** In contrast, after the precursor scan in DIA, every part of the *m/z* range (typically 400 to 1000 Da wide) is then sampled by stepping through the *m/z* range using windows of 10 to 20 Da, collecting all of the precursors in each window and fragmenting/analyzing them by MS/MS. This cycle of precursor/survey scan and MS/MS steps (typically ≤ 2 Hz) is repeated throughout the chromatographic run. Methods to improve the specificity of the DIA approach include overlapping windows and randomization of the MS/MS windows (not shown). For both methods, there is significant post–data acquisition analysis using software to determine the identity and abundance of peptides in the sample (not shown). (Figure courtesy Dr. Tim Collier, PhD. Used with permission, copyright Tim Collier, 2015.)

specific *m/z* values can be placed on an exclusion list for a fixed duration, which prevents further analysis until the exclusion criteria expire. Using this experimental strategy, the final data set contains information on the *m/z*, intensity, retention time of each individual precursor ion, and related fragmentation data.

Finally, data-independent acquisition (DIA) strategies use a full scan precursor survey, followed by the sequential isolation/fragmentation (MS/MS) of all ions in a small window of the precursor scan (eg, 5 to 20 Da) until the entire *m/z* range of the initial full scan is covered (see Fig. 6.4B). The survey scan and MS/MS scans are repeated throughout the chromatographic run. In contrast to DDA, DIA product ion spectra are composite spectra that include all fragment ions from all precursors isolated in the small window rather than just one precursor, as in DDA. To interpret the data, extensive processing deconvolutes precursor and fragment ion data to yield time, *m/z*, and intensity for each precursor peptide, as well as fragmentation data inferred from the composite spectra.

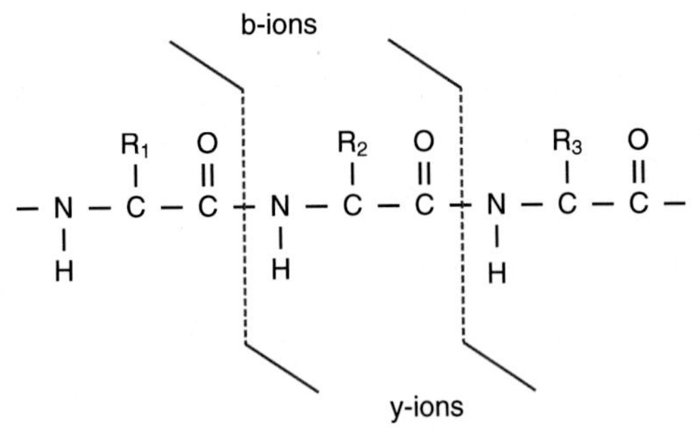

FIGURE 6.5 Fragmentation of peptides in the gas phase. Peptides are fragmented after being excited in the gas phase within the mass spectrometer. The most common ion fragments analyzed in triple-quadrupole mass spectrometers for targeted assays are *b-ions* and *y-ions*, respectively. Other ions are formed and may be more predominant in other types of mass analyzers (eg, a-, c-, x-, and z-ions). The characteristic fragmentation patterns of peptides make it possible to search databases for peptide identification from the fingerprint spectra.

For both DDA and DIA experiments, peptide identification relies on the ability of the mass spectrometer to create fragments in the gas phase to generate product ion spectra, also called MS/MS spectra. This fragmentation process occurs by imparting energy into a peptide by accelerating it and colliding it with an inert gas. Once it is at a higher energy state, a peptide will dissociate into two fragments in a thermodynamically probabilistic fashion, most commonly cleaving at specific points along the amide backbone to yield b-ions and y-ions, which respectively include the amino- and carboxyl-terminus of the peptide (Fig. 6.5). The mass differences between peaks in the spectrum, which are associated with fragmentation at each amino acid in the peptide, can then be used to construct the amino acid sequence of the peptide. The resulting product ion spectra are also called *fingerprint* spectra because each spectrum has features that can be associated with a specific peptide without deriving the amino acid sequence by manual inspection. Software algorithms are used to accomplish the process of matching mass spectra to peptide sequences in a database. The databases are generated by *in silico* digestion of all theoretical proteins that are derived from genomic data. In some cases, theoretical databases are supplemented with empirical MS data from a variety of sources, including public repositories representing hundreds of millions of processed mass spectra. Protein and peptide identification software such as Sequest,[43] MASCOT,[7] X! Tandem,[44] Andromeda,[45] and OMSSA[46] have been rigorously evaluated and are widely used by the MS community.

In each of these experimental workflows, isotope labels[47,48] and chemical tags[49,50] also have been used to allow the multiplex analysis of many samples simultaneously and/or to provide a reference against which both experimental sample types (eg, disease and healthy) can be compared to improve precision. For example, two samples, one from a diseased patient and one from a healthy patient, are chemically labeled during the preanalytical steps (ie, after proteolysis of the samples). In one case, the chemical label contains isotopes of natural abundance and the other is enriched in heavy isotopes. If the two samples are mixed after labeling, they will contain many identical peptides, but each peptide in the different samples will have a different mass because of the mass difference of the chemical label. Because this mixture of case and control is analyzed by LC-MS/MS, pairs of peaks will appear in the spectra, each representing the same peptide but at slightly different masses, which can be easily resolved by the mass spectrometer. These strategies help increase the number of technical and experimental replicates that can be achieved on any given instrument and also can facilitate the comparison of data acquired over weeks or months where instrument drift can make comparison of data sets more difficult.

Processing of Discovery Proteomics Data

Postprocessing of these data sets is required before final analysis. First, the precursor *m/z*, intensity, and retention time for each data set is arrayed in three dimensions and overlaid using software tools. These software tools can be used to adjust the data to correct for known experimental artifacts, such as drift in retention time. Second, the data are normalized to correct for intensity differences that arise from day-to-day changes in sample preparation efficiency and mass spectrometer performance. Third, the precursor mass and associated fragmentation data are used for database searching to identify *peptides*, which can be assembled to generate protein identifications. With appropriate data processing, relative abundance and identities of thousands of proteins can be achieved in a single discovery experiment. Once these steps are completed, an array of biostatistics tools is used to find proteins where significant abundance increases or decreases are observed between case and control samples.

Although there are commercial software packages that attempt to provide comprehensive solutions to these diverse workflows, many academic and industrial groups build data processing pipelines from custom-built, open-source, and commercial software packages.

Variations and Details

The experimental strategy that is ultimately selected is defined by the availability of specific instrumentation. There are subtle differences in each of the workflows that reflect tradeoffs in data depth versus breadth. For example, the greatest precision is generally observed in full-scan-only experiments, which yield no protein identifications without follow-up MS/MS analyses. The two drawbacks to this approach are (1) the reduced specificity of the method for low-abundance peaks, particularly when multiple peptides elute nearby and it becomes difficult to pick which peak corresponds to the peptide of interest, and (2) subsequent MS/MS analyses that can be helpful in identifying the peaks that differ between two samples or two groups of samples may have a shift in retention time between injections, which can make it difficult to assign MS/MS data to the particular peak of interest. DDA provides peptide and protein identification, but because of greater variability can have less power to discriminate abundance differences. The reason that DDA is more variable is because the automated process of picking peaks to analyze by MS/MS during the analysis is stochastic. More specifically, if more abundant peptides elute at the same time, the peptide of interest may not be selected for MS/MS analysis. DIA can provide the relative abundance of many peptides, but specificity is an issue because of overlapping peaks. In other words, the peptides that coelute with the analyte of interest will be fragmented at the same time and these fragments can make it difficult to tease out the signal of a single peptide from the signals present as a result of other peptides. Overall, significant recent improvements in instrumentation have enabled greater depth of proteome analysis with fewer compromises, but limits remain and the ability to exhaustively interrogate a proteome in a single LC-MS run is not yet achievable.

An alternative to these approaches is to use targeted mass spectrometric experiments to discover protein biomarkers.[51,52] With a list of predefined proteins of interest derived from previous proteomics experiments or biological insight, a tandem mass spectrometric (ie, MS/MS) method using a triple quadrupole or quadrupole–high mass accuracy analyzer hybrid instrument can be developed to detect only specific peptides in the sample and quantify them as chromatographic peak areas. The resulting peak areas are then normalized in some fashion to provide the relative abundance of the peptides of interest (ie, representing a subset of the total proteome). This type of discovery experiment is distinguished from fully quantitative experiments in that it lacks internal standards (ISs) to control for sample preparation and mass spectrometer performance and external calibrators to control for day-to-day digestion variability. It also typically lacks the quality control materials that are the cornerstone of longitudinal monitoring of assay quality in the clinical laboratory. Unlike DDA and DIA experiments, the decision to specifically target individual peptides places a limit on the total number of observations that can be achieved in a run. Although the precision of this approach is often good for research purposes (CV <25%), the breadth of protein coverage is limited to a few hundred peptides in a single run.

POINTS TO REMEMBER

Discovery Proteomics
- It is generally used to identify new biomarkers.
- Methods are less specific and precise compared to targeted quantitative methods.
- Gel electrophoresis was one of the first techniques used to examine the proteome.
- MALDI–time-of-flight (TOF) MS and LC-MS/MS replaced 2D gels.

Targeted Quantitative Experiment

After discovery, targeted quantitative mass spectrometric approaches are used in the development of specific, precise assays. These higher throughput assays, which are designed for improved specificity and precision, provide a mechanism to verify or validate initial proteomic discoveries by running hundreds of patient samples. During this stage, many candidates will fail. In some cases, proteins with marginal statistical significance will not hold up to further testing. In others, proteins will be removed because of poor performance characteristics (ie, stability, degradation, and modification), which would make them unsuitable for further assay development.

Targeted assays are typically performed using triple-quadrupole instruments after a well-developed technique known as stable isotope dilution (readers are referred to Chapter 5 for an in-depth discussion of the principles of isotope dilution assays). In this experiment, calibration materials with known concentrations of protein analyte are processed in parallel with patient samples where each sample and calibrator is spiked with an IS at a constant concentration. The IS is most often chemically and structurally identical to the analyte of interest, with the exception that stable (nonradioactive) heavy isotopes are incorporated. This modification changes only the mass of the peptide and not the chemical characteristics. Samples and calibrators are analyzed by MS to determine the intensity of signals specific for the analyte of interest and the IS, which are used in the determination of the peak area ratio (analyte peak area ÷ IS peak area). Using the calibration materials, a response curve is generated by plotting the peak area ratio versus concentration, which can be fit with a line by standard regression techniques. From this response curve, the concentration of the analyte of interest in an unknown sample can be determined from the peak area ratio.

The selection of MS to perform validation and clinical studies should be evaluated against the possibility of using commercially available immunoassays. Well-validated immunoassays with good sensitivity and specificity could alleviate the need for development and validation of high-quality MS-based assays, which can be time consuming (ie, 6- to 12-month development time for very-high-quality assays for low-abundance proteins is not unexpected). Frequently, though, discovery experiments yield proteins for which no well-validated immunoassay is commercially available, and in these cases the path through assay development can be significantly shorter with MS when compared to the development of an acceptable immunoassay.[35,36] Even when commercial kits are available, there is a growing appreciation of the substantial limitations afforded by immunoassays,[53] including lack of quality control for commercial research–grade enzyme-linked immunosorbent assay (ELISA) kits, nonspecific recognition of nontarget proteins, autoantibody and heterophilic antibody interference, poor concordance between kits for the same analyte, and saturation of sandwich assay reagents leading to falsely low results.[54] A well-designed mass spectrometric assay can avoid these issues, which makes it an attractive alternative to immunoassays.

POINTS TO REMEMBER

Targeted Proteomics
- Mass spectrometry (MS) is sometimes used without internal standards in discovery experiments.
- May or may not include proteolysis before liquid chromatography (LC).
- Without significant sample preparation, LC-MS or LC-MS/MS is more specific and precise than MALDI-TOF MS.

Bottom-Up and Top-Down Experiments

From the preceding sections, it may be apparent that proteins can be quantified by MS using two different approaches. In some cases, proteins are of sufficiently low molecular weight to be detected in their intact state, with or without dissociation in the gas phase in the mass spectrometer. This approach to quantification has been termed *top-down* proteomics to contrast with *bottom-up* proteomics, which relies on proteolysis and the quantification of surrogate peptides to determine protein concentration (Fig. 6.6). Both approaches are discussed in the following sections.

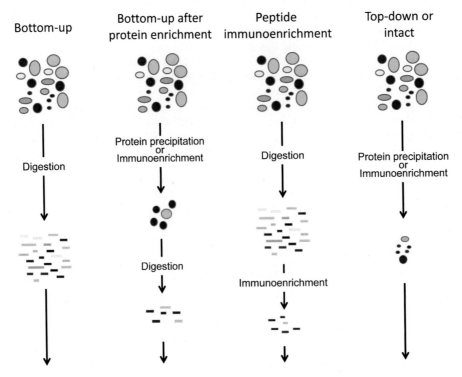

FIGURE 6.6 Common workflows for proteomic assays by mass spectrometry (MS). Bottom-up proteomics assays incorporate a proteolytic digestion step and use peptides detected by the mass spectrometer as surrogates for the protein contained in the sample. Intact or top-down proteomics assays require protein enrichment (either biochemical or immunoaffinity) before analysis of the protein by MS. Internal standard peptides (not shown) can be added before or after digestion (ideally before). Intact protein internal standards (not shown) are ideally added before protein enrichment.

BOTTOM-UP TARGETED PROTEOMICS

Workflow Overview

The quantification of proteins is predicated on the stable, reproducible liberation of specific peptides under carefully optimized digestion conditions, which may be different from the conditions used in the discovery experiment. Before digestion, proteins are chemically treated to make them more amenable to digestion. This often includes denaturation, reduction of disulfide bonds, and alkylation of reduced cysteine side chains to prevent reformation of disulfide bonds. Once prepared in this manner, the proteins in the sample can be digested using one or more of a number of different enzymes.[55] Each of these steps is optimized with the goal of maximizing specific rather than global peptide yield.

After digestion, it may be necessary to enrich the peptides of interest away from unwanted potential interferences in the sample to improve the robustness of the measurement. Salts, carbohydrates, and lipids may be removed using bulk chromatographic techniques such as solid-phase extraction. If the peptide of interest is very low in abundance and not detectable over the noise from other peptides, an antibody can be used to enrich the peptide.[56,57] During development of a bottom-up assay, peptides are selected, tested for performance, optimized, reevaluated, and reselected if necessary.[58,59] This cycle may be repeated iteratively until a set of suitable peptides has been identified that provides suitable performance. Once peptides are identified, ISs are synthesized and a calibration strategy is devised, both of which are discussed in more detail later. Before full validation, prevalidation experiments provide initial benchmarks that can be used to build confidence that the validation will be successful according to established guidelines (eg, Liquid Chromatography-Mass Spectrometry Methods [Clinical Laboratory Standards Institute, C62-A]). These prevalidation experiments are also described in C62-A and investigate precision, linearity, analytical sensitivity (ie, lower limit of quantification, lower limit of the measuring interval), stability, and matrix effects. A simple set of experiments also has been described in the literature.[60]

Peptide Selection

Development of a bottom-up assay begins with the selection of the peptides that will be surrogates for the protein measurement. Optimal peptides for quantitative protein analysis are freely liberated and unique within the human proteome, have good chromatographic and ionization properties, lack potentially complicating amino acid residues or modifications, present a linear and precise response, are free from interferences, and correlate with other peptides in the protein. Peptides that satisfy the first two criteria have been called *proteotypic* peptides,[44] with the other criteria arising from the basic requirements for a stable analyte suitable for quantitative analysis.

It is difficult to predict the peptides that will be well liberated from the protein during digestion and that have favorable chromatographic and ionization properties. Beyond data that are collected during discovery experiments, numerous algorithms exist to digest proteins *in silico* and predict which peptides will be observed in a typical proteomics experiment.[59,61-65] Several large public databases catalog peptides that have been observed in DDA proteomics experiments.[66] These resources provide spectrometric data and information on frequency of observation. More recently, a database with information regarding the basic assay characteristics (ie, limit of quantification, precision, accuracy) for many peptides under standard conditions has been made freely available.[65] Because of the enormous number of potential peptides that could be used to build an assay, heuristic rules, authentic data, curated MS databases, and protein sequence databases are often combined to prioritize peptides in a largely empirical process. However, a simplified approach using Skyline that evaluates all tryptic peptides in a targeted fashion and relies solely on empirical data is successful in identifying robust peptides.[67]

Any list of candidate peptides can be narrowed based on uniqueness and chemical stability. To evaluate whether a peptide is unique and therefore offers the best potential for specificity, database searching (eg, using the BLAST algorithm) of target peptides against the human genome can identify any other proteins that share sequence identity for the selected peptide.[68] Typically, peptides that are longer than 6 to 8 amino acids have the highest likelihood to be unique in the proteome. Some amino acids or amino acid combinations are not ideal residues for target peptides in bottom-up proteomics assays because of likely preanalytical modifications or heterogeneity that can lead to errors in quantitation. These amino acids are often reactive: methionine can be irreversibly oxidized to the sulfoxide form; glutamine and asparagine can be deamidated. Both of these modifications change the mass of the peptide, rendering them undetectable in an assay designed specifically for the unmodified form. Cysteines are often chemically alkylated during sample preparation, but accurate quantitation of a cysteine-containing peptide assumes stoichiometric conversion to the chemically modified form, which actually can be quite variable. Although the presence of reactive residues does not render them completely unsuitable for use, candidate peptides containing them do require extra attention during development.

A narrowed list of candidate peptides can be further honed empirically. Digestion time-courses, during which aliquots of the digest are removed and analyzed over the course of the digestion, help identify peptides that are liberated quickly from the protein and that are resistant to nonspecific degradation during the digestion.[67,69,70] Although these are ideal characteristics of peptides used in bottom-up assays, the actual rates of cleavage and degradation are strongly dependent on the denaturing conditions and the amount and quality of trypsin used in the digestion. The precision of the amount of peptide liberated (and degraded) is assessed by digesting several samples over several days. Linearity can be assessed using mixing studies of samples with different concentrations of endogenous protein or samples with a relevant protein-free matrix to help identify peptides and transitions that have interference.[58] Spiking purified or recombinant protein into human matrix also helps assess linearity. In addition to being interference free, robust peptides for the quantification of protein concentration often correlate very strongly with other peptides in the protein when measured across samples collected from multiple individuals.[71] Two peptides that correlate strongly suggest that they are liberated from the target protein similarly and when combined in the estimate of protein concentration (eg, averaged), may reduce the variability of the measurement. There are, of course, instances in which there is only one useful peptide that is liberated or when there is a particular posttranslational modification or polymorphism of interest.[67,72] In these cases, the chromatographic and mass spectrometric methods need extra attention to optimize performance, because there are no other peptides against which to compare. In these cases, alternative proteases with different cleavage specificity may be necessary.

Denaturation

Quaternary, tertiary, and secondary protein structural properties can interfere with digestion and can be minimized using chaotropic agents, organic solvents, and detergents (Table 6.1). This "linearization" of globular proteins

TABLE 6.1 Denaturants Used in Bottom-Up Proteomics Experiments

Reagent or Equipment	Notes
Urea	Must be prepared fresh and used at lower temperatures because of the potential for carbamylation of proteins and peptides.
Trifluoroethanol	Must be diluted to <5% for proteolysis with trypsin.
RapiGest, PPS	Compatible detergents. Hydrolyze under acidic conditions. Products are retained on reverse-phase columns but typically washed away at the end of the gradient.
Deoxycholate	Anionic, weak detergent. Precipitates on the addition of acid facilitating removal before LC-MS.
SDS (RIPA buffer)	Strong anionic detergent with very strong solubilizing activity. Removal before LC-MS is required.
CHAPS	Zwitterionic detergent. Used at <1%. Retained on reverse-phase columns but typically washed away at the end of the gradient.
NP-40	Nonionic detergent. Used at <1%. Removal before LC-MS is required.
Acetonitrile	Denatures proteins but causes precipitation at high concentrations.
Heat	Can be used alone or in combination with an additional denaturant (not urea).

LC-MS, Liquid chromatography–mass spectrometry.

improves reactivity toward reducing and alkylating agents as well as the efficiency and rate of proteolytic digestion. Some proteases retain activity under these conditions and can dramatically enhance digestion efficiency.

Typical denaturants, such as urea and guanidine, are used in concentrations up to 8 M, although urea predominates because it is less inhibitory to proteases. Trifluoroethanol, acetonitrile, methanol, and ethanol all have denaturing properties and are used as additives up to 80%. Detergents such as 3-[(3-cholamidopropyl)dimethylammonio]-1-propanesulfonate (CHAPS) and deoxycholate also have been used, but they are often avoided because of the potential for strong suppression of ionization. Novel detergents that degrade to noninterfering molecules on the addition of acid are commercially available. In general, the additives used during the digestion steps are generally incompatible with LC-MS/MS analysis and require significant dilution or removal by solid-phase extraction (SPE) or an alternative chromatographic step.

Reduction and Alkylation

In addition to denaturation, the reduction of protein disulfide bonds can help in the linearization process by further dissociating protein complexes and unfolding target proteins. The most common reducing agents used in protein biochemistry laboratories are 2-mercaptoethanol, dithiothreitol (DTT), and tris(2-carboxyethyl)phosphine (TCEP). Although any of these reducing agents are effective, DTT and TCEP have become popular in proteomics assays. The concentration needed to fully reduce disulfide bonds in each sample can be estimated from the protein concentration in the solution but generally ranges from 5 to 50 mM. When urea is not used as the denaturant, heat and agitation can be applied to accelerate the reaction.

After reduction, the free cysteine thiol groups can be covalently blocked by alkylation with iodoacetamide to form a carbamidomethyl cysteine. This easily detected modification blocks the re-formation of disulfide bonds and adds 57.03 Da to the mass of each cysteine (this is the monoisotopic mass; because of the presence of ^{13}C in iodoacetamide, the average mass added is 57.07 Da). The addition of a free amine group to the peptide adds another position to bind a proton and can make ionization of the peptide more favorable. As mentioned earlier, the reaction is variable and the alkylated cysteine is still sensitive to oxidation of sulfur. The usefulness of alkylation after reduction should be evaluated empirically. When choosing the amount of iodoacetamide to add to the reaction, it is important to remember that the number of reactive thiol groups in the sample is affected by the DTT added to the sample. A threefold excess of iodoacetamide (eg, 15 mM iodoacetamide after the addition of 5 mM DTT) provides sufficient alkylating reagent to react with DTT and the cysteines in the sample. Iodoacetamide is selective for cysteine but when present in large excess or at high pH, residues other than cysteine can be alkylated. In some protocols, additional DTT is added after alkylation to quench the remaining iodoacetamide to prevent undesirable side reactions. In other protocols, alkylation is avoided altogether.

Protease Selection and Activity

The most commonly used protease for bottom-up proteomics assays is trypsin, which cleaves C-terminally to arginine and lysine residues (except when proline is adjacent on the carboxyl side). Trypsin has been studied over many years and is readily available in highly purified and well-characterized forms from bovine, porcine, and recombinant sources at very reasonable costs. These preparations are also available in modified and covalently bound forms, which reduce autolysis and minimize nonspecific digestion activity. Trypsin frequently liberates peptides of suitable length for mass spectrometric analysis. The liberated peptides have a carboxyl-terminal lysine or arginine, which imparts a positive charge to the peptide at low pH (in addition to the amino terminus), aiding fragmentation. Finally, trypsin has limited secondary structure specificity, meaning that most lysines or arginines will be cleaved with similar efficiency.

Fidelity of digestion is essential to maximize peptide yield and minimize the generation of undesired peptides that complicate analysis. Preparation of trypsin from natural sources will often contain trace amounts of chymotrypsin, which has very different proteolytic sequence specificity compared to trypsin. This undesired activity can be essentially eliminated by treatment with tosyl phenylalanyl chloromethyl ketone (TPCK). Although very-high-purity preparations of various proteases are available, all show varying degrees of nonspecific activity, which arises during extended incubation and can be monitored by the loss of the desired peptide and appearance of nonspecific digestion products.[70]

The use of multiple proteases can enhance digestion. For sample preparation with 8 M urea, Lys-C (cleaving C-terminally of lysine) retains activity and can be used first for partial digestion. Subsequently, the digest can be diluted to 1 to 4 M urea and trypsin is added to complete the digestion.

There is no requirement for an assay to use trypsin. Other proteases can be used if proper attention is paid to selectivity and efficiency (Table 6.2). Their use can help resolve issues that arise from nonoptimal peptide length or difficulties in observing specific peptides.[55] However, less common proteases may have specific limitations such as poor availability, high cost, and unpredictable behavior.

Digestion Optimization

Although desirable, it is very unusual for proteolysis of complex mixtures to go to completion. Protein structure, protein sequence, protease autolysis, buffer conditions, and product inhibition can inhibit digestion. Given the many variables that contribute to incomplete digestion, optimization of digestion conditions for a specific protein aims to meet analytical performance goals.[73] Robustness testing with variable amounts of denaturants, reducing and alkylating reagents, trypsin, buffers, enzyme substrate ratios, temperature, cationic and anionic additives, as well as variable incubation times, will identify the most sensitive elements of the digestion process. The use of digest progress curves and extensive optimization have yielded an array of protocols that are capable of high-fidelity digestion in minutes to hours.

Internal Standards and Peptide Quality

Stable isotope-labeled ISs are added to the sample in quantitative bottom-up proteomics assays to control for variability in system performance (ie, volume of injection by the autosampler, retention characteristics of the chromatographic column, electrospray stability, and sensitivity of the

TABLE 6.2 Proteolytic Enzymes

Enzyme	Specificity	Notes
Trypsin	C side of lysine and arginine	May have contaminating enzymes (eg, chymotrypsin)
LysC	C side of lysine	Can be used in conjunction with trypsin. Active in urea up to 8 M
LysN	N side of lysine	Relatively new
AspN	N side of aspartic acid	Selectivity issues
GluC	C side of glutamic acid	Selectivity issues
ArgC	C side of arginine	Uncommon
Chymotrypsin	Preference at hydrophobic residues	Uncommonly used because of cleavage specificity
Elastase	Preference at hydrophobic residues	Uncommon
Pepsin	Preference at hydrophobic residues	Secondary structure is important; complicated digests with many overlapping peptides; generally not used for quantitative analysis

mass spectrometer), in ion suppression, and, increasingly, to control for digestion variability. Five types of ISs used in proteomics experiments include (1) isotope-labeled peptides with sequences identical to those of the target peptides, (2) isotope-labeled peptides with additional residues at the C- and N-termini of the peptide that get cleaved off during digestion, known as *winged* peptides,[74] (3) intact recombinant isotope-labeled protein,[75-78] (4) isotope-labeled concatenated peptides that liberate a labeled analog of the target peptide on digestion,[79] and (5) mutated peptides with a similar sequence but different mass to the target peptide.

Isotope-labeled peptides are most commonly used in the development of clinical assays because of their availability and acceptable performance when it is not necessary to control for the variability of digestion conditions. Recent experiments suggest that adding labeled peptides before digestion can be beneficial to overall assay performance.[70] Typically, amino acids with carbon-13 (^{13}C) and nitrogen-15 (^{15}N) incorporation (commonly Arg, Lys, Leu, Ile, Phe, Pro, and Val) are used in peptide synthesis. Deuterated amino acids and other amino acids are available in labeled forms but are used less frequently because of cost or subtle changes in chromatographic behavior. Winged peptides are synthesized in a similar fashion by including extra amino acids on each end of the tryptic analyte peptide that correspond to the native sequence of the protein. Recombinant protein expression systems are used to generate authentic protein or other proteins consisting of concatenated peptides labeled at specific amino acids or completely labeled with replacement of all ^{14}N with ^{15}N and/or all ^{12}C with ^{13}C. Synthetic peptides, winged peptides, and labeled proteins have all been used in clinical bottom-up proteomics assays.[72,74,77,80-84] Mutated peptides are generally not considered viable ISs because of shifts in chromatographic properties and the possibility of differential response to interferences.

To the extent that digestion efficiency is being recognized as a key element of assay precision, the utility of winged peptides, concatenated peptides, and fully isotope-labeled recombinant proteins to faithfully reflect the full complexity of digestion in matrix is under investigation. As described earlier, subtle differences in digestion efficiency associated with polypeptide structure are well-known and it is unlikely that any IS will behave identically to a native sample.

Once an IS is selected, adequate quality assurance during production ensures that IS protein and peptide quality is sufficient for reliable measurements. For each batch of ISs, four useful analyses to characterize the ISs before use include (1) amino acid analysis of the actual concentration of the IS, (2) HPLC with ultraviolet (UV) detection for peptides and proteins or sodium dodecyl sulfate polyacrylamide gel electrophoresis (SDS-PAGE) for proteins to give an estimate of purity, (3) MALDI MS or LC-MS/MS analysis to confirm sequence, and (4) correlation of the protein concentration in patient samples using the old versus the new IS (N determined by the precision of the assay). Peptides and proteins often adhere to storage containers, aggregating or precipitating, and becoming oxidized (cysteine, methionine, and tryptophan), deamidated (asparagine and glutamine), or degraded by light (tryptophan and phenylalanine), and these problems are exacerbated at low concentration.[84a] To help minimize these issues, stocks of ISs can be kept in an appropriate solvent at high concentration in amber vials under an inert gas below −60°C.

Calibration

There are many possible approaches to calibrating quantitative bottom-up proteomics assays, including (1) a simple calculation in which the ratio of peak area of the endogenous protein-derived peptide to the peak area of the IS peptide is multiplied by the concentration of the added IS, which is added at a known concentration; (2) calibration materials comprising peptides spiked into digested matrix; (3) calibration materials made by spiking purified or recombinant proteins into a relevant matrix; (4) native human samples with given concentrations (pooled or in singlicate); and (5) native human samples diluted into a relevant matrix. Other approaches are certainly conceivable.

Many research assays aimed at identifying novel protein biomarkers for disease use the first and second options for ease and availability of reagents. These options rely on a number of assumptions: the digestion goes to completion, the spiked peptide and IS are stable, and the relationship between analyte in calibrator matrix and experimental sample matrix is stable and linear. These approaches do not include the necessary controls for a validated clinical quantitative assay but can be useful in biomarker verification studies in which relative protein concentration is compared among samples.

Depending on the protein, spiking purified protein into a relevant matrix can be a very useful method for calibration of bottom-up proteomics methods.[77,81,85-88] It is important that the purified or recombinant protein be properly folded and be posttranslationally modified similarly to the endogenous protein in actual patient samples so that the proteins in the calibration materials are digested equivalently. For other proteins, such as those that exist in macromolecular complexes or those that are strongly bound to other molecules, it may be more difficult to generate useful calibrators if the native structure cannot be established.[71] In these cases, value-assigned human samples can be used. Calibrators made from spiked proteins or native samples are useful for improving between-day precision[71,73,89] and between-laboratory precision.[77] Importantly, the proper folding and posttranslational modifications of proteins in native samples and the near identity of the matrix between calibrator and patient samples allows the calibration material to behave as much like patient samples as possible. This helps control for the day-to-day differences in digestion that are the largest source of imprecision.

Preparing calibrators requires assignment of concentration. Comparison to certified or standard reference materials using reference measurement procedures would be ideal for this purpose, but these resources are not widely available for almost all protein targets. When available, certified or standard reference materials can be used to make matrix-matched calibrators. In the absence of reference materials, highly purified or recombinant proteins should be well-characterized by amino acid analysis, Karl-Fischer analysis (if crystalline), SDS-PAGE, HPLC-UV, and LC-MS to assign the concentration of the calibrator based on the mass of protein added to a blank matrix. A final option is to use an immunoassay to assign the concentration, but accuracy can suffer as a result of known issues with lot-to-lot variability and other well-known issues with immunoassay performance.

Protein or Peptide Enrichment

Similar to the issues encountered in discovery proteomics in which highly abundant proteins predominate the observable proteome, the sensitivity and selectivity of mass spectrometric measurements of proteins using bottom-up proteomics is limited by ion suppression and interference from other peptides. Generally speaking, without enrichment before LC-MS/MS, the lowest abundance serum and plasma proteins that can be reliably measured are in the low micromolar range.[71,72,90,91] Biochemical enrichment of peptides and small proteins can enhance sensitivities to the low nanomolar range.[75-77,92,93] Antibodies can be used to enrich proteins before digestion or specific peptides after digestion that can enable limits of quantitation in the low picomolar range.[37,74,80,81,86,94]

The biochemical enrichment of proteins and peptides relies on substantial differences in physical and chemical properties. Small proteins can be separated from the larger proteins in a sample by protein precipitation using organic solvents or polyethylene glycol. Ion-exchange SPE in any form (ie, strong vs. weak, cation vs. anion) can be an effective technique to achieve crude fractionation based on charge, pH, type, and concentration. Reverse-phase SPE enables separation by hydrophobicity. Finally, chemical affinity techniques such as immobilized metal affinity chromatography (IMAC) can selectively enrich acidic peptides, especially those that are phosphorylated. Using nonaffinity approaches, resolution of the separation is modest but can enrich proteins and peptides of interest more than 10-fold. In contrast, affinity techniques using antibodies or other highly selective chemistries can achieve enrichment of 1000- to 10,000-fold. The higher performance of antibody-based approaches is due to selectivity. Although differences in chemical properties are suitable to separate peptides and proteins by class (hydrophobic/hydrophilic, acidic/basic), the ability of antibodies to bind specific amino acid sequences allows for very selective purification even in samples in which target analytes are present in trace amounts. Antibodies, whether traditional immunoglobulins or immunoglobulin fragments,[95] can be conjugated to a solid phase and used in formats common in bioanalytical laboratories. Single-use and multiuse columns in 96-well format or inline flow through columns are available.[88] In particular, paramagnetic particles with conjugated antibody allow for easy automation of the binding, washing, and elution steps used. Both monoclonal and polyclonal antibodies have been used successfully, but success rates are higher with monoclonal antibodies, which can be selected to have higher affinity.

One of the best examples of a clinical assay that uses peptide immunoaffinity enrichment after trypsin digestion for protein quantification is the cancer biomarker thyroglobulin. As one of the best-studied biomarkers in laboratory medicine, particularly with respect to the sample-specific interferences that render the immunoassay measurement meaningless in a significant number of patients, the protein was an early target for MS. More specifically, in 25% to 30% of patients, autoantibodies can mask the epitopes recognized by reagent antibodies, leading to false results. Using the approaches described previously, peptides were selected and antibodies were generated in rabbits for the enrichment of specific peptides in a targeted bottom-up proteomics assay. The resulting workflow[81] and the iterations that came after it[74,80,86] laid the groundwork for other clinical assays using this technology. One of the assays developed for thyroglobulin actually uses rabbit antithyroglobulin antiserum to coat all of the analyte in a serum sample before protein precipitation of large proteins, digestion, and specific peptide enrichment.[74] The assay takes advantage of the fact that many patient samples already have autoantibodies coating the analyte of interest and that the binding of thyroglobulin to rabbit antibodies allows it to be separated from bulk protein more easily.

The coupling of the mass spectrometer to an antibody enrichment step is likely to provide the most sensitive and specific biomolecular detection available. Although antibodies can have significant issues with nonspecific interactions when used in immunoassays relying on secondary detection, the addition of mass and sequence information derived directly from the peptide in the mass spectrometer means that any nonspecific binding of protein or peptide to the antibody will not contribute to the signal of an assay.

Multiplexing

The ability to quantify many analytes in the same assay is as straightforward for peptides derived from proteins as it is for small molecules. Similar to small molecules, if preanalytical preparation techniques can recover peptides similarly, many proteins can be quantitated in a single assay. Development of multiplexed assays is much more complex and requires attention to all of the details mentioned earlier, as well as issues that can arise when attempting to optimize mutually exclusive assay parameters. However, good examples of assays that multiplex protein quantification have been described with and without peptide immunoaffinity enrichment.[71,73,89,94]

Quality control for a multiplexed assay, particularly when results from multiple proteins will be considered together, can be complicated and may involve decision trees or other approaches in addition to standard Westgard rules. For example, a peptide may have an IS peak area that is below the established cutoff, but the ratio of the peptide signal versus another peptide from the same protein may fall at the mean of the established acceptability criteria. These situations should be evaluated for their effects on the accuracy and precision of the assay during method development and evaluation.

Polymorphisms

Genetic polymorphisms can alter the amino acid sequence of proteins of interest. In some cases, changes are relevant to disease by affecting cellular localization, enzyme activity, or binding affinities. As a result, the identification of polymorphisms by MS can complement the quantification of protein concentration in the diagnosis of disease. One example of a bottom-up proteomics assay that quantifies the protein concentration and identifies important polymorphisms is an assay developed for α_1-antitrypsin.[72] In this assay, protein concentration is quantified by the peptide SASLHLPK, the S isoform (containing the E287V mutation) is detected by monitoring for the peptide LQHLVNELTHDIITK, and the Z isoform (containing the E365K mutation) is detected by monitoring for the peptide AVLTIDKK. The ability to quantify protein concentration and identify isoforms in the same assay eliminates the need for a second gel-based assay or a genetic assay for most patients.

When they are not important for the diagnosis of disease, amino acid changes secondary to genetic polymorphisms can interfere with protein quantification. A change in peptide mass as a result of amino acid substitution or the elimination of a cleavage site to produce a new peptide will reduce the amount of peptide detected in the digest. For this reason, databases of common polymorphisms should be considered before selecting a peptide for protein quantification. In addition, designing a quantitative assay with more than one peptide allows one to monitor for the heterozygous loss of a peptide (the ratio of two peptides should be constant from sample to sample and will deviate from the norm when a polymorphism is present in one of the peptides).

Chromatography

For high-throughput targeted clinical proteomics, the use of narrow-bore columns and flow rates approximately 100 to 500 µL/min are common. At these flow rates, injection-to-injection cycle times are minimized and robust operation is achievable. Whereas sensitivity is reduced compared with the nanoflow chromatographic separations and ionization methods common in discovery experiments, new ionization sources and high-efficiency ion transmission instrument designs now offer sensitivities near parity to nanospray techniques.

Summary

To summarize, an optimal workflow for a peptide immunoaffinity assay might begin with the denaturation and digestion of a human sample using an optimized digestion protocol, which includes reduction and alkylation, if necessary. IS is added before digestion (note that an intact protein or winged peptide would allow for normalization of digestion variability and account for the degradation of liberated peptide during the digestion). Matrix-matched calibration material is processed in parallel with patient samples so that the protein concentration can be calculated based on calibration curves rather than the simple ratio of the endogenous peak area to the IS peak area. Matrix-matched quality control materials with endogenous protein are included to evaluate the quality of the entire process. Monoclonal antibodies conjugated to paramagnetic beads are incubated with the digest. A magnet is used to isolate the beads to the side of the vessel, the beads are washed, and then the peptides are eluted with an acid (eg, 5% acetic acid). The eluted endogenous and IS peptides liberated during the digestion step are quantified by LC-MS/MS, and the peak area ratio is compared with the calibration curve to establish the protein concentration in each patient sample.

POINTS TO REMEMBER

Bottom-Up Quantitative Assays
- Internal standards help control for the variability in peptide liberation, peptide degradation, sample handling, and chromatograph and mass spectrometer performance.
- External calibration material greatly minimizes day-to-day variation.
- Native human samples can help reduce variability across laboratories.
- Bottom up quantitative assays are being used clinically in the detection and therapeutic management of disease.

TOP-DOWN PROTEOMICS

Workflow Overview

In proteomics, the nature of the sample is generally categorized based on whether the peptide or protein of interest has been proteolytically cleaved before analysis. For simplicity, polypeptides first might be categorized based arbitrarily on their molecular weight, with peptides being less than 3 kDa, small proteins being 3 to 35 kDa, and large proteins being greater than 35 kDa. The top-down approach to proteomic analysis refers to the introduction of undigested proteins into the mass spectrometer and often uses high-resolution, high mass accuracy instruments with or without dissociation or fragmentation.[96-99] This is in contrast to the bottom-up methods using proteolysis, as described previously. High-throughput, automated analysis of intact proteins is a recent development in the clinical proteomics field, and the high-performance mass spectrometers needed for these analyses represent significant advances over previous technologies, with sufficient resolution, mass accuracy, and sensitivity to be effective. The widespread commercial availability of these instruments has occurred only in the past several years.

The general approach to the analysis of intact proteins is no different than for other molecules. With the availability of a suitable IS, calibration material, separation strategy, and high-performance mass spectrometer, a quantitative analysis can be achieved. Similar to peptides, proteins also can be fragmented in the gas phase using collision-induced dissociation or electron transfer dissociation, which are both types of gas phase fragmentation. These techniques allow a mass spectrometric analysis analogous to selected reaction monitoring, with the detection of precursor-fragment pairs. Although this scheme is accurate in concept, it is very challenging to realize in practice. First, when proteins are ionized, they adopt a distribution of charge states that is roughly correlated with the number of basic amino acids. Whereas a tryptic peptide will typically contain 90% of its signal in a +2 charge state, a protein can spread its signal across 10 or more charge states, which reduces the total signal if only a single charge state is selected for analysis. Second, the fragmentation of very large molecules in the gas phase often yields few fragments with very low sensitivity, frequently making quantitative analysis impractical. Electron capture dissociation may have advantages over collision-induced dissociation (ie, what is used in triple-quadrupole instruments) for fragmentation efficiency, but the ability to perform selection reaction monitoring (also called multiple reaction monitoring when monitoring multiple fragments, see Chapter 2) is still largely experimental for quantitative analysis.[100] For applications in which quantitative analysis is not necessary, such as the determination of clinically relevant amino acid polymorphisms in hemoglobin, top-down analysis allows for unambiguous identification of sequence changes associated with disease.[101-103] The success of this application relies on the analysis of a very-high-abundance protein where quantitative information is generally not needed.

For quantitative analysis, accurate mass, high-resolution full-scan data can be used to quantify the peptide without any fragmentation in the gas phase and has been applied to insulin and insulin-like growth factor-1 (IGF-1).[75,104,105] Although this type of analysis might be better described as intact protein analysis, it is considered with top-down methods because of the similarities in sample preparation. This workflow is appealing because it does not require digestion or the laborious effort to produce, identify, and verify signature peptides. The collection of full-scan data also provides the opportunity for retrospective analysis of other signals in the spectra that were not originally analyzed.[75] Drawbacks include potentially lower sensitivity as a result of the presence of multiple charge states, poor ionization, and difficulties achieving chromatographic separations of the same resolution as those that are routinely achieved for small molecules and peptides.

Sample preparation for the analysis of intact proteins may be similar to that for intact peptides (eg, removal of large proteins by protein precipitation), and many examples of enrichment of small proteins and their analysis by high-resolution mass spectrometers also have been described or are available from reference laboratories.

Assay Specificity

In the absence of the selectivity afforded by the generation of highly specific fragment ions from a selected precursor, additional information from spectra is sought to ensure the integrity of the measurement. In a top-down analysis, high mass accuracy and isotope ratio information can be used to gain confidence in the specificity of measurement.[76] Although accurate mass determination alone cannot allow unequivocal confirmation of the molecular formula of large molecules, it provides a substantial constraint on the number of proteins in a proteome that could give rise to an observed signal. As mass accuracy is a function of resolution, a high-quality spectrum can be interrogated by inspection of very small m/z ranges reducing the possibility that any coeluting molecule will contribute to the signal. In addition, isotope ratios of the intact analyte are characteristic of the molecular formula and can be used to detect the presence of an interfering signal via perturbation of the expected ratio, similar to the use of fragment ion ratios in a selected reaction monitoring experiment.

Calibration

For proper calibration of a top-down protein assay, it is essential to have highly purified protein that is chemically and biologically identical to the human protein of interest. These proteins can be purified from pooled human specimens or obtained by heterologous expression and purification.

At present, commercial sources of proteins purified directly from human matrix are limited. Several organizations, such as the National Institute for Biological Standards and Control, offer reference materials (eg, 83/500 World Health Organization International Standard for Insulin) that are isolated from a human source. Other organizations dedicated to specific clinical diseases are also sources of reference materials. However, the difficulty of isolating large quantities of low-abundance proteins from very complex matrices makes commercial supply of highly purified material a challenging prospect.

Recombinant DNA techniques and related protein expression systems have made commercial availability of proteins more commonplace and are available through extensive searchable catalogs online. For proteins that are not found through commercial sources, the production of recombinant proteins can be outsourced. However, it is essential to carefully characterize these materials for identity, proper folding, posttranslational modifications, and purity before use in assay development. The quality of the quantitative information describing the amount of protein provided is generally insufficient (or nonexistent) to allow direct use of these proteins in clinical assays. Therefore further characterization using well-calibrated and quality-controlled amino acid analysis, Karl-Fischer analysis, spectrophotometry, and/or LC-MS analysis is needed to correctly assign calibrator concentrations.

Compared with bottom-up assays, it is even more important for top-down protein assays to use calibrators with the correct secondary and tertiary structure. Very small changes in α-helix content, β-sheet content, protein folding, and disulfide arrangement often lead to dramatic changes in chromatographic behavior. Thus, if a protein is enriched from human samples and chromatographed under conditions that retain elements of tertiary and secondary structure during analysis, these elements will need to be similarly conserved in proteins used for calibration. For recombinant proteins, correct folding and formation of disulfide bonding is not guaranteed and structural studies of commercially derived material (eg, functional assays, native gel electrophoresis or circular dichroism) may be required before successfully using the materials in a quantitative assay.

Internal Standards

Quantitative analysis of intact proteins by isotope dilution requires the availability of chemically or isotopically labeled proteins with a different mass but intact structure. Using ^{15}N ammonium chloride as the sole nitrogen source in cell culture, it is possible to replace all ^{14}N atoms during protein expression to yield a heavy labeled protein. A small number of proteins are now commercially available (eg, ApoA-I, ubiquitin, and IGF-1 from Cambridge Isotopes or Sigma). For novel targets, it is possible to outsource the production of labeled proteins, though this can be very costly. Although chemical and structural identity is the gold standard for an IS, reasonable compromises for proteins have been successfully accommodated. Homologous proteins with very closely related amino acid sequences have been used; similarly, recombinant proteins with conservative amino acid changes, insertions, or deletions can be used. However, extensive validation to demonstrate identical recovery, response, and chromatographic behavior to the native analyte is required.

Protein Enrichment

Quantitative top-down or intact mass spectrometric measurements of proteins require significant simplification of the matrix before being analyzed by MS. For some peptides and small proteins, this can be achieved simply by precipitating larger proteins using acidic solvents or other chemical treatment before chromatography. Alternatively, proteins can be enriched using immunoaffinity methods, which will necessarily purify any protein isoform that contains the epitope(s). This means that in one assay, all relevant posttranslationally modified protein isoforms could be identified and quantified. Many proteins have been evaluated using this technology including apolipoproteins,[106,107] serum amyloid A,[108] cell signaling molecules,[109] insulin,[110] IGF-1,[111] and parathyroid hormone.[112]

> **POINTS TO REMEMBER**
>
> **Top-Down Quantitative Assays**
> - Current methods rely on high-resolution instrumentation for intact peptides and small proteins.
> - Fragmentation of large proteins in the gas phase is still experimental.
> - It is absolutely imperative that internal standards and calibration materials be properly folded and contain the correct complement of posttranslational modifications.
> - Top-down assays are being used clinically for intact small proteins, including insulin and IGF-1.

PREANALYTICAL AND OTHER TECHNICAL CONSIDERATIONS

Specimen Collection Considerations

Obtaining high-quality specimens for discovery experiments is a crucial component of a successful biomarker pipeline. Although banked serum and plasma samples are the most readily available specimens for discovery studies, the suitability of such specimens has been called into question over a wide range of potential shortcomings, such as the lack of SOPs for collection protocols and collection devices, undocumented clot times for serum, unknown effects of storage time and temperature, and stability to freeze–thaw events.

Recommendations for SOPs for many clinical sample types have been published by the Human Proteome Organization, Early Detection Research Network, and Clinical Proteomic Tumor Analysis Consortium and provide a starting point for investigators evaluating previously banked specimens or embarking on a new collection effort.[113,114]

Urine as a Matrix for Proteomics Assays

For many pathologic conditions, biomarker discovery and clinical testing of urine could be the ideal matrix. As a body fluid, urine is generally much simpler than other specimens. Filtered in the glomerulus and extensively modified throughout its path to the collecting ducts, urine typically contains salts and small molecules diluted in a large volume of water. Depending on the hydration and metabolic status of the individual, urine can be significantly more concentrated and have extremes of pH. In addition, many different pathologic conditions can cause the release of serum proteins into the urine, particularly the conditions that would make urine a useful specimen. The large variability of the constituents in urine make it a very challenging matrix for proteomics experiments. As a general approach, proteins in urine are concentrated before analysis by ultrafiltration or protein precipitation; albumin and uromodulin (Tamm-Horsfall protein) are the most abundant proteins present. Immunodepletion of urine can be used to remove high-abundance serum proteins, with the associated costs and risk for depletion of the proteins of interest. Besides the measurement of albumin and uromodulin in urine,[115,116,116a] clinical proteomics assays in urine continue to be quite challenging.

Tissues

Clinical proteomic studies of tissues have been limited compared with traditional specimen types.[117] This is partially due to the infrequent use of quantitative diagnostic testing to diagnose disease in whole tissues. Much more commonly, the diagnosis of disease from surgically collected tissues is achieved by histologic and immunohistochemistry examination with the identification of morphologic elements and cellular organization that are associated with pathologic conditions.[118-120] In an effort to couple MS with microscopic analysis by a pathologist, laser capture microdissection has been used to isolate selected groups of cells for proteomic analysis.[121]

One challenge in the application of MS to tissue analysis has been the chemical treatments used to preserve and fix tissue before microscopy. Techniques such as formalin fixation and paraffin embedding (FFPE) dramatically alter proteins through cross-linking and profound dehydration. These processes yield proteins that are highly resistant to tryptic digestion and also heavily modified, which complicates any mass spectrometric analysis.[122] Chemical strategies for analysis of FFPE tissues have been reported that can improve the yield of tryptic peptides and quantitative analysis of selected cancer-related protein markers.

Application of MALDI MS to fresh tissue sections has been explored extensively. In pioneering work at Vanderbilt University, thin sections of frozen tissue collected on slides and overlaid with matrix were used to collect MS images directly from the surface of the tissue.[123] In the absence of any accompanying separation technology, tissue images are often predominated by highly abundant constituents of cells such as membrane lipids. Extensive development of this technology has improved image quality, resolution, and the ability to probe the metabolites and proteins with greater depth. The clinical applications of tissue imaging are promising[124,125] but have yet to make an impact on the diagnosis of disease or to be adopted by the pathology community.

Posttranslational Modifications

Up to this point in the chapter, methods describing the analysis of proteins and peptides have been in the context of ideal situations with individual, simple proteins. However, the number of known biologically relevant protein modifications is substantial and includes highly studied modifications such as phosphorylation and glycosylation. The presence of posttranslational modifications is key in regulating protein function and the aberrant modification of proteins may be central to the disease process, with temporal and abundance changes being relevant.

MS has played a vital role in the identification and analysis of posttranslational modifications, and it is reasonable to envision that selected reaction monitoring, along with chemically synthesized ISs and calibrators, would facilitate the facile development of clinical assays to detect and quantify posttranslationally modified proteins. In principle this is true, but the highly dynamic nature and low stoichiometry of many relevant modifications exacerbates the fundamental issues that challenge any measurement technology. For example, consider a protein present at 1 ng/mL in serum in which 0.1% of the protein is phosphorylated in the normal state and phosphorylated at 0.05% in a pathologic state. In this case, the investigator is faced with the challenge of trying to measure a small change under extremes of dynamic range.

Stability

As for any assay, an understanding of preanalytical stability of the analyte is extremely important. It is well known that many proteins are labile and that they must be quantified in specimens that are carefully drawn and processed.[126,127] Others are much more stable. Indeed, broad surveys of protein abundances suggest that under ideal storage conditions, few proteins experience dramatic changes.[128,129] However, previous comparisons of immunoassays and mass spectrometric assays in regular clinical samples demonstrate a positive bias in mass spectrometric assays,[72] suggesting the possibility that protein fragments are detectable by MS that are not detected by immunoassay. As a result, it is essential to validate that protein measurements by MS are unaffected by common specimen processing and handling procedures.

In addition to protein stability, it is important to consider peptide stability, particularly in bottom-up proteomics methods. After digestion, peptide recovery can suffer because of poor physiochemical properties, which leads to aggregation, peptide oxidation, and adsorption to surfaces of pipettes and vials. Validation studies must therefore include an assessment of peptide stability during analysis by LC-MS, including incubation in the autosampler.

Interferences

The challenges associated with interferences in LC-MS for clinical proteomics are similar to those faced in small molecule analysis (see Chapters 2 and 5)—signal suppression, signal enhancement, and presence of isobaric coeluting species (ie, molecules with the same retention time and precursor m/z). Signal suppression is commonly observed when salts, phospholipids, or other nonanalyte compounds coelute with the analyte of interest and inhibit the process of ionization. This results in loss of signal and in severe cases can fully interrupt the ionization process, leading to a complete loss of signal. Signal enhancement is the opposite effect wherein the interference causes an increase in observed signal. While absolute elimination of suppression is a goal of method development, it is seldom achieved and the effects of suppression on assay performance must be carefully evaluated. The use of a coeluting, isotopically labeled IS minimizes the influence of these effects. In cases of suppression and enhancement, the IS signal can be used to correct for a modest degree of perturbation because it will behave identically to the analyte of interest. Coelution of species with the same m/z values also can disrupt measurements by contributing signal where none is expected. Because fragmentation of molecules by collision-induced dissociation follows thermodynamic and kinetic rules specific to their structure, the abundance of various fragment ions remains relatively constant under fixed conditions. This property allows for the use of fragment ion ratios to contribute additional information about molecular identity. Coeluting compounds that interfere can contribute abundances that perturb ion ratios and allow for their presence to be identified.[58] The appearance of these types of interferences is not an uncommon occurrence in clinical proteomics, wherein the massive number of peptides in a complex digest leads to peptides with nearly equal precursor masses and shared fragment ion masses. Only with routine inspection of isotope ratios can these interferences be identified.

Instruments

The choice of instruments for discovery experiments, subsequent biomarker verification and validation experiments, and clinical implementation depends on how they will be used (Table 6.3). For discovery, instruments that collect high-resolution precursor scans and high-speed MS/MS

TABLE 6.3 Mass Spectrometers

Instruments	Notes
Single quadrupole	Not typically used for peptide analysis in the clinical laboratory
Triple quadrupole	Very common instrument for targeted methods, particularly isotope dilution
Ion trap	Commonly used for discovery experiments, but quantitative performance is limited because tandem-in-time operation
Time-of-flight	Can be used for quantitative analysis with adequate internal standards
Orbitrap	New high-resolution and high mass accuracy instrument that has the same limitations as an ion trap
Hybrid-orbitrap (eg, LTQ Orbitrap, Thermo Fisher Scientific, Waltham Massachusetts)	Commonly used for discovery experiments, lacks the precision needed for quantitative assays
Hybrid quadrupole high-resolution and high mass accuracy analyzer (eg, Q-TOF or Q-Orbitrap [Thermo Fisher)	An emerging competitor to triple quadrupole instruments, can be used for the quantitative analysis of intact proteins or peptides after digestion

fragmentation spectra in a data-dependent fashion are most desirable and include orbitrap hybrid instruments and high-performance TOF instruments. These instruments provide data of sufficient quality to achieve high-fidelity peptide identification and relative quantitation.

For biomarker verification and validation, as well as clinical assay implementation, instruments that are able to perform high-speed MS/MS data collection are common. Triple-quadrupole tandem mass spectrometers are ideal for this application, although hybrid quadrupole—high resolution and high mass accuracy instruments are emerging as alternatives.

Chapter 2 has detailed information describing instrumentation for use in proteomics experiments.

CONCLUSIONS

Current Clinical Proteomics Assays

Despite more than 20 years of work, the number of new clinical biomarkers derived purely from discovery proteomics is limited. So far, only two lung cancer diagnostics[130,131] and an ovarian cancer test[132] are commercially available. However, many single protein and peptide diagnostic markers measured by MS are offered by a number of reference laboratories. These include thyroglobulin,[74,80-82,84-87] IGF-1,[75,76,133] angiotensin-I (plasma renin activity),[92] insulin,[134] C-peptide, PTHrP, α_1-antitrypsin,[72] ADAMTS13,[135] and amyloidosis[136] and the identification of hemoglobin variants. The amyloidosis assay is particularly important to highlight as the first of its kind to employ a discovery proteomics—type platform in the regular clinical evaluation of surgical pathologic samples. The ADAMTS13 and plasma renin activity assay measure endogenous enzyme activity, and the IGF-1, insulin, C-peptide, and hemoglobin assays are all examples of top-down proteomics methods. This list of assays thus includes almost every imaginable workflow for the quantification of a protein by MS, which speaks to the versatility of MS for the measurement of proteins.

New Directions

In the coming years, MS will become a central method for protein and peptide quantification in the clinical laboratory. Improvements in instrumentation speed and sensitivity will complement the specificity inherent in the technique. It seems very likely that clinical methods will extend well beyond those currently used to measure proteins in blood, urine, and certain tissues to replace many immunoassays and immunohistochemical stains. The ability of immunoenrichment to achieve very low limits of detection will be instrumental in the application of the technique in these new areas. Advances in data processing software and the development of instruments to uniformly apply matrix to tissue samples also will help make MALDI imaging more acceptable to general pathology practices. The reality of protein MS for the care of patients is certainly exciting.

REFERENCES

1. Wasinger VC, Cordwell SJ, Cerpa-Poljak A, et al. Progress with gene-product mapping of the mollicutes: *Mycoplasma genitalium*. *Electrophoresis* 1995;**16**:1090—4.
2. Anderson L, Anderson NG. High resolution two-dimensional electrophoresis of human plasma proteins. *Proc Natl Acad Sci USA* 1977;**74**:5421—5.
3. Aebersold RH, Teplow DB, Hood LE, et al. Electroblotting onto activated glass: high efficiency preparation of proteins from analytical sodium dodecyl sulfate-polyacrylamide gels for direct sequence analysis. *J Biol Chem* 1986;**261**:4229—38.
4. Tanaka K, Waki H, Ido Y, et al. Protein and polymer analyses up to m/z 100,000 by laser ionization time-of-flight mass spectrometry. *Rapid Commun Mass Spectrom* 1988;**2**:151—3.
5. Yamashita M, Fenn JB. Electrospray ion source: another variation on the free-jet theme. *J Phys Chem* 1984;**88**:4451—9.
6. Eng JK, McCormack AL, Yates JR. An approach to correlate tandem mass spectral data of peptides with amino acid sequences in a protein database. *J Am Soc Mass Spectrom* 1994;**5**:976—89.
7. Perkins DN, Pappin DJ, Creasy DM, et al. Probability-based protein identification by searching sequence databases using mass spectrometry data. *Electrophoresis* 1999;**20**:3551—67.
8. Meehan KL, Holland JW, Dawkins HJS. Proteomic analysis of normal and malignant prostate tissue to identify novel proteins lost in cancer. *Prostate* 2002;**50**:54—63.

9. Shen J, Person MD, Zhu J, et al. Protein expression profiles in pancreatic adenocarcinoma compared with normal pancreatic tissue and tissue affected by pancreatitis as detected by two-dimensional gel electrophoresis and mass spectrometry. *Cancer Res* 2004;**64**:9018−26.
10. Li J, Zhang Z, Rosenzweig J, et al. Proteomics and bioinformatics approaches for identification of serum biomarkers to detect breast cancer. *Clin Chem* 2002;**48**:1296−304.
11. Lin Y, Goedegebuure PS, Tan MCB, et al. Proteins associated with disease and clinical course in pancreas cancer: a proteomic analysis of plasma in surgical patients. *J Proteome Res* 2006;**5**:2169−76.
12. Koo B-S, Lee D-Y, Ha H-S, et al. Comparative analysis of the tear protein expression in blepharitis patients using two-dimensional electrophoresis. *J Proteome Res* 2005;**4**:719−24.
13. Thongboonkerd V, McLeish KR, Arthur JM, et al. Proteomic analysis of normal human urinary proteins isolated by acetone precipitation or ultracentrifugation. *Kidney Int* 2002;**62**:1461−9.
14. Corthals GL, Wasinger VC, Hochstrasser DF, et al. The dynamic range of protein expression: a challenge for proteomic research. *Electrophoresis* 2000;**21**:1104−15.
15. Gygi SP, Corthals GL, Zhang Y, et al. Evaluation of two-dimensional gel electrophoresis-based proteome analysis technology. *Proc Natl Acad Sci USA* 2000;**97**:9390−5.
16. Hortin GL, Sviridov D, Anderson NL. High-abundance polypeptides of the human plasma proteome comprising the top 4 logs of polypeptide abundance. *Clin Chem* 2008;**54**:1608−16.
17. Zuo X, Speicher DW. A method for global analysis of complex proteomes using sample prefractionation by solution isoelectrofocusing prior to two-dimensional electrophoresis. *Anal Biochem* 2000;**284**:266−78.
18. Cox B, Emili A. Tissue subcellular fractionation and protein extraction for use in mass-spectrometry-based proteomics. *Nat Protoc* 2006;**1**:1872−8.
19. Chernokalskaya E, Gutierrez S, Pitt AM, et al. Ultrafiltration for proteomic sample preparation. *Electrophoresis* 2004;**25**:2461−8.
20. Greening DW, Simpson RJ. A centrifugal ultrafiltration strategy for isolating the low-molecular weight (<or=25K) component of human plasma proteome. *J Proteomics* 2010;**73**:637−48.
21. Hinerfeld D, Innamorati D, Pirro J, et al. Serum/plasma depletion with chicken immunoglobulin Y antibodies for proteomic analysis from multiple mammalian species. *J Biomol Tech* 2004;**15**:184−90.
22. Björhall K, Miliotis T, Davidsson P. Comparison of different depletion strategies for improved resolution in proteomic analysis of human serum samples. *Proteomics* 2005;**5**:307−17.
23. Hutchens TW, Yip T-T. New desorption strategies for the mass spectrometric analysis of macromolecules. *Rapid Commun Mass Spectrom* 1993;**7**:576−80.
24. Chapman K. The ProteinChip biomarker system from Ciphergen Biosystems: a novel proteomics platform for rapid biomarker discovery and validation. *Biochem Soc Trans* 2002;**30**:82−7.
25. Fung E. Ciphergen ProteinChip technology: a platform for protein profiling and biomarker identification. *Nat Genet* 2001;**27**:54.
26. Diamandis EP. Proteomic patterns in biological fluids: do they represent the future of cancer diagnostics? *Clin Chem* 2003;**49**:1272−5.
27. Fung ET, Yip T-T, Lomas L, et al. Classification of cancer types by measuring variants of host response proteins using SELDI serum assays. *Int J Cancer* 2005;**115**:783−9.
28. Li J, White N, Zhang Z, et al. Detection of prostate cancer using serum proteomics pattern in a histologically confirmed population. *J Urol* 2004;**171**:1782−7.
29. Petricoin EF, Ardekani AM, Hitt BA, et al. Use of proteomic patterns in serum to identify ovarian cancer. *Lancet* 2002;**359**:572−7.
30. Anonymous. Proteomic diagnostics tested. *Nature* 2004;**429**:487.
31. Baggerly KA, Coombes KR, Morris JS. Bias, randomization, and ovarian proteomic data: a reply to "producers and consumers." *Cancer Inform* 2005;**1**:9−14.
32. Baggerly KA, Morris JS, Coombes KR. Reproducibility of SELDI-TOF protein patterns in serum: comparing datasets from different experiments. *Bioinformatics* 2004;**20**:777−85.
33. Sorace JM, Zhan M. A data review and re-assessment of ovarian cancer serum proteomic profiling. *BMC Bioinformatics* 2003;**4**:24.
34. Washburn MP, Wolters D, Yates JR. Large-scale analysis of the yeast proteome by multidimensional protein identification technology. *Nat Biotechnol* 2001;**19**:242−7.
35. Paulovich AG, Whiteaker JR, Hoofnagle AN, et al. The interface between biomarker discovery and clinical validation: the tar pit of the protein biomarker pipeline. *Proteomics Clin Appl* 2008;**2**:1386−402.
36. Rifai N, Gillette MA, Carr SA. Protein biomarker discovery and validation: the long and uncertain path to clinical utility. *Nat Biotechnol* 2006;**24**:971−83.
37. Whiteaker JR, Lin C, Kennedy J, et al. A targeted proteomics-based pipeline for verification of biomarkers in plasma. *Nat Biotechnol* 2011;**29**:625−34.
38. Covey TR, Huang EC, Henion JD. Structural characterization of protein tryptic peptides via liquid chromatography/mass spectrometry and collision-induced dissociation of their doubly charged molecular ions. *Anal Chem* 1991;**63**:1193−200.
39. Chervet JP, Ursem M, Salzmann JP. Instrumental requirements for nanoscale liquid chromatography. *Anal Chem* 1996;**68**:1507−12.
40. Wilm MS, Mann M. Electrospray and Taylor-Cone theory, Dole's beam of macromolecules at last? *Int J Mass Spectrom Ion Process* 1994;**136**:167−80.
41. Loo JA, Udseth HR, Smith RD. Peptide and protein analysis by electrospray ionization-mass spectrometry and capillary electrophoresis-mass spectrometry. *Anal Biochem* 1989;**179**:404−12.
42. Strittmatter EF, Ferguson PL, Tang K, et al. Proteome analyses using accurate mass and elution time peptide tags with capillary LC time-of-flight mass spectrometry. *J Am Soc Mass Spectrom* 2003;**14**:980−91.
43. Yates 3rd JR, Eng JK, McCormack AL, et al. Method to correlate tandem mass spectra of modified peptides to amino acid sequences in the protein database. *Anal Chem* 1995;**67**:1426−36.
44. Craig R, Cortens JP, Beavis RC. The use of proteotypic peptide libraries for protein identification. *Rapid Commun Mass Spectrom* 2005;**19**:1844−50.
45. Cox J, Neuhauser N, Michalski A, et al. Andromeda: a peptide search engine integrated into the MaxQuant environment. *J Proteome Res* 2011;**10**:1794−805.

46. Geer LY, Markey SP, Kowalak JA, et al. Open mass spectrometry search algorithm. *J Proteome Res* 2004;**3**:958−64.
47. Geiger T, Velic A, Macek B, et al. Initial quantitative proteomic map of 28 mouse tissues using the SILAC mouse. *Mol Cell Proteomics* 2013;**12**:1709−22.
48. Jiang H, English AM. Quantitative analysis of the yeast proteome by incorporation of isotopically labeled leucine. *J Proteome Res* 2002;**1**:345−50.
49. Ross PL, Huang YN, Marchese JN, et al. Multiplexed protein quantitation in *Saccharomyces cerevisiae* using amine-reactive isobaric tagging reagents. *Mol Cell Proteomics* 2004;**3**:1154−69.
50. Thompson A, Schäfer J, Kuhn K, et al. Tandem mass tags: a novel quantification strategy for comparative analysis of complex protein mixtures by MS/MS. *Anal Chem* 2003;**75**:1895−904.
51. Anderson L, Hunter CL. Quantitative mass spectrometric multiple reaction monitoring assays for major plasma proteins. *Mol Cell Proteomics* 2006;**5**:573−88.
52. Ang C-S, Nice EC. Targeted in-gel MRM: a hypothesis-driven approach for colorectal cancer biomarker discovery in human feces. *J Proteome Res* 2010;**9**:4346−55.
53. Rifai N, Watson ID, Miller WG. Commercial immunoassays in biomarkers studies: researchers beware! *Clin Chem* 2012;**58**:1387−8.
54. Hoofnagle AN, Wener MH. The fundamental flaws of immunoassays and potential solutions using tandem mass spectrometry. *J Immunol Methods* 2009;**347**:3−11.
55. Tsiatsiani L, Heck AJR. Proteomics beyond trypsin. *FEBS J* 2015;**282**:2612−26.
56. Anderson NL, Anderson NG, Haines LR, et al. Mass spectrometric quantitation of peptides and proteins using stable isotope standards and capture by anti-peptide antibodies (SISCAPA). *J Proteome Res* 2004;**3**:235−44.
57. Becker JO, Hoofnagle AN. Replacing immunoassays with tryptic digestion-peptide immunoaffinity enrichment and LC−MS/MS. *Bioanalysis* 2012;**4**:281−90.
58. Abbatiello SE, Mani DR, Keshishian H, et al. Automated detection of inaccurate and imprecise transitions in peptide quantification by multiple reaction monitoring mass spectrometry. *Clin Chem* 2010;**56**:291−305.
59. Bereman MS, Maclean B, Tomazela DM, et al. The development of selected reaction monitoring methods for targeted proteomics via empirical refinement. *Proteomics* 2012;**12**:1134−41.
60. Grant RP, Hoofnagle AN. From lost in translation to paradise found: enabling protein biomarker method transfer by mass spectrometry. *Clin Chem* 2014;**60**:941−4.
61. de Graaf EL, Altelaar AF, van Breukelen B, et al. Improving SRM assay development: a global comparison between triple quadrupole, ion trap, and higher energy CID peptide fragmentation spectra. *J Proteome Res* 2011;**10**:4334−41.
62. Eyers CE, Lawless C, Wedge DC, et al. CONSeQuence: prediction of reference peptides for absolute quantitative proteomics using consensus machine learning approaches. *Mol Cell Proteomics* 2011;**10**. M110 003384.
63. Marx H, Lemeer S, Schliep JE, et al. A large synthetic peptide and phosphopeptide reference library for mass spectrometry-based proteomics. *Nat Biotechnol* 2013;**31**:557−64.
64. Qeli E, Omasits U, Goetze S, et al. Improved prediction of peptide detectability for targeted proteomics using a rank-based algorithm and organism-specific data. *J Proteomics* 2014;**108**:269−83.
65. Whiteaker JR, Halusa GN, Hoofnagle AN, et al. CPTAC Assay Portal: a repository of targeted proteomic assays. *Nat Methods* 2014;**11**:703−74.
66. Desiere F, Deutsch EW, King NL, et al. The PeptideAtlas project. *Nucleic Acids Res* 2006;**34**:D655−8.
67. Henderson CM, Lutsey PL, Misialek JR, et al. Measurement by a novel LC-MS/MS methodology reveals similar serum concentrations of vitamin D-binding protein in blacks and whites. *Clin Chem* 2016;**62**:179−87.
68. Altschul SF, Gish W, Miller W, et al. Basic local alignment search tool. *J Mol Biol* 1990;**215**:403−10.
69. Proc JL, Kuzyk MA, Hardie DB, et al. A quantitative study of the effects of chaotropic agents, surfactants, and solvents on the digestion efficiency of human plasma proteins by trypsin. *J Proteome Res* 2010;**9**:5422−37.
70. Shuford CM, Sederoff RR, Chiang VL, et al. Peptide production and decay rates affect the quantitative accuracy of protein cleavage isotope dilution mass spectrometry (PC-IDMS). *Mol Cell Proteomics* 2012;**11**:814−23.
71. Agger SA, Marney LC, Hoofnagle AN. Simultaneous quantification of apolipoprotein A-I and apolipoprotein B by liquid-chromatography-multiple-reaction-monitoring mass spectrometry. *Clin Chem* 2010;**56**:1804−13.
72. Chen Y, Snyder MR, Zhu Y, et al. Simultaneous phenotyping and quantification of alpha-1-antitrypsin by liquid chromatography-tandem mass spectrometry. *Clin Chem* 2011;**57**:1161−8.
73. van den Broek I, Romijn FPHTM, Smit NPM, et al. Quantifying protein measurands by peptide measurements: where do errors arise? *J Proteome Res* 2015;**14**:928−42.
74. Kushnir MM, Rockwood AL, Roberts WL, et al. Measurement of thyroglobulin by liquid chromatography-tandem mass spectrometry in serum and plasma in the presence of antithyroglobulin autoantibodies. *Clin Chem* 2013;**59**:982−90.
75. Bystrom C, Sheng S, Zhang K, et al. Clinical utility of insulin-like growth factor 1 and 2; determination by high resolution mass spectrometry. *PLoS ONE* 2012;**7**. e43457.
76. Bystrom CE, Sheng S, Clarke NJ. Narrow mass extraction of time-of-flight data for quantitative analysis of proteins: determination of insulin-like growth factor-1. *Anal Chem* 2011;**83**:9005−10.
77. Cox HD, Lopes F, Woldemariam Ga, et al. Interlaboratory agreement of insulin-like growth factor 1 concentrations measured by mass spectrometry. *Clin Chem* 2014;**60**:541−8.
78. Hoofnagle AN, Becker JO, Oda MN, et al. Multiple-reaction monitoring-mass spectrometric assays can accurately measure the relative protein abundance in complex mixtures. *Clin Chem* 2012;**58**:777−81.
79. Pratt JM, Simpson DM, Doherty MK, et al. Multiplexed absolute quantification for proteomics using concatenated signature peptides encoded by QconCAT genes. *Nat Protoc* 2006;**1**:1029−43.
80. Clarke NJ, Zhang Y, Reitz RE. A novel mass spectrometry-based assay for the accurate measurement of thyroglobulin from patient samples containing antithyroglobulin autoantibodies. *J Investig Med* 2012;**60**:1157−63.
81. Hoofnagle AN, Becker JO, Wener MH, et al. Quantification of thyroglobulin, a low-abundance serum protein, by immunoaffinity peptide enrichment and tandem mass spectrometry. *Clin Chem* 2008;**54**:1796−804.

82. Hoofnagle AN, Roth MY. Improving the measurement of serum thyroglobulin with mass spectrometry. *J Clin Endocrinol Metab* 2013;**98**:1343–52.
83. Kaiser P, Akerboom T, Ohlendorf R, et al. Liquid chromatography-isotope dilution-mass spectrometry as a new basis for the reference measurement procedure for hemoglobin A1c determination. *Clin Chem* 2010;**56**:750–4.
84. Kushnir MM, Rockwood AL, Straseski JA, et al. Comparison of LC-MS/MS to immunoassay for measurement of thyroglobulin in fine-needle aspiration samples. *Clin Chem* 2014;**60**:1452–3.
84a. Hoofnagle AN, Whiteaker JR, Carr SA, et al. Recommendations for the generation, quantification, storage, and handling of peptides used for mass spectrometry-based assays. *Clin Chem* 2016;**62**:48–69.
85. Netzel BC, Grant RP, Hoofnagle AN, et al. First steps toward harmonization of LC-MS/MS thyroglobulin assays. *Clin Chem* 2016;**62**:297–9.
86. Netzel BC, Grebe SK, Algeciras-Schimnich A. Usefulness of a thyroglobulin liquid chromatography-tandem mass spectrometry assay for evaluation of suspected heterophile interference. *Clin Chem* 2014;**60**:1016–8.
87. Netzel BC, Grebe SK, Carranza Leon BG, et al. Thyroglobulin (Tg) testing revisited: Tg assays, TgAb assays, and correlation of results with clinical outcomes. *J Clin Endocrinol Metab* 2015;**100**:E1074–83.
88. Neubert H, Gale J, Muirhead D. Online high-flow peptide immunoaffinity enrichment and nanoflow LC-MS/MS: assay development for total salivary pepsin/pepsinogen. *Clin Chem* 2010;**56**:1413–23.
89. van den Broek I, Nouta J, Razavi M, et al. Quantification of serum apolipoproteins A-I and B-100 in clinical samples using an automated SISCAPA–MALDI-TOF-MS workflow. *Methods* 2015:1–12.
90. Bondar OP, Barnidge DR, Klee EW, et al. LC-MS/MS quantification of Zn-alpha2 glycoprotein: a potential serum biomarker for prostate cancer. *Clin Chem* 2007;**53**:673–8.
91. Keshishian H, Addona T, Burgess M, et al. Quantitative, multiplexed assays for low abundance proteins in plasma by targeted mass spectrometry and stable isotope dilution. *Mol Cell Proteomics* 2007;**6**:2212–29.
92. Bystrom CE, Salameh W, Reitz R, et al. Plasma renin activity by LC-MS/MS: development of a prototypical clinical assay reveals a subpopulation of human plasma samples with substantial peptidase activity. *Clin Chem* 2010;**56**:1561–9.
93. Gerber SA, Rush J, Stemman O, et al. Absolute quantification of proteins and phosphoproteins from cell lysates by tandem MS. *Proc Natl Acad Sci USA* 2003;**100**:6940–5.
94. Kuhn E, Addona T, Keshishian H, et al. Developing multiplexed assays for troponin I and interleukin-33 in plasma by peptide immunoaffinity enrichment and targeted mass spectrometry. *Clin Chem* 2009;**55**:1108–17.
95. Whiteaker JR, Zhao L, Frisch C, et al. High-affinity recombinant antibody fragments (Fabs) can be applied in peptide enrichment immuno-MRM assays. *J Proteome Res* 2014;**13**:2187–96.
96. Kelleher NL. Top-down proteomics. *Anal Chem* 2004;**76**:197A–203 A.
97. Kellie JF, Higgs RE, Ryder JW, et al. Quantitative measurement of intact alpha-synuclein proteoforms from post-mortem control and Parkinson's disease brain tissue by intact protein mass spectrometry. *Sci Rep* 2014;**4**:5797.
98. Savaryn JP, Catherman AD, Thomas PM, et al. The emergence of top-down proteomics in clinical research. *Genome Med* 2013;**5**:53.
99. Sze SK, Ge Y, Oh H, et al. Top-down mass spectrometry of a 29-kDa protein for characterization of any posttranslational modification to within one residue. *Proc Natl Acad Sci USA* 2002;**99**:1774–9.
100. Ji QC, Rodila R, Gage EM, et al. A strategy of plasma protein quantitation by selective reaction monitoring of an intact protein. *Anal Chem* 2003;**75**:7008–14.
101. Edwards RL, Creese AJ, Baumert M, et al. Hemoglobin variant analysis via direct surface sampling of dried blood spots coupled with high-resolution mass spectrometry. *Anal Chem* 2011;**83**:2265–70.
102. Rai DK, Griffiths WJ, Landin B, et al. Accurate mass measurement by electrospray ionization quadrupole mass spectrometry: detection of variants differing by <6 Da from normal in human hemoglobin heterozygotes. *Anal Chem* 2003;**75**:1978–82.
103. Shackleton CHL, Falick AM, Green BN, et al. Electrospray mass spectrometry in the clinical diagnosis of variant hemoglobins. *J Chromatogr B Biomed Sci Appl* 1991;**562**:175–90.
104. Darby SM, Miller ML, Allen RO, et al. A mass spectrometric method for quantitation of intact insulin in blood samples. *J Anal Toxicol* 2001;**25**:8–14.
105. Mazur MT, Cardasis HL, Spellman DS, et al. Quantitative analysis of intact apolipoproteins in human HDL by top-down differential mass spectrometry. *Proc Natl Acad Sci USA* 2010;**107**:7728–33.
106. Niederkofler EE, Tubbs KA, Kiernan UA, et al. Novel mass spectrometric immunoassays for the rapid structural characterization of plasma apolipoproteins. *J Lipid Res* 2003;**44**:630–9.
107. Trenchevska O, Schaab MR, Nelson RW, et al. Development of multiplex mass spectrometric immunoassay for detection and quantification of apolipoproteins C-I, C-II, C-III and their proteoforms. *Methods* 2015;**81**:86–92.
108. Kiernan UA, Tubbs KA, Nedelkov D, et al. Detection of novel truncated forms of human serum amyloid A protein in human plasma. *FEBS Lett* 2003;**537**:166–70.
109. Oran PE, Sherma ND, Borges CR, et al. Intrapersonal and populational heterogeneity of the chemokine RANTES. *Clin Chem* 2010;**56**:1432–41.
110. Oran PE, Jarvis JW, Borges CR, et al. Mass spectrometric immunoassay of intact insulin and related variants for population proteomics studies. *Proteomics Clin Appl* 2011;**5**:454–9.
111. Oran PE, Trenchevska O, Nedelkov D, et al. Parallel workflow for high-throughput (>1,000 samples/day) quantitative analysis of human insulin-like growth factor 1 using mass spectrometric immunoassay. *PLoS ONE* 2014;**9**. e92801.
112. Lopez MF, Rezai T, Da Sarracino, et al. Selected reaction monitoring-mass spectrometric immunoassay responsive to parathyroid hormone and related variants. *Clin Chem* 2010;**56**:281–90.
113. Rai AJ, Ca Gelfand, Haywood BC, et al. HUPO Plasma Proteome Project specimen collection and handling: towards the standardization of parameters for plasma proteome samples. *Proteomics* 2005;**5**:3262–77.
114. Tuck MK, Chan DW, Chia D, et al. Standard operating procedures for serum and plasma collection: early detection research network consensus statement standard operating procedure integration working group. *J Proteome Res* 2009;**8**:113–7.

115. Beasley-Green A, Burris NM, Bunk DM, et al. Multiplexed LC-MS/MS assay for urine albumin. *J Proteome Res* 2014;**13**: 3930—9.
116. Singh R, Crow FW, Babic N, et al. A liquid chromatography-mass spectrometry method for the quantification of urinary albumin using a novel 15N-isotopically labeled albumin internal standard. *Clin Chem* 2007;**53**:540—2.
116a. Fu Q, Grote E, Zhu J, et al. An empirical approach to signature peptide choice for selected reaction monitoring: quantification of uromodulin in urine. *Clin Chem* 2016;**62**:198—207.
117. Schoenherr RM, Whiteaker JR, Zhao L, et al. Multiplexed quantification of estrogen receptor and HER2/Neu in tissue and cell lysates by peptide immunoaffinity enrichment mass spectrometry. *Proteomics* 2012;**12**:1253—60.
118. Azimzadeh O, Barjaktarovic Z, Aubele M, et al. Formalin-fixed paraffin-embedded (FFPE) proteome analysis using gel-free and gel-based proteomics. *J Proteome Res* 2010;**9**:4710—20.
119. Scicchitano MS, Dalmas DA, Boyce RW, et al. Protein extraction of formalin-fixed, paraffin-embedded tissue enables robust proteomic profiles by mass spectrometry. *J Histochem Cytochem* 2009;**57**:849—60.
120. Sprung RW, Brock JWC, Tanksley JP, et al. Equivalence of protein inventories obtained from formalin-fixed paraffin-embedded and frozen tissue in multidimensional liquid chromatography-tandem mass spectrometry shotgun proteomic analysis. *Mol Cell Proteomics* 2009;**8**:1988—98.
121. Guo T, Wang W, Rudnick PA, et al. Proteome analysis of microdissected formalin-fixed and paraffin-embedded tissue specimens. *J Histochem Cytochem* 2007;**55**:763—72.
122. Nirmalan NJ, Harnden P, Selby PJ, et al. Mining the archival formalin-fixed paraffin-embedded tissue proteome: opportunities and challenges. *Mol Biosyst* 2008;**4**:712—20.
123. Caldwell RL, Caprioli RM. Tissue profiling by mass spectrometry: a review of methodology and applications. *Mol Cell Proteomics* 2005;**4**:394—401.
124. Balluff B, Rauser S, Meding S, et al. MALDI imaging identifies prognostic seven-protein signature of novel tissue markers in intestinal-type gastric cancer. *Am J Pathol* 2011;**179**:2720—9.
125. Rauser S, Marquardt C, Balluff B, et al. Classification of HER2 receptor status in breast cancer tissues by MALDI imaging mass spectrometry. *J Proteome Res* 2010;**9**:1854—63.
126. Yi J, Kim C, Gelfand CA, et al. Inhibition of intrinsic proteolytic activities moderates preanalytical variability and instability of human plasma. *J Proteome Res* 2007;**6**: 1768—81.
127. Yi J, Liu Z, Craft D, et al. Intrinsic peptidase activity causes a sequential multi-step reaction (SMSR) in digestion of human plasma peptides. *J Proteome Res* 2008;**7**:5112—8.
128. Pasella S, Baralla A, Canu E, et al. Pre-analytical stability of the plasma proteomes based on the storage temperature. *Proteome Sci* 2013;**11**:10.
129. Zimmerman LJ, Li M, Yarbrough WG, et al. Global stability of plasma proteomes for mass spectrometry-based analyses. *Mol Cell Proteomics* 2012. M111.014340.
130. Gregorc V, Novello S, Lazzari C, et al. Predictive value of a proteomic signature in patients with non-small-cell lung cancer treated with second-line erlotinib or chemotherapy (PROSE): a biomarker-stratified, randomised phase 3 trial. *Lancet Oncol* 2014;**15**:713—21.
131. Vachani A, Pass HI, Rom WN, et al. Validation of a multi-protein plasma classifier to identify benign lung nodules. *J Thorac Oncol* 2015;**10**:629—37.
132. Kim KH, Alvarez RD. Using a multivariate index assay to assess malignancy in a pelvic mass. *Obstet Gynecol* 2012;**119**: 365—7.
133. Hines J, Milosevic D, Ketha H, et al. Detection of IGF-1 protein variants by use of LC-MS with high-resolution accurate mass in routine clinical analysis. *Clin Chem* 2015;**62**: 990—1.
134. Chen Z, Caulfield MP, McPhaul MJ, et al. Quantitative insulin analysis using liquid chromatography-tandem mass spectrometry in a high-throughput clinical laboratory. *Clin Chem* 2013;**59**:1349—56.
135. Jin M, Cataland S, Bissell M, et al. A rapid test for the diagnosis of thrombotic thrombocytopenic purpura using surface enhanced laser desorption/ionization time-of-flight (SELDI-TOF)-mass spectrometry. *J Thromb Haemost* 2006;**4**: 333—8.
136. Vrana JA, Gamez JD, Madden BJ, et al. Classification of amyloidosis by laser microdissection and mass spectrometry-based proteomic analysis in clinical biopsy specimens. *Blood* 2009;**114**:4957—9.

INDEX

Page numbers followed by *f* indicate figures, *t* indicate tables, and *b* indicate boxes.

A

Accuracy, LC-MS/MS and, 158—159, 163—164
Acid-base digestion, 75
Acylcarnitine detection, 52
Adsorption chromatography, 2, 15, 15f
Aerobic bacteria, 99—101
Affinity chromatography, 15f, 19—21, 19f
Alkylation, 190
AMDIS deconvolution program, 51
Anaerobic bacteria, MALDI-TOF-MS for identification of, 101
Anion-exchange chromatography, 17
Antimicrobial susceptibility testing (AST)
    MALDI-TOF-MS and, 104—105
Atmospheric pressure chemical ionization (APCI), 40
Atomic mass units, 33
Automation, mass spectrometry sample preparation and, 68—69

B

Bacteria
    aerobic, 99—101
    anaerobic, 94—98, 101
    MALDI-TOF-MS and typing of, 104—105
    PCR-ESI-MS and, 105
Band-broadening, 4—5
Baseline width, 2
Batch size, 175
Beam-type mass spectrometers, 43—46
Biomarker discovery pipeline. *See also* Genomics; Proteomics
    data acquisition strategies, 184—186, 185f
    data processing, 183
    discovery experiments, 183—184
    overview of, 182—183
    separations, 184
    targeted quantitative experiments, 187
    variations, 186—187
BLAST searches, 189
Blood samples. *See also* Dried blood spots; Plasma
    mass spectrometry sample preparation and, 75
Bonded phases, 8
Boronate affinity chromatography, 20
Bottom-up proteomics, 187, 188f
Bronchoalveolar lavage (BAL) fluids, 105
Bruker's Filamentous Fungi Laboratory 1.0, 102
Buffer exchange, 74

C

Calibration
    bottom-up targeted proteomics and, 191—192
    mass spectrometry and, 55—56
    top-down proteomics and, 194—195
Capillary columns, 9—10
Capillary microsampling, 85
Carnitine, 52
Carrier gas sources, 9
Carryover (sample), LC-MS/MS and, 135
Carryover blanks, 156—157
Cation-exchange chromatography, 17
Centrifugation, mass spectrometry sample preparation and, 70—71
Certified reference materials (CRM), LC-MS/MS and, 115
Charged aerosol detectors, 27
Chemical ionization (CI), 39—42
Chiral stationary phases, 20—21
Chromatograms, 11
Chromatography. *See also* Liquid chromatography-tandem mass spectrometry
    basic principles
        analyte identification and quantification, 6—7
        band-broadening and efficiency, 4—5
        resolution and peak capacity, 5—6, 5f—6f
        retention and selectivity, 2—4
        terminology and components of, 1—2, 2f
    bottom-up targeted proteomics and, 193
    gas. *See also* Gas chromatography-mass spectrometry
        gas-liquid, 8
        gas-solid, 8
        instrumentation, 8—14
        types of, 7—8
    liquid. *See also* High-performance liquid chromatography; Liquid chromatography-mass spectrometry
        adsorption chromatography, 15
        affinity, 19—21, 19f
        common myths of, 125t
        development of in LC-MS/MS, 121—131, 125t
        hydrophilic interaction and mixed-mode, 21—22
        instrumentation, 22—28, 22f
        ion-exchange, 17—18
        partition chromatography, 15—17
        size-exclusion, 18—19, 18f
        types of, 15—22, 15f
    mass spectrometry sample preparation and, 67
    multidimensional separations, 29—30
    planar, 1—2, 28—29, 29f
    supercritical fluid, 28
Collision energy (CE), 57, 57f, 118—119
Column bleed, 8
Column chromatography, 1—2
Column-switching, 82—85, 82f, 84f
Conjugation, phase II metabolic, 74
Constant neutral loss scans, 48—49, 49f
Continuous dynode electron multipliers (CDEM), 50
Coulometry, 26—27
Cryopumps, 42
Cytopathic effect, 109

D

Daltons, 33—34
Data-dependent acquisition (DDA), 51—52, 185, 185f
Data-independent acquisition (DIA), 185, 185f
Dead volumes, 126
Deconvolution protocols, 51
Denaturation, 189—190, 189t
Density gradient separations, 70—71
Derivatization
    gas chromatography and, 10
    LC-MS/MS and, 173—174
    liquid chromatography and, 23
    liquid chromatography-mass spectrometry and, 52
    mass spectrometry and, 77
Desalting, 74
Desorption electrospray ionization (DESI), 42
Dialysis, 73—74
Differential mobility analyzers, 50
Direct analysis in real time (DART), 42
Discrete dynode electron multipliers, 50, 50f
Dried blood spots (DBS), 85
Dried plasma spots (DPS), 85
Dual column mode, 83—85, 84f
Dye-ligand affinity chromatography, 20
Dynamic light-scattering detectors, 27

E

Effective force, 43—44
Efficiency, 4—5
Electron capture detectors, 12t, 13
Electron ionization (EI), 34, 38—39, 38f
Electrospray ionization (ESI)
    LC-MS/MS and, 118, 124f
    overview of, 34, 39—40, 39f—40f
Electrospray ionization mass spectrometry (ESI-MS), 105—109
Electrospray mass spectrometry-mass spectrometry, 52
Electrospray voltage, 57
Electrostatic repulsion hydrophilic liquid chromatography (ERLIC, eHILIC), 21
Elution time, 2
Elution volume, 2
Elutropic strength, 15
Endocarditis, 107
Endotracheal aspirates, 105
Enrichment, 76—77, 117, 192, 195
Enzymatic hydrolysis, 74—75
Evaporation, 76—77
Evaporative light-scattering detectors (ELSD), 27
Extracted ion chromatograms, 35, 36f

F

Faraday detection cups, 50
Fenn, John, 94, 105
Field asymmetric ion mobility spectrometry (FAIMS), 50
Fingerprints, 53
Fixed-wavelength absorbance detectors, 26
Flame ionization detectors (FID), 11, 12t
Flame photometric detectors (FPD), 12t, 13—14
Flow injection analysis (FIA), 118, 119f, 124f
Fluorescence detectors, 26
Forced error, 174
Fourier transform ion cyclotron resonance (FT-ICR), 42
Fourier transform ion cyclotron resonance-mass spectrometry (FT-ICR-MS), 47
Fragmentation, 34, 37f
Fungi
    MALDI-TOF-MS for identification of, 98, 101—102
    PCR-ESI-MS and, 105

G

Gas chromatography-mass spectrometry (GC-MS), 51
Gas chromatography/infrared spectroscopy (GC/IR), 13—14
Gas chromatography/mass spectrometry (GC-MS)
    overview of, 13, 51
Gas-liquid chromatography (GLC), 8, 8t
Gas-solid chromatography (GSC), 8
Gel filtration, 19
Genetic disorder screening
    liquid chromatography-mass spectrometry, 52
Genomics, mass spectrometry and, 54—55
Guidance for Bioanalytical Method Validation (FDA), 156

H

Half-height width, 4
Hard ionization, 34
Height equivalent of theoretical plate (HETP), 4
Hemoglobinopathies, 55
Hemolysis, LC-MS/MS and, 167, 169t

203

High-performance liquid chromatography (HPLC)
  biomarker discovery pipeline and, 184
  common myths of, 125t
  LC-MS/MS and, 121–131
High-throughput methods, 81
Hydrophilic interaction liquid chromatography (HILIC), 21–22, 121
Hydrophobic interaction chromatography (HIC), 17

I

Icterus, LC-MS/MS and, 167–168, 169t
Immobilized metal-ion affinity chromatography (IMAC), 20
Immunoaffinity chromatography (IAC), 20
Inductively coupled plasma (ICP) ionization, 38, 41
Inductively coupled plasma mass spectrophotometry (ICP-MS), 41, 41f, 54–55, 75
Information-dependent acquisition, 51–52
Infrared (IR) spectroscopy, 12t, 13–14
Infrared multiphoton dissociation, 40
Injection systems, 9–10, 23
Interferences, biomarker discovery pipeline and, 184f
Internal calibration, 6–7, 7f
Internal standards (IS), 116–117, 137t, 156–157, 165, 176, 187, 190–191
Ion cyclotron resonance mass spectrometry (ICR-MS), 47
Ion mirrors, 45–46
Ion mobility spectrometers (IMS), 49–50
Ion sources, 34, 38
Ion suppression, 58–59, 166
Ion-exchange chromatography (IEC), 2, 17–18
Ion-pair chromatography (IPC), 21
Ionization, 34, 41, 121
IRIDICA, 105
Isobaric tags for relative and absolute quantification (iTRAQ), 77
Isobars, 126–129, 140
Isotope dilution mass spectrometry (IDMS), 38
Isotope-coded affinity tags (iCAT), 77
Isotopes, 36–37
Isotopically labeled internal standards, 56

K

Karl-Fischer analysis, 116

L

Laser desorption ionization, 38
Limit of blank (LOB), 164
Limit of detection (LOD), 164
Linear ion traps, 43, 47
Linearity, 171, 189
Lipemia, LC-MS/MS and, 167, 169t
Liquid chromatography. See under Chromatography
Liquid chromatography-mass spectrometry (LC-MS), 27, 51–52
Liquid chromatography-tandem mass spectrometry (LC-MS/MS)
  history of proteomics and, 181–183
  practical considerations
    additional HPLC optimization, 116
    adsorptive loss evaluation, 131–132
    calibration standard and matrix, 116–117
    equipment setup and tuneup, 118–119
    final stage 2, 132–135
    internal standard selection and utility, 117, 117t
    liquid chromatography development, 121–131
    on-column detection limits, 132
    solvent chemistry optimization, 121
    prevalidation, 156–161, 157t–159t
  sample preparation
    calibration and carryover assessment, 147–150
    liquid-liquid extraction, 138–140
    method refinement (final), 155–156

Liquid chromatography-tandem mass spectrometry (LC-MS/MS) (Continued)
    online extraction, 144–147
    protein precipitation, 137–138
    selectivity, 151–155
    solid-phase extraction, 141–144
    supported liquid extraction, 140–141
    tools and considerations, 135–137
  validation, 161
    accuracy, 163–164
    analytical interferences, 165–166
    calibration curve fit and linearity, 171
    carryover, 164–165
    dilutional linearity, 171–172
    external quality assessment, certified reference materials, and reference intervals, 176–177
    general rules, 161–163, 163t
    hydrolysis and derivatization, 173–174
    interassay comparison and correlation, 175–176
    ion suppression and enhancement, 166
    long-term stability, 175
    matrix effects, 166–167
    postprocessing stability, 174
    precision, 164
    preprocessing stability, 172–173
    ruggedness and maximum batch size, 174–175
    selectivity, 169–171
    spike and recovery, 167–169
Liquid chromatography-tandem mass spectrometry (LC-MS/MS), 30, 51–52, 75, 183
Liquid chromatography/electrochemical detection (LC-EC), 26–27
Liquid-liquid extraction (LLE)
  LC-MS/MS and, 138, 139t, 140f
  mass spectrometry sample preparation and, 77, 78f
Lower limit of measurement interval (LLMI), 121
Lower limit of quantification (LLoQ), 121, 188

M

m/z analyzers, 34
Magnetic sector mass spectrometers, 43–44
MALDI Biotyper system, 98–102
Mass analyzers, 42–49
Mass spectrometers, classes of, 43–47
Mass spectrometry
  basic concepts and definitions, 33–38
  clinical applications
    gas chromatography-mass spectrometry, 51
    inductively coupled plasma mass spectrometry, 54
    liquid chromatography-mass spectrometry, 27, 51–52
    matrix-assisted laser desorption ionization mass spectrometry, 52–54, 53f, 53t
    in proteomics, genomics metabolomics, 54–55
  instrumentation, 38f
  optimization of instrument conditions, 56–60
  practical aspects, 55–56
  sample preparation for
    background, 67–69, 68b
    chromatographic techniques, 82–85
    enrichment techniques, 76–77
    evolving techniques, 85
    extraction techniques, 77–81, 78f
    general techniques, 69–71
    precipitation techniques, 75–76, 76f
    separation techniques, 71–75, 72t
Mass spectrums, 34–35, 34f
Mass transitions, 56–57
Mass-to-charge ratios, 33–34
Mathieu equation, 43–44
Matrix
  LC-MS/MS and, 116–117, 118t
  mass spectrometry and, 58–59
Matrix-assisted laser desorption ionization (MALDI), 41, 42f

Matrix-assisted laser desorption ionization mass spectrometry (MALDI-MS), 52–54, 53f, 53t
Matrix-assisted laser desorption ionization time-of-flight (MALDI-TOF), 52–53, 53f, 53t
Matrix-assisted laser desorption ionization time-of-flight mass spectrometry (MALDI-TOF-MS)
  for antimicrobial susceptibility testing, 103
  for direct testing of clinical samples, 103
  future perspectives, 105
  for identification of aerobic bacteria, 99–101
  for identification of anaerobic bacteria, 101
  for identification of fungi, 101–102
  for identification of mycobacteria, 101
  laboratory workflow, cost and, 102–103
  overview of, 93–98, 94f–95f
  for testing of positive blood culture bottles, 103–104
  for typing, 104
Matuszewski method, 58–59
Maurer Drug Library, 51
Metabolic conjugation, 74
Metabolomics, 55
Microdialysis, 73–74
Microflow separation, 58
Molecular ions, 34
Molecular mass, 33
Monolithic columns, 24–25, 74
Multichannel plate electron multipliers, 50
Multidrug resistant tuberculosis (MDR-TB), 108
Multiple reaction monitoring (MRM), 36, 48, 49f, 56
Multiplexing, 192–193
Multipliers, 50, 50f
Mycobacteria spp
  laboratory methods
    MALDI-TOF-MS, 93
    MALDI-TOF-MS for identification of, 101
    PCR-ESI-MS and, 105–106

N

Nanoflow chromatography, 23
NIST Mass Spectral Database, 51
Nitrogen-phosphorus detectors (NPD), 11–13, 12t
Normal-phase chromatography, 16
Nozzle-skimmer dissociation, 40
Nuclear magnetic resonance (NMR), 27

O

Oasis HLB, 80
Off-line solid phase extraction, 141
On-column detection limits, 132
Online solid phase extraction, 79, 81, 144–147
Orbitrap mass spectrometry, 47
Orthogonal acceleration TOF-MS (OA-TOF-MS), 45–46

P

Partition chromatography, 15–17, 15f
Pass bands, 43–44
Passive filtration, 72
Passive ultrafiltration, 72
Peak area, 6
Peak asymmetry, 129–131
Peak capacity, 5–6
Peak height, 6
Peak width, 129–131
Peptide
  bottom-up targeted proteomics and, 188–189
  mass fingerprinting, 52–53, 53t
Pfleger Drug Library, 51
Phase II metabolic conjugation, 74
Phospholipids, 125, 129
Photodiode array detectors, 26
Photoionization detectors (PID), 11, 12t
Planar chromatography, 28–29, 29f. See also Thin-layer chromatography
Plasma, mass spectrometry sample preparation and, 68

INDEX

Plate number, 4
PLEX-ID, 107—109
Poly(styrene divinylbenzene), 80, 80f
Polymerase chain reaction electrospray ionization mass spectrometry (PCR-ESI-MS)
    bacterial diagnostics and, 106—107
    fungal diagnostics and, 107—108
    limitations and future perspectives of, 109
    mycobacterial testing and, 108
    overview of, 105—106
    viral diagnostics and, 109
Polysiloxane, 8, 8f
Porous-layer open tubular (PLOT) columns, 10
Posttranslational modifications, biomarker discovery pipeline and, 51
Potassium, 1263—1264
Precursor ions
    LC-MS/MS and, 52
    tandem mass spectrometry and, 34, 84f
Prevalidation studies, 156—161, 163t
Product ion scans, 48
Product ions, 48—49, 119
Proficiency testing (external quality assessment) LC-MS/MS and, 176
Proteases, bottom-up targeted proteomics and, 188—193
Proteomics. *See also* Mass spectrometry
    biomarker pipeline
        data acquisition strategies, 184—186
        data processing, 186
        discovery experiments, 183—184
        overview of, 183—187
        separations, 184
        targeted quantitative experiments, 187
        variations, 186—187
    bottom-up targeted
        calibration, 191—192
        chromatography, 193
        denaturation, 189—190
        digestion optimization, 190
        internal standards and peptide quality, 190—191
        multiplexing, 192—193
        overview of, 188—193
        peptide selection, 188—189
        polymorphisms, 193
        protease selection and activity, 190
        protein or peptide enrichment, 192
        reduction and alkylation, 190
        workflow overview, 188
    current clinical assays, 197
    future of, 197
    history of, 181—183
    instruments for, 196—197, 197t
    interferences and, 196
    MALDI-TOF-MS and, 93—98
    mass spectrometry and, 33
    posttranslational modifications, 196
    specimen collection and processing and, 195
    stability and, 196
    tissues and, 195—196
    top-down
        calibration, 194—195
        internal standards, 195
        protein enrichment, 195
        specificity, 194
        workflow overview, 193—194
    urine as matrix, 195

Q

Quadrupole ion traps (QIT), 46, 46f
Quadrupole mass spectrometers, 43—44, 43f
Qualifying ions, 152—155, 169—170

R

Reference intervals, LC-MS/MS and, 176—177
Refractive index (RI) detectors, 27
Relative abundance, 35
Retention factor, 2—3
Retention time, 2—3
Retention volume, 2—3
Reverse-phase (liquid) chromatography, 16
Reverse-phase high-performance liquid chromatography (RPLC), 121, 126f—127f
Ribosomes, 70
Robotic liquid handling workstations, 70

S

Safety, laboratory
    gas chromatography and, 14
    liquid chromatography and, 28
Sample collection, biomarker discovery pipeline and, 183
Sample preparation
    for mass spectrometry
        background, 67—69
        chromatographic techniques, 82—85
        enrichment techniques, 76—77
        evolving techniques, 85
        extraction techniques, 77—81
        general techniques, 69—71
        precipitation techniques, 75—76
        separation techniques, 71—75
        stability and, 172—173
Secular frequency, 43—44
Selected ion monitoring (SIM), 13, 36
Selectivity factor, 3—4
Separation factor, 3—4
Sepsis, 106
Size-exclusion chromatography (SEC), 15f, 18—19, 18f
Soft ionization, 39
Solid-phase extraction (SPE)
    LC-MS/MS and, 141—144, 143t, 144f
    mass spectrometry sample preparation and, 79—81, 79f
    online, 144—147, 145t
Solid-supported liquid-liquid extraction (SS-LLE), 78—79
Spectrophotometry, liquid chromatography and, 26
Stable-isotope dimethyl labeling, 77
Standard deviation (SD), of peak, in chromatography, 4
Stereoisomers, 20—21
Strain typing, 103
Supercritical fluid chromatography (SCF), 28
Support-coated open tubular (SCOT) columns, 10
Surrogate matrices, 116—117
System suitability test (SST) injections, 135

T

Tanaka, Koichi, 94
Tandem mass spectrometry (MS/MS), 42—49, 187. *See also* Liquid chromatography-tandem mass spectrometry
Temperature
    gas chromatography and, 11
    liquid chromatography and, 25
Thermal conductivity detectors (TCD), 9, 12t
Thermally induced dissociation, 40
Thermionic selective detectors (TSD), 11—13, 12t
Thin-layer chromatography (TLC), 29. *See also* Planar chromatography
Thomson units, 42—43
Thyroglobulin (Tg), 55
Time-of-flight mass spectrometry (TOF-MS), 45—46, 45f
Top-down proteomics, 54
Total ion chromatograms, 35
Toxicology screening, liquid chromatography-mass spectrometry and, 51—52
Transition ratios, 155
Trapping mass spectrometers, 46—47
Trypsin (TRY), bottom-up targeted proteomics and, 189, 191t
Tswett, Mikhail, 1
Tuberculosis, PCR-ESI-MS and, 105
Turbomolecular pumps, 42
Turbulent flow chromatography (TFC), 82, 144—147
Two-dimensional (2D) chromatographic separation, 57—58
Two-dimensional (2D) electrophoresis, 182

U

Ultra-high performance liquid chromatography (UPLC, UHPLC), 23
Ultrafiltration, 72
Unified atomic mass units, 33
Upper limit of measurement interval (ULMI), 147—150
Upper limit of quantification (UloQ), 147—150
Urea, 189t, 190
Urine specimens
    liquid-liquid extraction and, 138—139
    MALDI-TOF-MS and, 103
    mass spectrometry sample preparation and, 67, 69
    as matrix for proteomics assays, 195

V

Validation, LC-MS/MS and, 161, 162t
van Deemter equation, 5
Viral infections, PCR-ESI-MS and, 109
Vitek systems, 98, 100

W

Wall-coated open tubular (WCOT) columns, 10
Weber Drug Library, 51
Wiley Registry of Mass Spectral Data, 51
Wüthrich, Kurt, 94

Y

Yamashita, Masamichi, 105
Yeasts, MALDI-TOF-MS for identification of, 94, 99—100

Z

Zwitterionic hydrophilic interaction liquid chromatography (ZIC-HILIC), 21

Printed in the United States
By Bookmasters